Basic Mathematics for Economis

Economics students will welcome the new edition of this excellent textbook. Given that many students come into economics courses without having studied mathematics for a number of years, this clearly written book will help to develop quantitative skills in even the least numerate student up to the required level for a general Economics or Business Studies course. All explanations of mathematical concepts are set out in the context of applications in economics.

This new edition incorporates several new features, including new sections on:

- financial mathematics
- continuous growth
- matrix algebra

Improved pedagogical features, such as learning objectives and end of chapter questions, along with an overall example-led format and the use of Microsoft Excel for relevant applications mean that this textbook will continue to be a popular choice for both students and their lecturers.

Mike Rosser is Principal Lecturer in Economics in the Business School at Coventry University.

Basic Mathematics for Economists

Second Edition

Mike Rosser

Routledge
Taylor & Francis Group

LONDON AND NEW YORK

First edition published 1993
by Routledge
This edition published 2003
by Routledge
11 New Fetter Lane, London EC4P 4EE

Simultaneously published in the USA and Canada
by Routledge
29 West 35th Street, New York, NY 10001

Routledge is an imprint of the Taylor & Francis Group

© 1993, 2003 Mike Rosser

Typeset in Times New Roman by
Newgen Imaging Systems (P) Ltd, Chennai, India
Printed and bound in Great Britain by
TJ International Ltd, Padstow, Cornwall

British Library Cataloguing in Publication Data
A catalogue record for this book is available from the British Library

Library of Congress Cataloging in Publication Data
A catalog record for this book has been requested

ISBN 0–415–26783–8 (hbk)
ISBN 0–415–26784–6 (pbk)

Contents

Preface

Over half of the students who enrol on economics degree courses have not studied mathematics beyond GCSE or an equivalent level. These include many mature students whose last encounter with algebra, or any other mathematics beyond basic arithmetic, is now a dim and distant memory. It is mainly for these students that this book is intended. It aims to develop their mathematical ability up to the level required for a general economics degree course (i.e. one not specializing in mathematical economics) or for a modular degree course in economics and related subjects, such as business studies. To achieve this aim it has several objectives.

First, it provides a revision of arithmetical and algebraic methods that students probably studied at school but have now largely forgotten. It is a misconception to assume that, just because a GCSE mathematics syllabus includes certain topics, students who passed examinations on that syllabus two or more years ago are all still familiar with the material. They usually require some revision exercises to jog their memories and to get into the habit of using the different mathematical techniques again. The first few chapters are mainly devoted to this revision, set out where possible in the context of applications in economics.

Second, this book introduces mathematical techniques that will be new to most students through examples of their application to economic concepts. It also tries to get students tackling problems in economics using these techniques as soon as possible so that they can see how useful they are. Students are not required to work through unnecessary proofs, or wrestle with complicated special cases that they are unlikely ever to encounter again. For example, when covering the topic of calculus, some other textbooks require students to plough through abstract theoretical applications of the technique of differentiation to every conceivable type of function and special case before any mention of its uses in economics is made. In this book, however, we introduce the basic concept of differentiation followed by examples of economic applications in Chapter 8. Further developments of the topic, such as the second-order conditions for optimization, partial differentiation, and the rules for differentiation of composite functions, are then gradually brought in over the next few chapters, again in the context of economics application.

Third, this book tries to cover those mathematical techniques that will be relevant to students' economics degree programmes. Most applications are in the field of microeconomics, rather than macroeconomics, given the increased emphasis on business economics within many degree courses. In particular, Chapter 7 concentrates on a number of mathematical techniques that are relevant to finance and investment decision-making.

Given that most students now have access to computing facilities, ways of using a spreadsheet package to solve certain problems that are extremely difficult or time-consuming to solve manually are also explained.

Although it starts at a gentle pace through fairly elementary material, so that the students who gave up mathematics some years ago because they thought that they could not cope with A-level maths are able to build up their confidence, this is *not* a watered-down 'mathematics without tears or effort' type of textbook. As the book progresses the pace is increased and students are expected to put in a serious amount of time and effort to master the material. However, given the way in which this material is developed, it is hoped that students will be motivated to do so. Not everyone finds mathematics easy, but at least it helps if you can see the reason for having to study it.

Preface to Second Edition

The approach and style of the first edition have proved popular with students and I have tried to maintain both in the new material introduced in this second edition. The emphasis is on the introduction of mathematical concepts in the context of economics applications, with each step of the workings clearly explained in all the worked examples. Although the first edition was originally aimed at less mathematically able students, many others have also found it useful, some as a foundation for further study in mathematical economics and others as a helpful reference for specific topics that they have had difficulty understanding.

The main changes introduced in this second edition are a new chapter on matrix algebra (Chapter 15) and a rewrite of most of Chapter 14, which now includes sections on differential equations and has been retitled 'Exponential functions, continuous growth and differential equations'. A new section on part-year investment has been added and the section on interest rates rewritten in Chapter 7, which is now called 'Financial mathematics – series, time and investment'. There are also new sections on the reduced form of an economic model and the derivation of comparative static predictions, in Chapter 5 using linear algebra, and in Chapter 9 using calculus. All spreadsheet applications are now based on Excel, as this is now the most commonly used spreadsheet program. Other minor changes and corrections have been made throughout the rest of the book.

The Learning Objectives are now set out at the start of each chapter. It is hoped that students will find these useful as a guide to what they should expect to achieve, and their lecturers will find them useful when drawing up course guides. The layout of the pages in this second edition is also an improvement on the rather cramped style of the first edition.

I hope that both students and their lecturers will find these changes helpful.

Mike Rosser
Coventry

Acknowledgements

Microsoft® Windows and Microsoft® Excel® are registered trademarks of the Microsoft Corporation. Screen shot(s) reprinted by permission from Microsoft Corporation.

I am still grateful to those who helped in the production of the first edition of this book, including Joy Warren for her efficiency in typing the final manuscript, Mrs M. Fyvie and Chandrika Chauhan for their help in typing earlier drafts, and Mick Hayes for his help in checking the proofs.

The comments I have received from those people who have used the first edition have been very helpful for the revisions and corrections made in this second edition. I would particularly like to thank Alison Johnson at the Centre for International Studies in Economics, SOAS, London, and Ray Lewis at the University of Adelaide, Australia, for their help in checking the answers to the questions. I am also indebted to my colleague at Coventry, Keith Redhead, for his advice on the revised chapter on financial mathematics, to Gurpreet Dosanjh for his help in checking the second edition proofs, and to the two anonymous publisher's referees whose comments helped me to formulate this revised second edition.

Last, but certainly not least, I wish to acknowledge the help of my students in shaping the way that this book was originally developed and has since been revised. I, of course, am responsible for any remaining errors or omissions.

1 Introduction

Learning objective

After completing this chapter students should be able to:

- Understand why mathematics is useful to economists.

1.1 Why study mathematics?

Economics is a social science. It does not just describe what goes on in the economy. It attempts to explain how the economy operates and to make predictions about what may happen to specified economic variables if certain changes take place, e.g. what effect a crop failure will have on crop prices, what effect a given increase in sales tax will have on the price of finished goods, what will happen to unemployment if government expenditure is increased. It also suggests some guidelines that firms, governments or other economic agents might follow if they wished to allocate resources efficiently. Mathematics is fundamental to any serious application of economics to these areas.

Quantification

In introductory economic analysis predictions are often explained with the aid of sketch diagrams. For example, supply and demand analysis predicts that in a competitive market if supply is restricted then the price of a good will rise. However, this is really only common sense, as any market trader will tell you. An economist also needs to be able to say by how much price is expected to rise if supply contracts by a specified amount. This quantification of economic predictions requires the use of mathematics.

Although non-mathematical economic analysis may sometimes be useful for making qualitative predictions (i.e. predicting the direction of any expected changes), it cannot by itself provide the quantification that users of economic predictions require. A firm needs to know how much quantity sold is expected to change in response to a price increase. The government wants to know how much consumer demand will change if it increases a sales tax.

Simplification

Sometimes students believe that mathematics makes economics more complicated. Algebraic notation, which is essentially a form of shorthand, can, however, make certain concepts much clearer to understand than if they were set out in words. It can also save a great deal of time and effort in writing out tedious verbal explanations.

For example, the relationship between the quantity of apples consumers wish to buy and the price of apples might be expressed as: 'the quantity of apples demanded in a given time period is 1,200 kg when price is zero and then decreases by 10 kg for every 1p rise in the price of a kilo of apples'. It is much easier, however, to express this mathematically as: $q = 1,200 - 10p$ where q is the quantity of apples demanded in kilograms and p is the price in pence per kilogram of apples.

This is a very simple example. The relationships between economic variables can be much more complex and mathematical formulation then becomes the only feasible method for dealing with the analysis.

Scarcity and choice

Many problems dealt with in economics are concerned with the most efficient way of allocating limited resources. These are known as 'optimization' problems. For example, a firm may wish to maximize the output it can produce within a fixed budget for expenditure on inputs. Mathematics must be used to obtain answers to these problems.

Many economics graduates will enter employment in industry, commerce or the public sector where very real resource allocation decisions have to be made. Mathematical methods are used as a basis for many of these decisions. Even if students do not go on to specialize in subjects such as managerial economics or operational research where the applications of these decision-making techniques are studied in more depth, it is essential that they gain an understanding of the sort of resource allocation problems that can be tackled and the information that is needed to enable them to be solved.

Economic statistics and estimating relationships

As well as using mathematics to work out predictions from economic models where the relationships are already quantified, one also needs mathematics in order to estimate the parameters of the models in the first place. For example, if the demand relationship in an actual market is described by the economic model $q = 1,200 - 10p$ then this would mean that the parameters (i.e. the numbers 1,200 and 10) had been estimated from statistical data.

The study of how the parameters of economic models can be estimated from statistical data is known as econometrics. Although this is not one of the topics covered in this book, you will find that a knowledge of several of the mathematical techniques that are covered is necessary to understand the methods used in econometrics. Students using this book will probably also study an introductory statistics course as a prerequisite for econometrics, and here again certain basic mathematical tools will come in useful.

Mathematics and business

Some students using this book may be on courses that have more emphasis on business studies than pure economics. Two criticisms of the material covered that these students sometimes make are as follows.

(a) These simple models do not bear any resemblance to the real-world business decisions that have to be made in practice.
(b) Even if the models are relevant to business decisions there is not always enough actual data available on the relevant variables to make use of these mathematical techniques.

Criticism (a) should be answered in the first few lectures of your economics course when the methodology of economic theory is explained. In summary, one needs to start with a simplified model that can explain how firms (and other economic agents) behave in general before looking at more complex situations only relevant to specific firms.

Criticism (b) may be partially true, but a lack of complete data does not mean that one should not try to make the best decision using the information that is available. Just because some mathematical methods can be difficult to understand to the uninitiated, this does not mean that efficient decision-making should be abandoned in favour of guesswork, rule of thumb and intuition.

1.2 Calculators and computers

Some students may ask, 'what's the point in spending a great deal of time and effort studying mathematics when nowadays everyone uses calculators and computers for calculations?' There are several answers to this question.

Rubbish in, rubbish out

Perhaps the most important point which has to be made is that calculators and computers can only calculate what they are told to. They are machines that can perform arithmetic computations much faster than you can do by hand, and this speed does indeed make them very useful tools. However, if you feed in useless information you will get useless information back – hence the well-known phrase 'rubbish in, rubbish out'.

At a very basic level, consider what happens when you use a pocket calculator to perform some simple operations. Get out your pocket calculator and use it to answer the problem

$$16 - 3 \times 4 - 1 = ?$$

What answer did you get? 3? 7? 51? 39? It all depends on which order you perform the calculations and the type of calculator you use.

There are set rules for the order in which basic arithmetic operations should be performed, which are explained in Chapter 2. Nowadays, these are programmed into most calculators but not some older basic calculators. If you only have an old basic calculator then it cannot help you. It is you who must tell the calculator in which order to perform the calculations. (The correct answer is 3, by the way.)

For another example, consider the demand relationship

$$q = 1,200 - 10p$$

referred to earlier. What would quantity demanded be if price was 150? A computer would give the answer -300, but this is clearly nonsense as you cannot have a negative quantity of apples. It only makes sense for the above mathematical relationship to apply to positive values of p and q. Therefore if price is 120, quantity sold will be zero, and if any price higher than 120 is charged, such as 130, quantity sold will still be zero. This case illustrates why you must take care to interpret mathematical answers sensibly and not blindly assume that any numbers produced by a computer will always be correct even if the 'correct' numbers have been fed into it.

Algebra

Much economic analysis involves algebraic notation, with letters representing concepts that are capable of taking on different values (see Chapter 3). The manipulation of these algebraic expressions cannot usually be carried out by calculators and computers.

Rounding errors

Despite the speed of operation of calculators and computers it can sometimes be quicker and more accurate to solve a problem manually. To illustrate this point, if you have an old basic calculator, use it to answer the problem

$$\frac{10}{3} \times 3 = ?$$

You may get the answer 9.9999999. However, if you use a modern mathematical calculator you will have obtained the correct answer of 10. So why do some calculators give a slightly inaccurate answer?

All calculators and computers have a limited memory capacity. This means that numbers have to be rounded off after a certain number of digits. Given that 10 divided by 3 is 3.3333333 recurring, it is difficult for basic calculators to store this number accurately in decimal form. Although modern computers have a vast memory they still perform many computations through a series of algorithms, which are essentially a series of arithmetic operations. At various stages numbers can be rounded off and so the final answer can be slightly inaccurate. More accuracy can often be obtained by using simple 'vulgar fractions' and by limiting the number of calculator operations that round off the answers. Modern calculators and computer programs are now designed to try to minimize inaccuracies due to rounding errors.

When should you use calculators and computers?

Obviously pocket calculators are useful for basic arithmetic operations that take a long time to do manually, such as long division or finding square roots. If you only use a basic calculator, care needs to be taken to ensure that individual calculations are done in the correct order so that the fundamental rules of mathematics are satisfied and needless inaccuracies through rounding are avoided.

However, the level of mathematics in this book requires more than these basic arithmetic functions. It is recommended that all students obtain a mathematical calculator that has at least the following function keys:

$$[y^x] \quad [\sqrt[x]{y}] \quad [\text{LOG}] \quad [10^x] \quad [\text{LN}] \quad [e^x]$$

The meaning and use of these functions will be explained in the following chapters.

Most of you who have recently left school will probably have already used this type of calculator for GCSE mathematics, but mature students may only currently possess an older basic calculator with only the basic square root $[\sqrt{\ }]$ function. The modern mathematical calculators, in addition to having more mathematical functions, are a great advance on these basic calculators and can cope with most rounding errors and sequences of operations in multiple calculations. In some sections of the book, however, calculations that could be done on a mathematical calculator are still explained from first principles to ensure that all students fully understand the mathematical method employed.

Most students on economics degree courses will have access to computing facilities and be taught how to use various computer program packages. Most of these will probably be used for data analysis as part of the statistics component of your course. The facilities and programs available to students will vary from institution to institution. Your lecturer will advise whether or not you have access to computer program packages that can be used to tackle specific types of mathematical problems. For example, you may have access to a graphics package that tells you when certain lines intersect or solves linear programming problems (see Chapter 5). Spreadsheet programs, such as Excel, can be particularly useful, especially for the sort of financial problems covered in Chapter 7 and for performing the mathematical operations on matrices explained in Chapter 15.

However, even if you do have access to computer program packages that can solve specific types of problem you will still need to understand the method of solution so that you will understand the answer that the computer gives you. Also, many economic problems have to be set up in the form of a mathematical problem before they can be fed into a computer program package for solution.

Most problems and exercises in this book can be tackled without using computers although in some cases solution only using a calculator would be very time-consuming. Some students may not have easy access to computing facilities. In particular, part-time students who only attend evening classes may find it difficult to get into computer laboratories. These students may find it worthwhile to invest a few more pounds in a more advanced calculator. Many of the problems requiring a large number of calculations are in Chapter 7 where methods of solution using the Excel spreadsheet program are suggested. However, financial calculators are now available that have most of the functions and formulae necessary to cope with these problems.

As Excel is probably the spreadsheet program most commonly used by economics students, the spreadsheet suggested solutions to certain problems are given in Excel format. It is assumed that students will be familiar with the basic operational functions of this program (e.g. saving files, using the copy command etc.), and the solutions in this book only suggest a set of commands necessary to solve the set problems.

1.3 Using the book

Most students using this book will be on the first year of an economics degree course and will not have studied A-level mathematics. Some of you will be following a mathematics course specifically designed for people without A-level mathematics whilst others will be mixed in with more mathematically experienced students on a general quantitative methods course. The book starts from some very basic mathematical principles. Most of these you will already have covered for GCSE mathematics (or O-level or CSE for some mature students). Only you can judge whether or not you are sufficiently competent in a technique to be able to skip some of the sections.

It would be advisable, however, to start at the beginning of the book and work through all the set problems. Many of you will have had at least a two-year break since last studying mathematics and will benefit from some revision. If you cannot easily answer all the questions in a section then you obviously need to work through the topic. You should find that a lot of material is familiar to you although more applications of mathematics to economics are introduced as the book progresses.

It is assumed that students using this book will also be studying an economic analysis course. The examples in the first few chapters only use some basic economic theory, such as supply and demand analysis. By the time you get to the later chapters it will be assumed that you have covered additional topics in economic analysis, such as production and cost theory. If you come across problems that assume a knowledge of economics topics that you have not yet covered then you should leave them until you understand these topics, or consult your lecturer.

In some instances the basic analysis of certain economic concepts is explained before the mathematical application of these concepts, but this should not be considered a complete coverage of the topic.

Practise, practise

You will not learn mathematics by reading this book, or any other book for that matter. The only way you will learn mathematics is by practising working through problems. It may be more hard work than just reading through the pages of a book, but your effort will be rewarded when you master the different techniques. As with many other skills that people acquire, such as riding a bike or driving a car, a book can help you to understand how something is supposed to be done, but you will only be able to do it yourself if you spend time and effort practising.

You cannot acquire a skill by sitting down in front of a book and hoping that you can 'memorize' what you read.

Group working

Your lecturer will make it clear to you which problems you must do by yourself as part of your course assessment and which problems you may confer with others over. Asking others for help makes sense if you are absolutely stuck and just cannot understand a topic. However, you should make every effort to work through all the problems that you are set before asking your lecturer or fellow students for help. When you do ask for help it should be to find out *how* to tackle a problem.

Some students who have difficulty with mathematics tend to copy answers off other students without really understanding what they are doing, or when a lecturer runs through an answer in class they just write down a verbatim copy of the answer given without asking for clarification of points they do not follow.

They are only fooling themselves, however. The point of studying mathematics in the first year of an economics degree course is to learn how to be able to apply it to various economics topics. Students who pretend that they have no difficulty with something they do not properly understand will obviously not get very far.

What is important is that you understand the *method* of solving different types of problems. There is no point in having a set of answers to problems if you do not understand how these answers were obtained.

Don't give up!

Do not get disheartened if you do not understand a topic the first time it is explained to you. Mathematics can be a difficult subject and you will need to read through some sections several times before they become clear to you. If you make the effort to try all the set problems and consult your lecturer if you really get stuck then you will eventually master the subject.

Because the topics follow on from each other, each chapter assumes that students are familiar with material covered in previous chapters. It is therefore very important that you keep up-to-date with your work. You cannot 'skip' a topic that you find difficult and hope to get through without answering examination questions on it, as it is sometimes possible to do in other subjects.

About half of all students on economics degree courses gave up mathematics at school at the age of 16, many of them because they thought that they were not good enough at mathematics to take it for A-level. However, most of them usually manage to complete their first-year mathematics for economics course successfully and go on to achieve an honours degree. There is no reason why you should not do likewise if you are prepared to put in the effort.

2 Arithmetic

Learning objectives

After completing this chapter students should be able to:

- Use again the basic arithmetic operations taught at school, including: the use of brackets, fractions, decimals, percentages, negative numbers, powers, roots and logarithms.
- Apply some of these arithmetic operations to simple economic problems.
- Calculate arc elasticity of demand values by dividing a fraction by another fraction.

2.1 Revision of basic concepts

Most students will have previously covered all, or nearly all, of the topics in this chapter. They are included here for revision purposes and to ensure that everyone is familiar with basic arithmetical processes before going on to further mathematical topics. Only a fairly brief explanation is given for most of the arithmetical rules set out in this chapter. It is assumed that students will have learned these rules at school and now just require something to jog their memory so that they can begin to use them again.

As a starting point it will be assumed that all students are familiar with the basic operations of addition, subtraction, multiplication and division, as applied to whole numbers (or integers) at least. The notation for these operations can vary but the usual ways of expressing them are as follows.

Example 2.1

Addition (+): $\qquad$ $24 + 204 = 228$
Subtraction (−): $\qquad$ $9{,}089 - 393 = 8{,}696$
Multiplication (× or ·): $\quad$ $12 \times 24 = 288$
Division (÷ or /): $\qquad$ $4{,}448 \div 16 = 278$

The sign '·' is sometimes used for multiplication when using algebraic notation but, as you will see from Chapter 2 onwards, there is usually no need to use any multiplication sign to signify that two algebraic variables are being multiplied together, e.g. *A* times *B* is simply written *AB*.

Most students will have learned at school how to perform these operations with a pen and paper, even if their long multiplication and long division may now be a bit rusty. However, apart from simple addition and subtraction problems, it is usually quicker to use a pocket calculator for basic arithmetical operations. If you cannot answer the questions below then you need to refer to an elementary arithmetic text or to see your lecturer for advice.

Test Yourself, Exercise 2.1

1. $323 + 3{,}232 =$
2. $1{,}012 - 147 =$
3. $460 \times 202 =$
4. $1{,}288/56 =$

2.2 Multiple operations

Consider the following problem involving only addition and subtraction.

Example 2.2

A bus leaves its terminus with 22 passengers aboard. At the first stop 7 passengers get off and 12 get on. At the second stop 18 get off and 4 get on. How many passengers remain on the bus?

Most of you would probably answer this by saying $22 - 7 = 15$, $15 + 12 = 27$, $27 - 18 = 9$, $9 + 4 = 13$ passengers remaining, which is the correct answer.

If you were faced with the abstract mathematical problem

$$22 - 7 + 12 - 18 + 4 = ?$$

you should answer it in the same way, i.e. working from left to right. If you performed the addition operations first then you would get $22 - 19 - 22 = -19$ which is clearly not the correct answer to the bus passenger problem!

If we now consider an example involving only multiplication and division we can see that the same rule applies.

Example 2.3

A restaurant catering for a large party sits 6 people to a table. Each table requires 2 dishes of vegetables. How many dishes of vegetables are required for a party of 60?

Most people would answer this by saying $60 \div 6 = 10$ tables, $10 \times 2 = 20$ dishes, which is correct.

If this is set out as the calculation $60 \div 6 \times 2 =$? then the left to right rule must be used. If you did *not* use this rule then you might get

$$60 \div 6 \times 2 = 60 \div 12 = 5$$

which is incorrect.

Thus the general rule to use when a calculation involves several arithmetical operations and

(i) only addition and subtraction are involved or
(ii) only multiplication and division are involved

is that the operations should be performed by working from left to right.

Example 2.4

(i) $48 - 18 + 6 = 30 + 6 = 36$

(ii) $6 + 16 - 7 = 22 - 7 = 15$

(iii) $68 + 5 - 32 - 6 + 14 = 73 - 32 - 6 + 14$
$$= 41 - 6 + 14$$
$$= 35 + 14 = 49$$

(iv) $22 \times 8 \div 4 = 176 \div 4 = 44$

(v) $460 \div 5 \times 4 = 92 \times 4 = 368$

(vi) $200 \div 25 \times 8 \times 3 \div 4 = 8 \times 8 \times 3 \div 4$
$$= 64 \times 3 \div 4$$
$$= 192 \div 4 = 48$$

When a calculation involves both addition/subtraction and multiplication/division then the rule is: multiplication and division calculations must be done before addition and subtraction calculations (except when brackets are involved – see Section 2.3).

To illustrate the rationale for this rule consider the following simple example.

Example 2.5

How much change do you get from £5 if you buy 6 oranges at 40p each?

Solution

All calculations must be done using the same units and so, converting the £5 to pence,

$$\text{change} = 500 - 6 \times 40 = 500 - 240 = 260\text{p} = £2.60$$

Clearly the multiplication must be done before the subtraction in order to arrive at the correct answer.

Test Yourself, Exercise 2.2

1. $962 - 88 + 312 - 267 =$
2. $240 - 20 \times 3 \div 4 =$
3. $300 \times 82 \div 6 \div 25 =$
4. $360 \div 4 \times 7 - 3 =$
5. $6 \times 12 \times 4 + 48 \times 3 + 8 =$
6. $420 \div 6 \times 2 - 64 + 25 =$

2.3 Brackets

If a calculation involves brackets then the operations within the brackets must be done first. Thus brackets take precedence over the rule for multiple operations set out in Section 2.2.

Example 2.6

A firm produces 220 units of a good which cost an average of £8.25 each to produce and sells them at a price of £9.95. What is its profit?

Solution

$$\text{profit per unit} = £9.95 - £8.25$$
$$\text{total profit} = 220 \times (£9.95 - £8.25)$$
$$= 220 \times £1.70$$
$$= £374$$

In a calculation that only involves addition or subtraction the brackets can be removed. However, you must remember that if there is a minus sign before a set of brackets then all the terms within the brackets must be multiplied by -1 if the brackets are removed, i.e. all $+$ and $-$ signs are reversed. (See Section 2.7 if you are not familiar with the concept of negative numbers.)

Example 2.7

$$(92 - 24) - (20 - 2) = ?$$

Solution

$$68 - 18 = 50 \text{ using brackets}$$

or

$$92 - 24 - 20 + 2 = 50 \text{ removing brackets}$$

Test Yourself, Exercise 2.3

1. $(12 \times 3 - 8) \times (44 - 14) =$
2. $(68 - 32) - (100 - 84 + 3) =$
3. $60 + (36 - 8) \times 4 =$
4. $4 \times (62 \div 2) - 8 \div (12 \div 3) =$
5. If a firm produces 600 units of a good at an average cost of £76 and sells them all at a price of £99, what is its total profit?
6. $(124 + 6 \times 81) - (42 - 2 \times 15) =$
7. How much net (i.e. after tax) profit does a firm make if it produces 440 units of a good at an average cost of £3.40 each, and pays 15p tax to the government on each unit sold at the market price of £3.95, assuming it sells everything it produces?

2.4 Fractions

If computers and calculators use decimals when dealing with portions of whole numbers why bother with fractions? There are several reasons:

1. Certain operations, particularly multiplication and division, can sometimes be done more quickly by fractions if one can cancel out numbers.
2. When using algebraic notation instead of actual numbers one cannot use calculators, and operations on formulae have to be performed using the basic principles for operations on fractions.
3. In some cases fractions can give a more accurate answer than a calculator owing to rounding error (see Example 2.15 below).

A fraction is written as

$$\frac{\text{numerator}}{\text{denominator}}$$

and is just another way of saying that the numerator is divided by the denominator. Thus

$$\frac{120}{960} = 120 \div 960$$

Before carrying out any arithmetical operations with fractions it is best to simplify individual fractions. Both numerator and denominator can be divided by any whole number that they are both a multiple of. It therefore usually helps if any large numbers in a fraction are 'factorized', i.e. broken down into the smaller numbers that they area multiple of.

Example 2.8

$$\frac{168}{104} = \frac{21 \times 8}{13 \times 8} = \frac{21}{13}$$

In this example it is obvious that the 8s cancel out top and bottom, i.e. the numerator and denominator can both be divided by 8.

Example 2.9

$$\frac{120}{960} = \frac{12 \times 10}{12 \times 8 \times 10} = \frac{1}{8}$$

Addition and subtraction of fractions is carried out by converting all fractions so that they have a common denominator (usually the largest one) and then adding or subtracting the different quantities with this common denominator. To convert fractions to the common (largest) denominator, one multiplies both top and bottom of the fraction by whatever number it is necessary to get the required denominator. For example, to convert 1/6 to a fraction with 12 as its denominator, one simply multiplies top and bottom by 2. Thus

$$\frac{1}{6} = \frac{2 \times 1}{2 \times 6} = \frac{2}{12}$$

Example 2.10

$$\frac{1}{6} + \frac{5}{12} = \frac{2}{12} + \frac{5}{12} = \frac{2+5}{12} = \frac{7}{12}$$

It is necessary to convert any numbers that have an integer (i.e. a whole number) in them into fractions with the same denominator before carrying out addition or subtraction operations involving fractions. This is done by multiplying the integer by the denominator of the fraction and then adding.

Example 2.11

$$1\frac{3}{5} = \frac{1 \times 5}{5} + \frac{3}{5} = \frac{5}{5} + \frac{3}{5} = \frac{8}{5}$$

Example 2.12

$$2\frac{3}{7} - \frac{24}{63} = \frac{17}{7} - \frac{8}{21} = \frac{51-8}{21} = \frac{43}{21} = 2\frac{1}{21}$$

Multiplication of fractions is carried out by multiplying the numerators of the different fractions and then multiplying the denominators.

Example 2.13

$$\frac{3}{8} \times \frac{5}{7} = \frac{15}{56}$$

The exercise can be simplified if one first cancels out any whole numbers that can be divided into both the numerator and the denominator.

Example 2.14

$$\frac{20}{3} \times \frac{12}{35} \times \frac{4}{5} = \frac{(4 \times 5) \times (4 \times 3) \times 4}{3 \times 35 \times 5} = \frac{4 \times 4 \times 4}{35} = \frac{64}{35}$$

The usual way of performing this operation is simply to cross through numbers that cancel

$$\frac{\overset{4}{\cancel{20}}}{\underset{1}{\cancel{3}}} \times \frac{\overset{4}{\cancel{12}}}{\underset{7}{\cancel{35}}} \times \frac{4}{5} = \frac{64}{35}$$

Multiplying out fractions may provide a more accurate answer than the one you would get by working out the decimal value of a fraction with a calculator before multiplying. However, nowadays if you use a modern mathematical calculator and store the answer to each part you should avoid rounding errors.

Example 2.15

$$\frac{4}{7} \times \frac{7}{2} = ?$$

Solution

$$\frac{4}{7} \times \frac{7}{2} = \frac{4}{2} = 2 \text{ using fractions}$$

$$0.5714285 \times 3.5 = 1.9999997 \text{ using a basic calculator}$$

Using a modern calculator, if you enter the numbers and commands

4 [÷] 7 [×] 7 [÷] 2 [=]

you should get the correct answer of 2.

However, if you were to perform the operation 4 [÷] 7, note the answer of 0.5714286 and then re-enter this number and multiply by 3.5, you would get the slightly inaccurate answer of 2.0000001.

To divide by a fraction one simply multiplies by its inverse.

Example 2.16

$$3 \div \frac{1}{6} = 3 \times \frac{6}{1} = 18$$

Example 2.17

$$\frac{44}{7} \div \frac{8}{49} = \frac{44}{7} \times \frac{49}{8} = \frac{11}{1} \times \frac{7}{2} = \frac{77}{2} = 38\frac{1}{2}$$

Test Yourself, Exercise 2.4

1. $\dfrac{1}{6} + \dfrac{1}{7} + \dfrac{1}{8} =$

2. $\dfrac{3}{7} + \dfrac{2}{9} - \dfrac{1}{4} =$

3. $\dfrac{2}{5} \times \dfrac{60}{7} \times \dfrac{21}{15} =$

4. $\dfrac{4}{5} \div \dfrac{24}{19} =$

5. $4\dfrac{2}{7} - 1\dfrac{2}{3} =$

6. $2\dfrac{1}{6} + 3\dfrac{1}{4} - \dfrac{4}{5} =$

7. $3\dfrac{1}{4} + 4\dfrac{1}{3} =$

8. $8\dfrac{1}{2} \div 2\dfrac{1}{6} =$

9. $20\dfrac{1}{4} - \dfrac{3}{5} \times 2\dfrac{1}{8} =$

10. $6 - \dfrac{2}{3} \div \dfrac{1}{12} + 3\dfrac{1}{3} =$

2.5 Elasticity of demand

The arithmetic operation of dividing a fraction by a fraction is usually the first technique that students on an economics course need to brush up on if their mathematics is a bit rusty. It is needed to calculate 'elasticity' of demand, which is a concept you should encounter fairly early in your microeconomics course, where its uses should be explained. Price elasticity of demand is a measure of the responsiveness of demand to changes in price. It is usually defined as

$$e = (-1)\frac{\%\ \text{change in quantity demanded}}{\%\ \text{change in price}}$$

The (-1) in this definition ensures a positive value for elasticity as either the change in price or the change in quantity will be negative. When there are relatively large changes in price and quantity it is best to use the concept of 'arc elasticity' to measure elasticity along a section of a demand schedule. This takes the changes in quantity and price as percentages of the averages of their values before and after the change. Thus arc elasticity is usually defined as

$$\text{arc } e = (-1)\frac{\dfrac{\text{change in quantity}}{0.5 \,(\text{1st quantity} + \text{2nd quantity})} \times 100}{\dfrac{\text{change in price}}{0.5 \,(\text{1st price} + \text{2nd price})} \times 100}$$

Although a positive price change usually corresponds to a negative quantity change, and vice versa, it is easier to treat the changes in both price and quantity as positive quantities. This allows the (-1) to be dropped from the formula. The 0.5 and the 100 will always cancel top and bottom in arc elasticity calculations. Thus we are left with

$$\text{arc } e = \frac{\dfrac{\text{change in quantity}}{(\text{1st quantity} + \text{2nd quantity})}}{\dfrac{\text{change in price}}{(\text{1st price} + \text{2nd price})}}$$

as the formula actually used for calculating price arc elasticity of demand.

Example 2.18

Calculate the arc elasticity of demand between points A and B on the demand schedule shown in Figure 2.1.

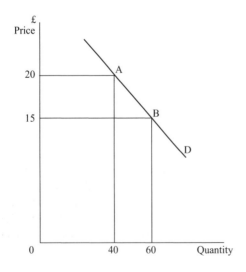

Figure 2.1

Solution

Between points A and B price falls by 5 from 20 to 15 and quantity rises by 20 from 40 to 60. Using the formula defined above

$$\text{arc } e = \frac{\dfrac{20}{40+60}}{\dfrac{5}{20+15}} = \frac{\dfrac{20}{100}}{\dfrac{5}{35}} = \frac{20}{100} \times \frac{35}{5} = \frac{1}{5} \times \frac{7}{1} = \frac{7}{5}$$

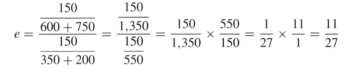

Example 2.19

When the price of a product is lowered from £350 to £200 quantity demanded increases from 600 to 750 units. Calculate the elasticity of demand over this section of its demand schedule.

Solution

Price fall is £150 and quantity rise is 150. Therefore using the concept of arc elasticity

$$e = \frac{\dfrac{150}{600+750}}{\dfrac{150}{350+200}} = \frac{\dfrac{150}{1,350}}{\dfrac{150}{550}} = \frac{150}{1,350} \times \frac{550}{150} = \frac{1}{27} \times \frac{11}{1} = \frac{11}{27}$$

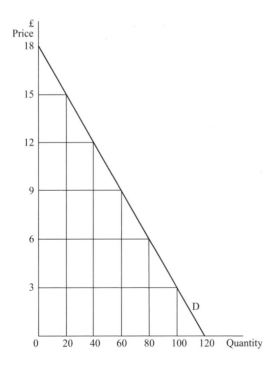

Figure 2.2

Test Yourself, Exercise 2.5

1. With reference to the demand schedule in Figure 2.2 calculate the arc elasticity of demand between the prices of (a) £3 and £6, (b) £6 and £9, (c) £9 and £12, (d) £12 and £15, and (e) £15 and £18.
2. A city bus service charges a uniform fare for every journey made. When this fare is increased from 50p to £1 the number of journeys made drops from 80,000 a day to 40,000. Calculate the arc elasticity of demand over this section of the demand schedule for bus journeys.
3. Calculate the arc elasticity of demand between (a) £5 and £10, and (b) between £10 and £15, for the demand schedule shown in Figure 2.3.

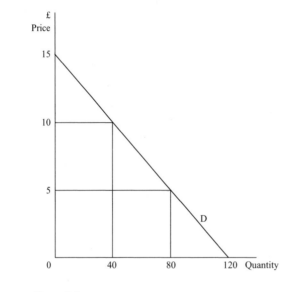

Figure 2.3

4. The data below show the quantity demanded of a good at various prices. Calculate the arc elasticity of demand for each £5 increment along the demand schedule.

Price	£40	£35	£30	£25	£20	£15	£10	£5	£0
Quantity	0	50	100	150	200	250	300	350	400

2.6   Decimals

Decimals are just another way of expressing fractions.

$$0.1 = 1/10$$

$$0.01 = 1/100$$

$$0.001 = 1/1,000 \text{ etc.}$$

Thus 0.234 is equivalent to 234/1,000.

Most of the time you will be able to perform operations involving decimals by using a calculator and so only a very brief summary of the manual methods of performing arithmetic operations using decimals is given here.

Addition and subtraction

When adding or subtracting decimals only 'like terms' must be added or subtracted. The easiest way to do this is to write any list of decimal numbers to be added so that the decimal points are all in a vertical column, in a similar fashion to the way that you may have been taught in primary school to add whole numbers by putting them in columns for hundreds, tens and units. You then add all the numbers that are the same number of digits away from the decimal point, carrying units over to the next column when the total is more than 9.

Example 2.20

$$1.345 + 0.00041 + 0.20023 = ?$$

Solution

$$
\begin{array}{r}
1.345 \ + \\
0.00041 \ + \\
0.20023 \ + \\
\hline
1.54564
\end{array}
$$

Multiplication

To multiply two numbers involving decimal fractions one can ignore the decimal points, multiply the two numbers in the usual fashion, and then insert the decimal point in the answer by counting the total number of digits to the right of the decimal point in both the numbers that were multiplied.

Example 2.21

$$2.463 \times 0.38 = ?$$

Solution

Removing the decimal places and multiplying the whole numbers remaining gives

$$
\begin{array}{r}
2{,}463 \ \times \\
38 \\
\hline
19{,}704 \\
73{,}890 \\
\hline
93{,}954
\end{array}
$$

There were a total of 5 digits to the right of the decimal place in the two numbers to be multiplied and so the answer is 0.93594.

Division

When dividing by a decimal fraction one first multiplies the fraction by the multiple of 10 that will convert it into a whole number. Then the number that is being divided is multiplied by the same multiple of 10 and the normal division operation is applied.

Example 2.22

$$360.54 \div 0.04 = ?$$

Solution

Multiplying both terms by 100 the problem becomes

$$36,054 \div 4 = 9,013.5$$

Given that actual arithmetic operations involving decimals can usually be performed with a calculator, perhaps one of the most common problems you are likely to face is how to express quantities as decimals before setting up a calculation.

Example 2.23

Express 0.01p as a decimal fraction of £1.

Solution

$$1p = £0.01$$

Therefore

$$0.01p = £0.0001$$

In mathematics a decimal format is often required for a value that is usually specified as a percentage in everyday usage. For example, interest rates are usually specified as percentages. A percentage format is really just another way of specifying a decimal fraction, e.g.

$$62\% = \frac{62}{100} = 0.62$$

and so percentages can easily be converted into decimal fractions by dividing by 100.

Example 2.24

$$22\% = 0.22 \qquad 0.24\% = 0.0024$$
$$24.56\% = 0.2456 \quad 0.02\% = 0.0002 \quad 2.4\% = 0.024$$

You will need to convert interest rate percentages to their decimal equivalent when you learn about investment appraisal methods and other aspects of financial mathematics, which are topics that we shall return to in Chapter 7.

Because some fractions cannot be expressed exactly in decimals, one may need to 'round off' an answer for convenience. In many of the economic problems in this book there is not much point in taking answers beyond two decimal places. Where this is done then the note '(to 2 dp)' is normally put after the answer. For example, 1/7 as a percentage is 14.29% (to 2 dp).

Test Yourself, Exercise 2.6

(Try to answer these without using a calculator.)

1. $53.024 - 16.11 =$
2. $44.2 \times 17 =$
3. $602.025 + 34.1006 - 201.016 =$
4. $432.984 \div 0.012 =$
5. $64.5 \times 0.0015 =$
6. $18.3 \div 0.03 =$
7. How many pencils costing 30p each can be bought for £42.00?
8. What is 1 millimetre as a decimal fraction of

    (a) 1 centimetre   (b) 1 metre   (c) 1 kilometre?

9. Specify the following percentages as decimal fractions:

    (a) 45.2%   (b) 243.15%

    (c) 7.5%   (d) 0.2%

2.7   Negative numbers

There are numerous instances where one comes across negative quantities, such as temperatures below zero or bank overdrafts. For example, if you have £35 in your bank account and withdraw £60 then your bank balance becomes −£25. There are instances, however, where it is not usually possible to have negative quantities. For example, a firm's production level cannot be negative.

To add negative numbers one simply subtracts the number after the negative sign, which is known as the absolute value of the number. In the examples below the negative numbers are written with brackets around them to help you distinguish between the addition of negative numbers and the subtraction of positive numbers.

Example 2.25

$$45 + (-32) + (-6) = 45 - 32 - 6 = 7$$

If it is required to subtract a negative number then the two negatives will cancel out and one adds the absolute value of the number.

Example 2.26

$$0.5 - (-0.45) - (-0.1) = 0.5 + 0.45 + 0.1 = 1.05$$

The rules for multiplication and division of negative numbers are:

● A negative multiplied (or divided) by a positive gives a negative.
● A negative multiplied (or divided) by a negative gives a positive.

Example 2.27

Eight students each have an overdraft of £210. What is their total bank balance?

Solution

$$\text{total balance} = 8 \times (-210) = -£1,680$$

Example 2.28

$$\frac{24}{-5} \div \frac{-32}{-10} = \frac{24}{-5} \times \frac{-10}{-32} = \frac{3}{1} \times \frac{2}{-4} = \frac{6}{-4} = -\frac{3}{2}$$

Test Yourself, Exercise 2.7

1. Subtract -4 from -6.
2. Multiply -4 by 6.
3. $-48 + 6 - 21 + 30 =$
4. $-0.55 + 1.0 =$
5. $1.2 + (-0.65) - 0.2 =$
6. $-26 \times 4.5 =$
7. $30 \times (4 - 15) =$
8. $(-60) \times (-60) =$

9. $\dfrac{-1}{4} \times \dfrac{9}{7} - \dfrac{4}{5} =$

10. $(-1)\dfrac{\dfrac{30 + 34}{-2}}{16 + 18} =$

2.8   Powers

We have all come across terms such as 'square metres' or 'cubic capacity'. A square metre is a rectangular area with each side equal to 1 metre. If a square room had all walls 5 metres long then its area would be $5 \times 5 = 25$ square metres.

When we multiply a number by itself in this fashion then we say we are 'squaring' it. The mathematical notation for this operation is the superscript 2. Thus '12 squared' is written 12^2.

Example 2.29

$2.5^2 = 2.5 \times 2.5 = 6.25$

We find the cubic capacity of a room, in cubic metres, by multiplying length $\times$ width $\times$ height. If all these distances are equal, at 3 metres say (i.e. the room is a perfect cube) then cubic capacity is $3 \times 3 \times 3 = 27$ cubic metres. When a number is cubed in this fashion the notation used is the superscript 3, e.g. 12^3.

These superscripts are known as 'powers' and denote the number of times a number is multiplied by itself. Although there are no physical analogies for powers other than 2 and 3, in mathematics one can encounter powers of any value.

Example 2.30

$12^4 = 12 \times 12 \times 12 \times 12 = 20{,}736$

$12^5 = 12 \times 12 \times 12 \times 12 \times 12 = 248{,}832$ etc.

To multiply numbers which are expressed as powers of the same number one adds all the powers together.

Example 2.31

$3^3 \times 3^5 = (3 \times 3 \times 3) \times (3 \times 3 \times 3 \times 3 \times 3) = 3^8 = 6{,}561$

To divide numbers in terms of powers of the same base number, one subtracts the superscript of the denominator from the numerator.

Example 2.32

$$\frac{6^6}{6^3} = \frac{6 \times 6 \times 6 \times 6 \times 6 \times 6}{6 \times 6 \times 6} = 6 \times 6 \times 6 = 6^3 = 216$$

In the two examples above the multiplication and division processes are set out in full to illustrate how these processes work with exponents. In practice, of course, one need not do this and it is just necessary to add or subtract the indices.

Any number to the power of 1 is simply the number itself. Although we do not normally write in the power 1 for single numbers, we must not forget to include it in calculation involving powers.

Example 2.33

$$4.6 \times 4.6^3 \times 4.6^2 = 4.6^6 = 9,474.3 \text{ (to 1 dp)}$$

In the example above, the first term 4.6 is counted as 4.6^1 when the powers are added up in the multiplication process.

Any number to the power of 0 is equal to 1. For example, $8^2 \times 8^0 = 8^{(2+0)} = 8^2$ so 8^0 must be 1.

Powers can also take negative values or can be fractions (see Section 2.9). A negative superscript indicates the number of times that one is dividing by the given number.

Example 2.34

$$3^6 \times 3^{-4} = \frac{3^6}{3^4} = \frac{3 \times 3 \times 3 \times 3 \times 3 \times 3}{3 \times 3 \times 3 \times 3} = 3^2$$

Thus multiplying by a number with a negative power (when both quantities are expressed as powers of the same number) simply involves adding the (negative) power to the power of the number being multiplied.

Example 2.35

$$8^4 \times 8^{-2} = 8^2 = 64$$

Example 2.36

$$14^7 \times 14^{-9} \times 14^6 = 14^4 = 38,416$$

The evaluation of numbers expressed as exponents can be time-consuming without a calculator with the function $[y^x]$, although you could, of course, use a basic calculator and put the number to be multiplied in memory and then multiply it by itself the required number of times. (This method would only work for whole number exponents though.)

To evaluate a number using the $[y^x]$ function on your calculator you should read the instruction booklet, if you have not lost it. The usual procedure is to enter y, the number to be multiplied, then hit the $[y^x]$ function key, then enter x, the exponent, and finally hit the $[=]$ key. For example, to find 14^4 enter $14\,[y^x]\,4\,[=]$ and you should get 38,416 as your answer.

If you do not, then you have either pressed the wrong keys or your calculator works in a slightly different fashion. To check which of these it is, try to evaluate the simpler answer to *Example 2.35* (8^2 which is obviously 64) by entering $8\,[y^x]\,2\,[=]$. If you do not get 64 then you need to find your calculator instructions.

Most calculators will not allow you to use the $[y^x]$ function to evaluate powers of negative numbers directly. Remembering that a negative multiplied by a positive gives a negative number, and a negative multiplied by a negative gives a positive, we can work out that if a negative number has an even whole number exponent then the whole term will be positive.

Example 2.37

$$(-3)^4 = (-3)^2 \times (-3)^2 = 9 \times 9 = 81$$

Similarly, if the exponent is an odd number the term will be negative.

Example 2.38

$$(-3)^5 = 3^5 \times (-1)^5 = 243 \times (-1) = -243$$

Therefore, when using a calculator to find the values of negative numbers taken to powers, one works with the absolute value and then puts in the negative sign if the power value is an odd number.

Example 2.39

$$(-19)^6 = 19^6 = 47,045,881$$

Example 2.40

$$(-26)^5 = -(26^5) = -11,881,376$$

Example 2.41

$$(-2)^{-2} \times (-2)^{-1} = (-2)^{-3} = \frac{1}{(-2)^3} = \frac{1}{-8} = -0.125$$

Test Yourself, Exercise 2.8

1. $4^2 \div 4^3 =$
2. $123^7 \times 123^{-6} =$
3. $6^4 \div (6^2 \times 6) =$
4. $(-2)^3 \times (-2)^3 =$
5. $1.42^4 \times 1.42^3 =$
6. $9^5 \times 9^{-3} \times 9^4 =$
7. $8.673^3 \div 8.673^6 =$
8. $(-6)^5 \times (-6)^{-3} =$
9. $(-8.52)^4 \times (-8.52)^{-1} =$
10. $(-2.5)^{-8} + (0.2)^6 \times (0.2)^{-8} =$

2.9   Roots and fractional powers

The *square root* of a number is the quantity which when squared gives the original number. There are different forms of notation. The square root of 16 can be written

$$\sqrt{16} = 4 \quad \text{or} \quad 16^{0.5} = 4$$

We can check this exponential format of $16^{0.5}$ using the rule for multiplying powers.

$$(16^{0.5})^2 = 16^{0.5} \times 16^{0.5} = 16^{0.5+0.5} = 16^1 = 16$$

Even most basic calculators have a square root function and so it is not normally worth bothering with the rather tedious manual method of calculating square roots when the square root is not obvious, as it is in the above example.

Example 2.42

$$2246^{0.5} = \sqrt{2,246} = 47.391982 \quad \text{(using a calculator)}$$

Although the positive square root of a number is perhaps the most obvious one, there will also be a negative square root. For example,

$$(-4) \times (-4) = 16$$

and so (-4) is a square root of 16, as well as 4. The negative square root is often important in the mathematical analysis of economic problems and it should not be neglected. The usual convention is to use the sign $\pm$ which means 'plus or minus'. Therefore, we really ought to say

$$\sqrt{16} = \pm 4$$

There are other roots. For example, $\sqrt[3]{27}$ or $27^{1/3}$ is the number which when multiplied by itself three times equals 27. This is easily checked as

$$(27^{1/3})^3 = 27^{1/3} \times 27^{1/3} \times 27^{1/3} = 27^1 = 27$$

When multiplying roots they need to be expressed in the form with a superscript, e.g. $6^{0.5}$, so that the rules for multiplying powers can be applied.

Example 2.43

$$47^{0.5} \times 47^{0.5} = 47$$

Example 2.44

$$15 \times 9^{0.75} \times 9^{0.75} = 15 \times 9^{1.5}$$
$$= 15 \times 9^{1.0} \times 9^{0.5}$$
$$= 15 \times 9 \times 3 = 405$$

These basic rules for multiplying numbers with powers as fractions will prove very useful when we get to algebra in Chapter 3.

Roots other than square roots can be evaluated using the $[\sqrt[x]{y}]$ function key on a calculator.

Example 2.45

To evaluate $\sqrt[5]{261}$ enter

$$261\,[\sqrt[x]{y}]\,5\,[=]$$

which should give 3.0431832.

Not all fractional powers correspond to an exact root in this sense, e.g. $6^{0.625}$ is not any particular root. To evaluate these other fractional powers you can use the $[y^x]$ function key on a calculator.

Example 2.46

To evaluate $452^{0.85}$ most calculators require you to enter

$$452\ [y^x]\ 0.85\ [=]$$

which should give you the answer 180.66236.

If you do not have this function key then you can use logarithms to evaluate these fractional powers (see Section 2.10). Roots and other powers less than 1 cannot be evaluated for negative numbers on calculators. A negative number cannot be the product of two positive or two negative numbers, and so the square root of a negative number cannot exist. Some other roots for negative numbers do exist, e.g. $\sqrt[3]{-1} = -1$, but you are not likely to need to find them.

In Chapter 7 some applications of these rules to financial problems are explained. For the time being we shall just work through a few more simple mathematical examples to ensure that the rules for working with powers are fully understood.

Example 2.47

$$24^{0.45} \times 24^{-1} = 24^{-0.55} = 0.1741341$$

Note that you must use the $[+/-]$ key on your calculator after entering 0.55 when evaluating this power. Alternatively you could have calculated

$$\frac{1}{24^{0.55}} = \frac{1}{5.7427007} = 0.1741341$$

Example 2.48

$$20 \times 8^{0.3} \times 8^{0.25} = 20 \times 8^{0.55} = 20 \times 3.1383364 = 62.766728$$

Sometimes it may help to multiply together two numbers with a common power. Both numbers can be put inside brackets with the common power outside the brackets.

Example 2.49

$$18^{0.5} \times 2^{0.5} = (18 \times 2)^{0.5} = 36^{0.5} = 6$$

Test Yourself, Exercise 2.9

Put the answers to the questions below as powers and then evaluate.

1. $\sqrt{625} =$
2. $\sqrt[3]{8} =$
3. $5^{0.5} \times 5^{-1.5} =$
4. $(7)^{0.5} \times (7)^{0.5} =$
5. $6^{0.3} \times 6^{-0.2} \times 6^{0.4} =$
6. $12 \times 4^{0.8} \times 4^{0.7} =$
7. $20^{0.5} \times 5^{0.5} =$
8. $16^{0.4} \times 16^{0.2} =$
9. $462^{-0.83} \times 462^{0.48} \div 462^{-0.2} =$
10. $76^{0.62} \times 18^{0.62} =$

2.10   Logarithms

Many people thought that logarithms went out of the window when pocket calculators became widely available. In the author's schooldays logarithms were used as a short-cut method for awkward long multiplication and long division calculations. Although pocket calculators have indeed now made log tables redundant for this purpose, they are still useful for some economic applications. For example, Chapters 7 and 14 show how logarithms can help calculate growth rates on investments. So, for those of you who have never seen log tables, or have forgotten what they are for, what are these mysterious logarithms?

The most commonly encountered logarithm is the base 10 logarithm. What this means is that the logarithm of any number is the power to which 10 must be raised to equal that number. The usual notation for logarithms to base 10 is 'log'.

Thus the logarithm of 100 is 2 since $100 = 10^2$. This is written as $\log 100 = 2$. Similarly

$$\log 10 = 1$$

and

$$\log 1,000 = 3$$

The square root of 10 is $3.1622777 = 10^{0.5}$ and so we know that $\log 3.1622777 = 0.5$.

The above logarithms are obvious. For the logarithms of other numbers you can use the [LOG] function key on a calculator or refer to a printed set of log tables.

If two numbers expressed as powers of 10 are multiplied together then we know that the indices are added, e.g.

$$10^{0.5} \times 10^{1.5} = 10^2$$

Therefore, to use logs to multiply numbers, one simply adds the logs, as they are just the powers to which 10 is taken. The resulting log answer is a power of 10. To transform it back to a normal number one can use the $[10^x]$ function on a calculator or 'antilog' tables if the answer is not obvious, as it is above.

Although you can obviously do the calculations more quickly by using the relevant function keys on a calculator, the following examples illustrate how logarithms can solve some multiplication, division and power evaluation problems so that you can see how they work. You will then be able to understand how logarithms can be applied to some problems encountered in economics.

Example 2.50

Evaluate $4,632.71 \times 251.07$ using logs.

Solution
Using the [LOG] function key on a calculator

$$\log 4,632.71 = 3.6658351$$
$$\log 251.07 = 2.3997948$$

Thus

$$4,632.71 \times 251.07 = 10^{3.6658351} \times 10^{2.3997948}$$
$$= 10^{6.0656299} = 1,163,134.5$$

using the $[10^x]$ function key.

The principle is therefore to put all numbers to be multiplied together in log form, add the logs, and then evaluate.

To divide, one index is subtracted from the other, e.g.

$$10^{2.5} \div 10^{1.5} = 10^{2.5-1.5} = 10^1 = 10$$

and so logs are subtracted.

Example 2.51

Evaluate $56,200 \div 3,484$ using logs.

Solution
From log tables

$$\log 56,200 = 4.7497363$$
$$\log 3,484 = 3.5420781$$

To divide, we subtract the log of the denominator since

$$56{,}200 \div 3{,}484 = 10^{4.7497363} \div 10^{3.5420781}$$

$$= 10^{4.7497363 - 3.5420781}$$

$$= 10^{1.2076582}$$

$$= 1.6130884$$

Note that when you use the [LOG] function key on a calculator to obtain the logs of numbers less than 1 you get a negative sign, e.g.

$$\log 0.31 = -0.5086383$$

Logarithms can also be used to work out powers and roots of numbers.

Example 2.52

Calculate $1{,}242.67^6$ using logs.

Solution

$$\log 1{,}242.67 = 3.0943558$$

This means

$$1{,}242.67 = 10^{3.0943558}$$

If this is taken to the power of 6, it means that this index of 10 is multiplied 6 times. Therefore

$$\log 1{,}242.67^6 = 6 \log 1{,}242.67 = 18.566135$$

Using the $[10^x]$ function to evaluate this number gives

$$3.6824 \times 10^{18} = 3{,}682{,}400{,}000{,}000{,}000{,}000$$

Example 2.53

Use logs to find $\sqrt[8]{226.34}$.

Solution

Log 226.34 must be divided by 8 to find the log of the number which when multiplied by itself 8 times gives 226.34, i.e. the eighth root. Thus

$$\log 226.34 = 2.3547613$$

$$\tfrac{1}{8} \log 226.34 = 0.2943452$$

Therefore $\sqrt[8]{(226.34)} = 10^{0.2943452} = 1.9694509.$

To summarize, the rules for using logs are as follows.

Multiplication:	add logs
Division:	subtract logs
Powers:	multiply log by power
Roots:	divide log by root

The answer is then evaluated by finding 10^x where x is the resulting value of the log.

Having learned how to use logarithms to do some awkward calculations which you could have almost certainly have done more quickly on a calculator, let us now briefly outline some of their economic applications. It can help in the estimation of the parameters of non-linear functions if they are specified in logarithmic format. This application is explained further in Section 4.9. Logarithms can also be used to help solve equations involving unknown exponent values.

Example 2.54

If $460(1.08)^n = 925$, what is n?

Solution

$$460(1.08)^n = 925$$
$$(1.08)^n = 2.0108696$$

Putting in log form

$$n \log 1.08 = \log 2.0108696$$
$$n = \frac{\log 2.0108696}{\log 1.08}$$
$$= \frac{0.3033839}{0.0334238}$$
$$= 9.0768945$$

We shall return to this type of problem in Chapter 7 when we consider for how long a sum of money needs to be invested at any given rate of interest to accumulate to a specified sum.

Although logarithms to the base 10 are perhaps the easiest ones to use, logarithms can be based on any number. In Chapter 14 the use of logarithms to the base $e = 2.7183$, known as natural logarithms, is explained (and also why such an odd base is used).

Test Yourself, Exercise 2.10

Use logs to answer the following.

1. $424 \times 638.724 =$
2. $6,434 \div 29.12 =$
3. $22.43^7 =$
4. $9.612^{8.34} =$
5. $\sqrt[36]{5,200} =$
6. $14^{3.2} \times 6.2^4 \times 81^{0.2} =$
7. If $(1.06)^n = 235$ what is n?
8. If $825(1.22)^n = 1,972$ what is n?
9. If $4,350(1.14)^n = 8,523$ what is n?

3 Introduction to algebra

Learning objectives

After completing this chapter students should be able to:

- Construct algebraic expressions for economic concepts involving unknown values.
- Simplify and reformulate basic algebraic expressions.
- Solve single linear equations with one unknown variable.
- Use the summation sign Σ.
- Perform basic mathematical operations on algebraic expressions that involve inequality signs.

3.1 Representation

Algebra is basically a system of shorthand. Symbols are used to represent concepts and variables that are capable of taking different values.

For example, suppose that a biscuit manufacturer uses the following ingredients for each packet of biscuits produced: 0.2 kg of flour, 0.05 kg of sugar and 0.1 kg of butter. One way that we could specify the total amount of flour used is: '0.2 kg times the number of packets of biscuits produced'. However, it is much simpler if we let the letter q represent the number of packets of biscuits produced. The amount of flour required in kilograms will then be 0.2 times q, which we write as $0.2q$.

Thus we can also say

amount of sugar required $= 0.05q$ kilograms

amount of butter required $= 0.1q$ kilograms

Sometimes an algebraic expression will have several terms in it with different algebraic symbols representing the unknown quantities of different variables.

Consider the total expenditure on inputs by the firm in the example above. Let the price (in £) per kilogram of flour be denoted by the letter a. The total cost of the amount of flour the firm uses will therefore be $0.2q$ times a, written as $0.2qa$.

If the price per kilogram of sugar is denoted by the letter b and the price per kilogram of butter is c then the total expenditure (in £) on inputs for biscuit production will be

$$0.2qa + 0.05qb + 0.1qc$$

When two algebraic symbols are multiplied together it does not matter in which order they are written, e.g. $xy = yx$. This of course, is, the same rule that applies when multiplying numbers. For example:

$$5 \times 7 = 7 \times 5$$

Any operation that can be carried out with numbers (such as division or deriving the square root) can be carried out with algebraic symbols. The difference is that the answer will also be in terms of algebraic symbols rather than numbers. An algebraic expression cannot be evaluated until values have been given to the variables that the symbols represent (see Section 3.2).

For example, an expression for the length of fencing (in metres) needed to enclose a square plot of land of as yet unknown size can be constructed as follows:

The length of a side will be $\sqrt{A}$ for a square that has area A square metres.
All squares have four sides.
Therefore the length of fencing $= 4 \times$ (length of one side) $= 4\sqrt{A}$.

Without information on the value of A we cannot say any more. Once the value of A is specified then we can simply work out the value of the expression using basic arithmetic.

For example, if the area is 100 square metres, then we just substitute 100 for A and so

$$\text{length of fencing} = 4\sqrt{A} = 4\sqrt{(100)} = 4 \times 10 = 40 \text{ metres}$$

One of the uses of writing an expression in an algebraic form is that it is not necessary to work out a solution for every different value of the unknown variable that one is faced with. The different values can just be substituted into the algebraic expression. In this section we start with some fairly simple expressions but later, as more complex relationships are dealt with, the usefulness of algebraic representation will become more obvious.

Example 3.1

You are tiling a bathroom with 10 cm square tiles. The number of square metres to be tiled is as yet unknown and is represented by q. Because you may break some tiles and will have to cut some to fit around corners etc. you work to the rule of thumb that you should buy enough tiles to cover the specified area plus 10%. Derive an expression for the number of tiles to be bought in terms of q.

Solution

One hundred 10 cm square tiles will cover 1 square metre and 110% written as a decimal is 1.1. Therefore the number of tiles required is

$$100q \times 1.1 = 110q$$

Test Yourself, Exercise 3.1

1. An engineering firm makes metal components. Each component requires 0.01 tonnes of steel, 0.5 hours of labour plus 0.5 hours of machine time. Let the number of components produced be denoted by x. Derive algebraic expressions for:

    (a) the amount of steel required;
    (b) the amount of labour required;
    (c) the amount of machine time required.

2. If the price per tonne of steel is given by r, the price per hour of labour is given by w and the price per hour of machine time is given by m, then derive an expression for the total production costs of the firm in question 1 above.
3. The petrol consumption of your car is 12 miles per litre. Let x be the distance you travel in miles and p the price per litre of petrol in pence. Write expressions for (a) the amount of petrol you use and (b) your expenditure on petrol.
4. Suppose that you are cooking a dinner for a number of people. You only know how to cook one dish, and this requires you to buy 0.1 kg of meat plus 0.3 kg of potatoes for each person. (Assume you already have a plentiful supply of any other ingredients.) Define your own algebraic symbols for relevant unknown quantities and then write expressions for:

    (a) the amount of meat you need to buy;
    (b) the amount of potatoes you need to buy;
    (c) your total shopping bill.

5. You are cooking again! This time it's a turkey. The cookery book recommends a cooking time of 30 minutes for every kilogramme weight of the turkey plus another quarter of an hour. Write an expression for the total cooking time (in hours) for your turkey in terms of its weight.
6. Make up your own algebraic expression for the total profit of a firm in terms of the amount of output sold, the price of its product and the average cost of production per unit.
7. Someone is booking a meal in a restaurant for a group of people. They are told that there is a set menu that costs £9.50 per adult and £5 per child, and there is also a fixed charge of £1 per head for each meal served. Derive an expression for the total cost of the meal, in pounds, if there are x adults and y children.
8. A firm produces a good which it can sell any amount of at £12 per unit. Its costs are a fixed outlay of £6,000 plus £9 in variable costs for each unit produced. Write an expression for the firm's profit in terms of the number of units produced.

3.2   Evaluation

An expression can be evaluated when the variables represented by algebraic symbols are given specific numerical values.

Example 3.2

Evaluate the expression $6.5x$ when $x = 8$.

Solution

$$6.5x = 6.5(8) = 52$$

Example 3.3

A firm's total costs are given by the expression

$$0.2qa + 0.05qb + 0.1qc$$

where q is output and a, b and c are the per unit costs (in £) of the three different inputs used. Evaluate these costs if $q = 1,000$, $a = 0.6$, $b = 1.3$ and $c = 2.1$.

Solution

$$
\begin{aligned}
0.2qa + 0.05qb + 0.1qc &= 0.2(1,000 \times 0.6) + 0.05(1,000 \times 1.3) + 0.1(1,000 \times 2.1) \\
&= 0.2(600) + 0.05(1,300) + 0.1(2,100) \\
&= 120 + 65 + 210 = 395
\end{aligned}
$$

Therefore the total cost is £395.

Example 3.4

Evaluate the expression $(3^x + 4)y$ when $x = 2$ and $y = 6$.

Solution

$$(3^x + 4)y = (3^2 + 4)6 = (9 + 4)6 = 13 \times 6 = 78$$

Example 3.5

A Bureau de Change will sell euros at an exchange rate of 1.62 euros to the pound and charges a flat rate commission of £2 on all transactions.

 (i) Write an expression for the number of euros that can be bought for £x (any given quantity of sterling), and
(ii) evaluate it for $x = 250$.

Solution

(i) Number of euros bought for £x = $1.62(x - 2)$.

(ii) £250 will therefore buy

$$1.62(250 - 2) = 1.62(248) = €401.76$$

Test Yourself, Exercise 3.2

1. Evaluate the expression $2x^3 + 4x$ when $x = 6$.
2. Evaluate the expression $(6x + 2y)y^2$ when $x = 4.5$ and $y = 1.6$.
3. When the UK government privatized the Water Authorities in 1989 it decided that annual percentage price increases for water would be limited to the rate of inflation plus z, where z was a figure to be determined by the government. Write an algebraic expression for the maximum annual percentage price increase for water and evaluate it for an inflation rate of 6% and a z factor of 3.
4. Make up your own values for the unknown variables in the expressions you have written for Test Yourself, Exercise 3.1 above and then evaluate.
5. Evaluate the expression $1.02^x + x^{-3.2}$ when $x = 2.8$.
6. A firm's average production costs (AC) are given by the expression

$$AC = 450q^{-1} + 0.2q^{1.5}$$

where q is output. What will AC be when output is 175?
7. A firm's profit (in £) is given by the expression $7.5q - 1650$. What profit will it make when q is 500?
8. If income tax is levied at a rate of 22% on annual income over £5,400 then:

    (a) write an expression for net monthly salary in terms of gross monthly salary (assumed to be greater than £450), and
    (b) evaluate it if gross monthly salary is £2,650.

3.3 Simplification: addition and subtraction

Simplifying an expression means rearranging the terms in it so that the expression becomes easier to work with. Before setting out the different rules for simplification let us work through an example.

Example 3.6

A businesswoman driving her own car on her employer's business gets paid a set fee per mile travelled for travelling expenses. During one week she records one journey of 234 miles, one of 166 miles and one of 90 miles. Derive an expression for total travelling expenses.

Solution

If the rate per mile is denoted by the letter M then her expenses will be $234M$ for the first journey and $166M$ and $90M$ for the second and third journeys respectively. Total travelling expenses for the week will thus be

$$234M + 166M + 90M$$

We could, instead, simply add up the total number of miles travelled during the week and then multiply by the rate per mile. This would give

$$(234 + 166 + 90) \times \text{rate per mile} = 490 \times \text{rate per mile} = 490M$$

It is therefore obvious that, as both methods should give the same answer, then

$$234M + 166M + 90M = 490M$$

In other words, in an expression with different terms all in the same format of

$$(\text{a number}) \times M$$

all the terms can be added together.

The general rule is that like terms can be added or subtracted to simplify an expression. 'Like terms' have the same algebraic symbol or symbols, usually multiplied by a number.

Example 3.7

$$3x + 14x + 7x = 24x$$

Example 3.8

$$45A - 32A = 13A$$

It is important to note that only terms that have exactly the same algebraic notation can be added or subtracted in this way. For example, the terms x, y^2 and xy are all different and cannot be added together or subtracted from each other.

Example 3.9

Simplify the expression $5x^2 + 6xy - 32x + 3yx - x^2 + 4x$.

Solution

Adding/subtracting all the terms in x^2 gives $4x^2$.
Adding/subtracting all the terms in x gives $-28x$.
Adding/subtracting all the terms in xy gives $9xy$.
(Note that the terms in xy and yx can be added together since $xy = yx$.)

Putting all these terms together gives the simplified expression

$$4x^2 - 28x + 9xy$$

In fact all the basic rules of arithmetic (as set out in Chapter 2) apply when algebraic symbols are used instead of actual numbers. The difference is that the simplified expression will still be in a format of algebraic terms.

The example below illustrates the rule that if there is a negative sign in front of a set of brackets then the positive and negative signs of the terms within the brackets are reversed if the brackets are removed.

Example 3.10

Simplify the expression

$$16q + 33q - 2q - (15q - 6q)$$

Solution

Removing brackets, the above expression becomes

$$16q + 33q - 2q - 15q + 6q = 38q$$

Test Yourself, Exercise 3.3

1. Simplify the expression $6x - (6 - 24x) + 10$.
2. Simplify the expression $4xy + (24x - 13y) - 12 + 3yx - 5y$.
3. A firm produces two goods, X and Y, which it sells at prices per unit of £26 and £22 respectively. Good X requires an initial outlay of £400 and then an expenditure of £16 on labour and £4 on raw materials for each unit produced. Good Y requires a fixed outlay of £250 plus £14 labour and £3 of raw material for each unit. If the quantities produced of X and Y are x and y, respectively, write an expression in terms of x and y for the firm's total profit and then simplify it.
4. A worker earns £6 per hour for the first 40 hours a week he works and £9 per hour for any extra hours. Assuming that he works at least 40 hours, write an expression for his gross weekly wage in terms of H, the total hours worked per week, and then simplify it.

3.4   Simplification: multiplication

When a set of brackets containing different terms is multiplied by a symbol or a number it may be possible to simplify an expression by multiplying out, i.e. multiplying each term within the brackets by the term outside. In some circumstances, though, it may be preferable to leave brackets in the expression if it makes it clearer to work with.

Example 3.11

$$x(4 + x) = 4x + x^2$$

Example 3.12

$$5(7x^2 - x) - 3(3x^2 + 6x) = 35x^2 - 5x - 9x^2 - 18x$$
$$= 26x^2 - 23x$$

Example 3.13

$$6y(8 + 3x) - 2xy + 12y = 48y + 18xy - 2xy + 12y$$
$$= 60y + 16xy$$

Example 3.14

The basic hourly rate for a weekly paid worker is £8 and any hours above 40 are paid at £12. Tax is paid at a rate of 25% on any earnings above £80 a week. Assuming hours worked per week (H) exceed 40, write an expression for net weekly wage in terms of H and then simplify it.

Solution

$$\text{gross wage} = 40 \times 8 + (H - 40)12$$
$$= 320 + 12H - 480$$
$$= 12H - 160$$
$$\text{net wage} = 0.75(\text{gross wage} - 80) + 80$$
$$= 0.75(12H - 160 - 80) + 80$$
$$= 9H - 120 - 60 + 80$$
$$= 9H - 100$$

If you are not sure whether the expression you have derived is correct, you can try to check it by substituting numerical values for unknown variables. In the above example, if 50 hours per week were worked, then

$$\text{gross pay} = (40 \text{ hours @ £8}) + (10 \text{ hours @ £12})$$
$$= £320 + £120$$
$$= £440$$

$$\text{tax payable} = (£440 - £80)0.25 = (£360)0.25 = £90$$

Therefore

$$\text{net pay} = £440 - £90 = £350$$

Using the expression derived in Example 3.14, if $H = 50$ then

$$\text{net pay} = 9H - 100 = 9(50) - 100 = 450 - 100 = £350$$

This checks out with the answer above and so we know our expression works.

It is rather more complicated to multiply pairs of brackets together. One method that can be used is rather like the long multiplication that you probably learned at school, but instead of keeping all units, tens, hundreds etc. in the same column it is the same algebraic terms that are kept in the same column during the multiplying process so that they can be added together.

Example 3.15

Simplify $(6 + 2x)(4 - 2x)$.

Solution

Writing this as a long multiplication problem:

$$
\begin{array}{lr}
 & 6 + 2x \quad \times \\
 & 4 - 2x \\
\text{Multiplying } (6 + 2x) \text{ by } -2x & -12x - 4x^2 \\
\text{Multiplying } (6 + 2x) \text{ by } 4 & 24 + 8x \\
\text{Adding together gives the answer} & 24 - 4x - 4x^2 \\
\end{array}
$$

One does not have to use the long multiplication format for multiplying out sets of brackets. The basic principle is that each term in one set of brackets must be multiplied by each term in the other set. Like terms can then be collected together to simplify the resulting expression.

Example 3.16

Simplify $(3x + 4y)(5x - 2y)$.

Solution

Multiplying the terms in the second set of brackets by $3x$ gives:

$$15x^2 - 6xy \tag{1}$$

Multiplying the terms in the second set of brackets by $4y$ gives:

$$20xy - 8y^2 \tag{2}$$

Therefore, adding (1) and (2) the whole expression is

$$15x^2 - 6xy + 20xy - 8y^2 = 15x^2 + 14xy - 8y^2$$

Example 3.17

Simplify $(x + y)^2$.

Solution

$$(x + y)^2 = (x + y)(x + y) = x^2 + xy + yx + y^2 = x^2 + 2xy + y^2$$

The above answer can be checked by referring to Figure 3.1. The area enclosed in the square with sides of length $x + y$ can be calculated by squaring the lengths of the sides, i.e. finding $(x + y)^2$. One can also see that this square is made up of the four rectangles A, B, C and D whose areas are x^2, xy, xy and y^2 respectively – in other words, $x^2 + 2xy + y^2$, which is the answer obtained above.

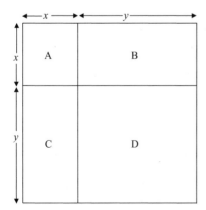

Figure 3.1

Example 3.18

Simplify $(6 - 5x)(10 - 2x + 3y)$.

Solution

Multiplying out gives

$$60 - 12x + 18y - 50x + 10x^2 - 15xy = 60 - 62x + 18y + 10x^2 - 15xy$$

Some expressions may be best left with the brackets still in.

Example 3.19

If a sum of £x is invested at an interest rate of $r\%$ write an expression for the value of the investment at the end of 2 years.

Solution

After 1 year the investment's value (in £) is $x\left(1 + \dfrac{r}{100}\right)$

After 2 years the investment's value (in £) is $x\left(1 + \dfrac{r}{100}\right)^2$

One could multiply out but in this particular case the expression is probably clearer, and also easier to evaluate, if the brackets are left in. The next section explains how in some cases, some expressions may be 'simplified' by reformatted into two expressions in brackets multiplied together. This is called 'factorization'.

Test Yourself, Exercise 3.4

Simplify the following expressions:
1. $6x(x - 4)$
2. $(x + 3)^2 - 2x$
3. $(2x + y)(x + 3)$
4. $(6x + 2y)(7x - 8y) + 4y + 2y$
5. $(4x - y + 7)(2y - 3) + (9x - 3y)(5 + 6y)$
6. $(12 - x + 3y + 4z)(10 + x + 2y)$
7. A good costs a basic £180 a unit but if an order is made for more than 10 units this price is reduced by a discount of £2 for every one unit increase in the size of an order (up to a maximum of 60 units purchased), i.e. if the order size is 11, price is £178; if it is 12, price is £176 etc. Write an expression for the total cost of an order in terms of order size and simplify it. Assume order size is between 10 and 60 units.

8. A holiday excursion costs £8 per person for transport plus £5 per adult and £3 per child for meals. Write an expression for the total cost of an excursion for x adults and y children and simplify it.

9. A firm is building a car park for its employees. Assume that a car park to accommodate x cars must have a length (in metres) of $4x + 10$ and a width of $2x + 10$. If 24 square metres will be specifically allocated for visitors' cars, write an expression for the amount of space available for the cars of the workforce in terms of x if x is the planned capacity of the car park.

10. A firm buys a raw material that costs £220 a tonne for the first 40 tonnes, £180 a tonne for the next 40 tonnes and £150 for any further quantities. Write an expression for the firm's total expenditure on this input in terms of the total amount used (which can be assumed to be greater than 80 tonnes), and simplify.

3.5 Simplification: factorizing

For some purposes (see, e.g. Section 3.6 below and Chapter 6 on quadratic equations) it may be helpful if an algebraic expression can be simplified into a format of two sets of brackets multiplied together. For example

$$x^2 + 4x + 4 = (x + 2)(x + 2)$$

This is rather like the arithmetical process of factorizing a number, which means finding all the prime numbers which when multiplied together equal that number, e.g.

$$126 = 2 \times 3 \times 3 \times 7$$

If an expression has only one unknown variable, x, and it is possible to factorize it into two sets of brackets that do not contain terms in x to a power other than 1, the expression must be in the form

$$ax^2 + bx + c$$

However, not all expressions in this form can be factorized into sets of brackets that only involve integers, i.e. whole numbers. There are no set rules for working out if and how an expression may be factorized. However, if the term in x^2 does not have a number in front of it (i.e. $a = 1$) then the expression can be factorized if there are two numbers which

(i)   give c when multiplied together, and
(ii) give b when added together.

Example 3.20

Attempt to factorize the expression $x^2 + 6x + 9$.

Solution

In this example $a = 1$, $b = 6$ and $c = 9$.
Since $3 \times 3 = 9$ and $3 + 3 = 6$, it can be factorized, as follows

$$x^2 + 6x + 9 = (x + 3)(x + 3)$$

This can be checked as

$$
\begin{array}{r}
x + 3 \quad \times \\
x + 3 \\
\hline
3x + 9 \\
x^2 + 3x \\
\hline
x^2 + 6x + 9 \\
\hline
\end{array}
$$

Example 3.21

Attempt to factorize the expression $x^2 - 2x - 80$.

Solution

Since $(-10) \times 8 = -80$ and $(-10) + 8 = -2$ then the expression can be factorized and

$$x^2 - 2x - 80 = (x - 10)(x + 8)$$

Check this answer yourself by multiplying out.

Example 3.22

Attempt to factorize the expression $x^2 + 3x + 11$.

Solution

There are no two numbers which when multiplied together give 11 and when added together give 3. Therefore this expression cannot be factorized.

It is sometimes possible to simplify an expression before factorizing if all the terms are divisible by the same number.

Example 3.23

Attempt to factorize the expression $2x^2 - 10x + 12$.

Solution

$$2x^2 - 10x + 12 = 2(x^2 - 5x + 6)$$

The term in brackets can be simplified as

$$x^2 - 5x + 6 = (x - 3)(x - 2)$$

Therefore

$$2x^2 - 10x + 12 = 2(x - 3)(x - 2)$$

In expressions in the format $ax^2 + bx + c$ where a is not equal to 1, then one still has to find two numbers which multiply together to give c. However, one also has to find two numbers for the coefficients of the two terms in x within the two sets of brackets that when multiplied together equal a, and allow the coefficient b to be derived when multiplying out.

Example 3.24

Attempt to factorize the expression $30x^2 + 52x + 14$.

Solution

If we use the results that $6 \times 5 = 30$ and $2 \times 7 = 14$ we can try multiplying

$$
\begin{array}{r}
5x + 7 \quad \times \\
6x + 2 \\
\hline
10x + 14 \\
30x^2 + 42x \\
\hline
30x^2 + 52x + 14
\end{array}
$$

This gives $\quad 30x^2 + 52x + 14$

Thus

$$30x^2 + 52x + 14 = (5x + 7)(6x + 2)$$

Similar rules apply when one attempts to factorize an expression with two unknown variables, x and y. This may be in the format

$$ax^2 + bxy + cy^2$$

where a, b and c are specified parameters.

Example 3.25

Attempt to factorize the expression $x^2 - y^2$.

Solution

In this example $a = 1, b = 0$ and $c = -1$. The two numbers -1 and 1 give -1 when multiplied together and 0 when added. Thus

$$x^2 - y^2 = (x - y)(x + y)$$

To check this, multiply out:

$$(x - y)(x - y) = x^2 - xy + y^2 - yx = x^2 - y^2$$

Example 3.26

Attempt to factorize the expression $3x^2 + 8x + 23$.

Solution

As 23 is a positive prime number, the only pairs of positive integers that could possibly be multiplied together to give 23 are 1 and 23. Thus, whatever permutations of combinations with terms in x that we try, the term in x when brackets are multiplied out will be at least $24x$, e.g. $(3x + 23)(x + 1) = 3x^2 + 26x + 23$, whereas the given expression contains the term $8x$. It is therefore **not** possible to factorize this expression.

Unfortunately, it is not always so obvious whether or not an expression can be factorized.

Example 3.27

Attempt to factorize the expression $3x^2 + 24 + 16$.

Solution

Although the numbers look promising, if you try various permutations you will find that this expression does not factorize.

There is no easy way of factorizing expressions and it is just a matter of trial and error. Do not despair though! As you will see later on, factorizing may help you to use short-cut methods of solving certain problems. If you spend ages trying to factorize an expression then this will defeat the object of using the short-cut method. If it is not obvious how an expression can be factorized after a few minutes of thought and experimentation with some potential possible solutions then it is usually more efficient to forget factorization and use some other method of solving the problem. We shall return to this topic in Chapter 6.

Test Yourself, Exercise 3.5

Attempt to factorize the following expressions:

1. $x^2 + 8x + 16$
2. $x^2 - 6xy + 9y^2$
3. $x^2 + 7x + 22$
4. $8x^2 - 10x + 33$
5. Make up your own expression in the format $ax^2 + bx + c$ and attempt to factorize it. Check your answer by multiplying out.

3.6 Simplification: division

To divide an algebraic expression by a number one divides every term in the expression by the number, cancelling where appropriate.

Example 3.28

$$\frac{15x^2 + 2xy + 90}{3} = 5x^2 + \frac{2}{3}xy + 30$$

To divide by an unknown variable the same rule is used although, of course, where the numerator of a fraction does not contain that variable it cannot be simplified any further.

Example 3.29

$$\frac{2x^2}{x} = 2x$$

Example 3.30

$$\frac{4x^3 - 2x^2 + 10x}{x} = \frac{x(4x^2 - 2x + 10)}{x}$$
$$= 4x^2 - 2x + 10$$

Example 3.31

$$\frac{16x + 120}{x} = 16 + \frac{120}{x}$$

Example 3.32

A firm's total costs are $25x + 2x^2$, where x is output. Write an expression for average cost.

Solution

Average cost is total cost divided by output. Therefore

$$AC = \frac{25x + 2x^2}{x} = 25 + 2x$$

If one expression is divided by another expression with more than one term in it then terms can only be cancelled top and bottom if the numerator and denominator are both multiples of the same factor.

Example 3.33

$$\frac{x^2 + 2x}{x + 2} = \frac{x(x + 2)}{x + 2} = x$$

Example 3.34

$$\frac{x^2 + 5x + 6}{x + 3} = \frac{(x + 3)(x + 2)}{x + 3} = x + 2$$

Example 3.35

$$\frac{x^2 + 5x + 6}{x^2 + x - 2} = \frac{(x + 2)(x + 3)}{(x + 2)(x - 1)} = \frac{x + 3}{x - 1}$$

Test Yourself, Exercise 3.6

1. Simplify

    $$\frac{6x^2 + 14x - 40}{2x}$$

2. Simplify

    $$\frac{x^2 + 12x + 27}{x + 3}$$

3. Simplify

$$\frac{8xy + 2x^2 + 24x}{2x}$$

4. A firm has to pay fixed costs of £200 and then £16 labour plus £5 raw materials for each unit produced of good X. Write an expression for average cost and simplify.

5. A firm sells 40% of its output at £200 a unit, 30% at £180 and 30% at £150. Write an expression for the average revenue received on each unit sold and then simplify it.

6. You have all come across this sort of party trick: Think of a number. Add 3. Double it. Add 4. Take away the number you first thought of. Take away 3. Take away the number you thought of again. Add 2. Your answer is 9. Show how this answer can be derived by algebraic simplification by letting x equal the number first thought of.

7. Make up your own 'think of a number' trick, writing down the different steps in the form of an algebraic expression that checks out the answer.

3.7   Solving simple equations

We have seen that evaluating an expression means calculating its value when one is given specific values for unknown variables. This section explains how it is possible to work backwards to discover the value of an unknown variable when the total value of the expression is given.

When an algebraic expression is known to equal a number, or another algebraic expression, we can write an equation, i.e. the two concepts are written on either side of an equality sign. For example

$$45 = 24 + 3x$$

In this chapter we have already written some equations when simplifying algebraic expressions. However, the ones we have come across so far have usually not been in a format where the value of the unknown variables can be worked out. Take, for example, the simplification exercise

$$3x + 14x - 5x = 12x$$

The expressions on either side of the equality sign are equal, but x cannot be calculated from the information given.

Some equations are what are known as '*identities*', which means that they must always be true. For example, a firm's total costs (TC) can be split into the two components total fixed costs (TFC) and total variable costs (TVC). It must therefore always be the case that

$$TC = TFC + TVC$$

Identities are sometimes written with the three bar equality sign '$\equiv$' instead of '$=$', but usually only when it is necessary to distinguish them from other forms of equations, such as functions.

A *function* is a relationship between two or more variables such that a unique value of one variable is determined by the values taken by the other variables in the function. (Functions are explained more fully in Chapter 4.) For example, statistical analysis may show that a demand function takes the form

$$q = 450 - 3p$$

where p is price and q is quantity demanded. Thus the expected quantity demanded can be predicted for any given value of p, e.g. if $p = 60$ then

$$q = 450 - 3(60) = 450 - 180 = 270$$

In this section we shall not distinguish between equations that are identities and those that relate to specific values of functions, since the method of solution is the same for both. We shall also mainly confine the analysis to linear equations with one unknown variable whose value can be deduced from the information given. A *linear equation* is one where the unknown variable does not take any powers other than 1, e.g. there may be terms in x but not x^2, x^{-1} etc.

Before setting out the formal rules for solving single linear equations let us work through some simple examples.

Example 3.36

You go into a foreign exchange bureau to buy US dollars for your holiday. You exchange £200 and receive $343. When you get home you discover that you have lost your receipt. How can you find out the exchange rate used for your money if you know that the bureau charges a fixed £4 fee on all transactions?

Solution

After allowing for the fixed fee, the amount actually exchanged into dollars will be

$$£200 - £4 = £196$$

Let x be the exchange rate of pounds into dollars. Therefore

$$343 = 196x$$
$$\frac{343}{196} = x$$
$$1.75 = x$$

Thus the exchange rate is $1.75 to the pound.

This example illustrates the fundamental principle that one can divide both sides of an equation by the same number.

Example 3.37

If $62 = 34 + 4x$ what is x?

Solution

Subtracting 34 from both sides gives

$$28 = 4x$$

then dividing both sides by 4 gives the solution

$$7 = x$$

This example illustrates the principle that one can subtract the same amount from both sides of an equation.

The *basic principles for solving equations* are that all the terms in the unknown variable have to be brought together by themselves on one side of the equation. In order to do this one can add, subtract, multiply or divide both sides of an equation by the same number or algebraic term. One can also perform other arithmetical operations, such as finding the square root of both sides of an equality sign.

Once the equation is in the form

$$ax = b$$

where a and b are numbers, then x can be found by dividing b by a.

Example 3.38

Solve for x if $16x - 4 = 68 + 7x$.

Solution

Subtracting $7x$ from both sides

$$9x - 4 = 68$$

Adding 4 to both sides

$$9x = 72$$

Dividing both sides by 9 gives the solution

$$x = 8$$

Example 3.39

Solve for x if

$$4 = \frac{96}{x}$$

Solution
Multiplying both sides by x

$$4x = 96$$

Dividing both sides by 4 gives the solution

$$x = 24$$

Example 3.40

Solve for x if $6x^2 + 12 = 162$.

Solution
Subtracting 12 from both sides

$$6x^2 = 150$$

Dividing through by 6

$$x^2 = 25$$

Taking square roots gives the solution

$$x = 5 \text{ or } -5$$

Example 3.41

A firm has to pay fixed costs of £1,500 plus another £60 for each unit produced. How much can it produce for a budget of £4,800?

Solution

$$\text{budget} = \text{total expenditure on production}$$

Therefore if x is output level

$$4,800 = 1,500 + 60x$$

Subtracting 1,500 from both sides

$$3,300 = 60x$$

Dividing by 60 gives the solution

$$55 = x$$

Thus the firm can produce 55 units for a budget of £4,800.

Example 3.42

You sell 500 shares in a company via a stockbroker who charges a flat £20 commission rate on all transactions under £1,000. Your bank account is credited with £692 from the sale of the shares. What price (in pence) were your shares sold at?

Solution

Let price per share be x. Therefore, working in pence,

$$69,200 = 500x - 2,000$$

Adding 2,000 to both sides

$$71,200 = 500x$$

Dividing both sides by 500 gives the solution

$$142.4 = x$$

Thus the share price is 142.4p.

Test Yourself, Exercise 3.7

1. Solve for x when $16x = 2x + 56$.
2. Solve for x when

$$14 = \frac{6 + 4x}{5x}$$

3. Solve for x when $45 = 24 + 3x$.
4. Solve for x if $5x^2 + 20 = 1,000$.
5. If $q = 560 - 3p$ solve for p when $q = 314$.
6. You get paid travelling expenses according to the distance you drive in your car plus a weekly sum of £21. You put in a claim for 420 miles travelled and receive an expenses payment of £105. What is the payment rate per mile?
7. In one module that you are studying, the overall module mark is calculated on the basis of a $30:70$ weighting between coursework and examination marks. If you have scored 57% for coursework, what examination mark do you need to get to achieve an overall mark of 40%?

8. You sell 900 shares via your broker who charges a flat rate of commission of £20 on all transactions of less than £1,000. Your bank account is credited with £340 from the share sale. What price were your shares sold at?

9. Your net monthly salary is £1,950. You know that National Insurance and pension contributions take 15% of your gross salary and that income tax is levied at a rate of 25% on gross annual earnings above a £5,400 exemption limit. What is your gross monthly salary?

10. You have 64 square paving stones and wish to lay them to form a square patio in your garden. If each paving stone is 0.5 metres square, what will the length of a side of your patio be?

11. A firm faces the marginal revenue schedule MR $= 80 - 2q$ and the marginal cost schedule MC $= 15 + 0.5q$ where q is quantity produced. You know that a firm maximizes profit when MC $=$ MR. What will the profit-maximizing output be?

3.8   The summation sign $\sum$

The summation sign $\sum$ can be used in certain circumstances as a shorthand means of expressing the sum of a number of different terms added together. (Σ is a Greek letter, pronounced 'sigma'.) There are two ways in which it can be used.

The first is when one variable increases its value by 1 in each successive term, as the example below illustrates.

Example 3.43

A new firm sells 30 units in the first week of business. Sales then increase at the rate of 30 units per week. If it continues in business for 5 weeks, its total cumulative sales will therefore be

$$(30 \times 1) + (30 \times 2) + (30 \times 3) + (30 \times 4) + (30 \times 5)$$

You can see that the number representing the week is increased by 1 in each successive term. This is rather a cumbersome expression to work with. We can instead write

$$\text{sales revenue} = \sum_{i=1}^{5} 30i$$

This means that one is summing all the terms $30i$ for values of i from 1 to 5.

If the number of weeks of business n was not known we could instead write

$$\text{sales revenue} = \sum_{i=1}^{n} 30i$$

To evaluate an expression containing a summation sign, one may still have to calculate the value of each term separately and then add up. However, spreadsheets can be used to do tedious calculations and in some cases short-cut formulae may be used (see Chapter 7).

Example 3.44

Evaluate

$$\sum_{i=3}^{n} (20 + 3i) \quad \text{for } n = 6$$

Solution

Note that in this example i starts at 3. Thus

$$\sum_{i=3}^{6} (20 + 3i) = (20 + 9) + (20 + 12) + (20 + 15) + (20 + 18)$$

$$= 29 + 32 + 35 + 38$$

$$= 134$$

The second way in which the summation sign can be used requires a set of data where observations are specified in numerical order.

Example 3.45

Assume that a researcher finds a random group of twelve students and observes their weight and height as shown in Table 3.1.

If we let H_i represent the height and W_i represent the weight of student i, then the total weight of the first six students can be specified as $\sum_{i=1}^{6} W_i$.

In this method i refers to the number of the observation and so the value of i is *not* incorporated into the actual calculations.

Staying with the same example, the average weight of the first n students could be specified as

$$\frac{1}{n} \sum_{i=1}^{n} W_i$$

When no superscript or subscripts are shown with the $\sum$ sign it usually means that all possible values are summed. For example, a price index is constructed by working out how much a weighted average of prices rises over time. One method of measuring how much, on

Table 3.1

Student no.	1	2	3	4	5	6	7	8	9	10	11	12
Height (cm)	178	175	170	166	168	185	169	189	175	181	177	180
Weight (kg)	72	68	58	52	55	82	55	86	70	71	65	68

average, prices rise between year 0 and year 1 is to use the Laspeyre price index formula

$$\frac{\sum p_i^1 x_i}{\sum p_i^0 x_i}$$

where p_i^1 is the price of good i in year 1, p_i^0 is the price of good i in year 0 and x_i is the percentage of consumer expenditure on good i in year 0. If all goods are in the index then $\sum x_i = 100$ by definition.

Example 3.46

Given the figures in Table 3.2 for prices and expenditure proportions, calculate the rate of inflation between year 0 and year 1 and compare the price rise of food with the weighted average price rise.

Solution

Note that in this example we are just assuming one price for each category of expenditure. In reality, of course, the prices of several individual goods are included in a price index. It must be stressed that these are *prices* not measures of expenditure on these goods and services.

The weighted average price increase will be

$$\frac{\sum p_i^1 x_i}{\sum p_i^0 x_i} = \frac{1,944 + 1,666 + 1,012 + 910 + 2,160 + 781 + 1,176 + 1,500}{1,800 + 1,360 + 770 + 840 + 2,120 + 682 + 1,128 + 1,300}$$

$$= \frac{11,149}{10,000} = 1.115$$

This means that, on average, prices in year 1 are 111.5% of prices in year 0, i.e. the inflation rate is 11.5%. The price of food went up from 80 to 98, i.e. by 22.5%, which is greater than

Table 3.2

	Percentage of expenditure (x_i)	Prices, year 0 (p_i^0)	Prices, year 1 (p_i^1)
Durable goods	9	200	216
Food	17	80	98
Alcohol and tobacco	11	70	92
Footwear and clothing	7	120	130
Energy	8	265	270
Other goods	11	62	71
Rent, rates, water	12	94	98
Other services	25	52	60
	100		

Note
All prices are in £.

the inflation rate. Adjusted for inflation, the real price increase for food is thus

$$100 \left(\frac{1.225}{1.115} - 1 \right) = 100(1.099 - 1) = 9.9\%$$

Test Yourself, Exercise 3.8

1. Refer to Table 3.1 above and write an expression for the average height of the first n students observed and evaluate for $n = 6$.
2. Evaluate

$$\sum_{i=1}^{5} (4 + i)$$

3. Evaluate

$$\sum_{i=2}^{5} (2^i)$$

   (Note that i is an exponent in this question.)
4. A firm sells 6,000 tonnes of its output in its first year of operation. Sales then decrease each year by 10% of the previous year's sales figure. Write an expression for the firm's total sales over n years and evaluate for $n = 3$.
5. Observations of a firm's sales revenue (in £'000) per month are as follows:

Month	1	2	3	4	5	6	7	8	9	10	11	12
Revenue	4.5	4.2	4.6	4.4	5.0	5.3	5.2	4.9	4.7	5.4	5.3	5.8

   (a) Write an expression for average monthly sales revenue for the first n months and evaluate for $n = 4$.
   (b) Write an expression for average monthly sales revenue over the preceding 3 months for any given month n, assuming that n is not less than 4. Evaluate for $n = 10$.

6. Assume that the expenditure and price data given in Example 3.46 above all still hold except that the price of alcohol and tobacco rises to £108 in year 1. Work out the new inflation rate and the new real price increase in the price of food.

3.9 Inequality signs

As well as the equality sign ($=$), the following four inequality signs are used in algebra:

> $>$ which means 'is always greater than'
> $<$ which means 'is always less than'
> $\geq$ which means 'is greater than or equal to'
> $\leq$ which means 'is less than or equal to'

The last two are sometimes called 'weak inequality' signs.

Example 3.47

If we let the number of days in any given month be represented by N, then whatever month is chosen it must be true that

$$N > 27$$
$$N < 32$$
$$N \geq 28$$
$$N \leq 31$$

Special care has to be taken when using inequality signs if unknown variables can take negative values. For example, the inequality $2x < 3x$ only holds if $x > 0$.

If x took a negative value, then the inequality would be reversed. For example, if $x = -5$, then $2x = -10$ and $3x = -15$ and so $2x > 3x$.

When considering inequality relationships, it can be useful to work in terms of the absolute value of a variable x. This is written $|x|$ and is defined as

$$|x| = x \text{ when } x \geq 0$$
$$|x| = -x \text{ when } x < 0$$

i.e. the absolute value of a positive number is the number itself and the absolute value of a negative number is the same number but with the negative sign removed.

If an inequality is specified in terms of the absolute value of an unknown variable, then the inequality will not be reversed if the variable takes on a negative value. For example

$$|2x| < |3x| \quad \text{for all positive and negative non-zero values of } x.$$

In economic applications, the unknown variable in an algebraic expression often represents a concept (such as quantity produced or price) that cannot normally take on a negative value. In these cases, the use of inequality signs is therefore usually more straightforward than in cases where negative values are possible.

It is possible to simplify an inequality relationship by performing the same arithmetical operation on both sides of the inequality sign. However, the rules for doing this differ from those that apply when manipulating both sides of an ordinary equality sign.

One can add any number to or subtract any number from both sides of an inequality sign.

Example 3.48

If $x + 6 > y + 2$ then $x + 4 > y$

One can multiply or divide both sides of an inequality sign by a positive number,

Example 3.49

If $x < y$ then $8x < 8y$ (multiplying through by 8).

However, if one multiplies or divides by a negative number then the direction of the inequality is reversed.

Example 3.50

If $3x < 18y$ then $-x > -6y$ (dividing both sides by -3)

If both sides of an inequality sign are squared, the same inequality sign only holds if both sides are initially positive values. This is because a negative number squared becomes a positive number.

Example 3.51

If $x + 3 < y$ then $(x + 3)^2 < y^2$ if $(x + 3) \geq 0$ and $y > 0$

Example 3.52

$-6 < -4$ but $(-6)^2 > (-4)^2$ since $36 > 16$

If both sides of an inequality sign are positive and are raised to the same negative power, then the direction of the inequality will be reversed.

Example 3.53

If $x > y$ then $x^{-1} < y^{-1}$ if x and y are positive,

Example 3.54

Two leisure park owners A and B have the same weekly running costs of £8,000. The numbers of customers visiting the two parks are x and y respectively. If $x > y$, what can be said about comparative average costs per customer?

Solution

Since $x > y$

then $x^{-1} < y^{-1}$

thus $\dfrac{£8,000}{x} < \dfrac{£8,000}{y}$

and so average cost for A $<$ average cost for B.

Test Yourself, Exercise 3.9

1. You are studying a subject which is assessed by coursework and examination with the total mark for the course being calculated on a $30:70$ weighting between these two components. Assuming you score 60% in coursework, insert the appropriate inequality sign between your possible overall mark for the course and the percentage figures below.

    (a) 18% ? overall mark   (b) 16% ? overall mark
    (c) 88% ? overall mark   (d) 90% ? overall mark

2. If $x \geq 1$, insert the appropriate inequality sign between:

    (a) $(x + 2)^2$ and 3   (b) $(x + 2)^2$ and 9
    (c) $(x + 2)^2$ and $3x$ (d) $(x + 2)^2$ and $6x$

3. If Q_1 and Q_2 represent positive production levels of a good and the equality $Q_2 = Z^n Q_1$ always holds where $Z > 1$, what can be said about the relationship between Q_1 and Q_2 if

    (a) $n > 0$ (b) $n = 0$ (c) $n < 0$?

4. If a monopolist can operate price discrimination and charge separate prices P_1 and P_2 in two different markets, it can be proved that for profit maximization the monopolist should choose values for P_1 and P_2 that satisfy the equation

    $$P_1 \left(1 - \frac{1}{e_1} \right) = P_2 \left(1 - \frac{1}{e_2} \right)$$

    where e_1 and e_2 are elasticities of demand in the two markets. In which market should price be higher if $|e_1| > |e_2|$?

4 Graphs and functions

Learning objectives

After completing this chapter students should be able to:

- Interpret the meaning of functions and inverse functions.
- Draw graphs that correspond to linear, non-linear and composite functions.
- Find the slopes of linear functions and tangents to non-linear function by graphical analysis.
- Use the slope of a linear demand function to calculate point elasticity.
- Show what happens to budget constraints when parameters change.
- Interpret the meaning of functions with two independent variables.
- Deduce the degree of returns to scale from the parameters of a Cobb–Douglas production function.
- Construct an Excel spreadsheet to plot the values of different functional formats.
- Sum marginal revenue and marginal cost functions horizontally to help find solutions to price discrimination and multi-plant monopoly problems.

4.1 Functions

Suppose that average weekly household expenditure on food (C) depends on average net household weekly income (Y) according to the relationship

$$C = 12 + 0.3Y$$

For any given value of Y, one can evaluate what C will be. For example

if $Y = 90$ then $C = 12 + 27 = 39$

Whatever value of Y is chosen there will be one unique corresponding value of C. This is an example of a function.

A relationship between the values of two or more variables can be defined as a *function* when *a unique value of one of the variables is determined by the value of the other variable* or variables.

If the precise mathematical form of the relationship is not actually known then a function may be written in what is called a *general form*. For example, a general form demand function is

$$Q_d = f(P)$$

This particular general form just tells us that quantity demanded of a good (Q_d) depends on its price (P). The 'f' is not an algebraic symbol in the usual sense and so f(P) means 'is a function of P' and not 'f multiplied by P'. In this case P is what is known as the '*independent variable*' because its value is given and is not dependent on the value of Q_d, i.e. it is exogenously determined. On the other hand Q_d is the '*dependent variable*' because its value depends on the value of P.

Functions may have more than one independent variable. For example, the general form production function

$$Q = f(K,L)$$

tells us that output (Q) depends on the values of the two independent variables capital (K) and labour (L).

The *specific form of a function* tells us exactly how the value of the dependent variable is determined from the values of the independent variable or variables. A specific form for a demand function might be

$$Q_d = 120 - 2P$$

For any given value of P the specific function allows us to calculate the value of Q_d. For example

when $P = 10$ then $Q_d = 120 - 2(10) = 120 - 20 = 100$

when $P = 45$ then $Q_d = 120 - 2(45) = 120 - 90 = 30$

In economic applications of functions it may make sense to restrict the 'domain' of the function, i.e. the range of possible values of the variables. For example, variables that represent price or output may be restricted to positive values. Strictly speaking the *domain limits the values of the independent variables* and the *range governs the possible values of the dependent variable*.

For more complex functions with more than one independent variable it may be helpful to draw up a table to show the relationship of different values of the independent variables to the value of the dependent variable. Table 4.1 shows some possible different values for the specific form production function $Q = 4K^{0.5}L^{0.5}$. (It is implicitly assumed that Q, $K^{0.5}$ and $L^{0.5}$ only take positive values.)

Table 4.1

K	L	$K^{0.5}$	$L^{0.5}$	Q
1	1	1	1	4
4	1	2	1	8
9	25	3	5	60
7	11	2.64575	3.31662	35.0998

When defining the specific form of a function it is important to make sure that only *one unique value of the dependent variable is determined* from each given value of the independent variable(s). Consider the equation

$$y = 80 + x^{0.5}$$

This does *not* define a function because any given value of x corresponds to two possible values for y. For example, if $x = 25$, then $25^{0.5} = 5$ or -5 and so $y = 75$ or 85. However, if we define

$$y = 80 + x^{0.5} \quad \text{for } x^{0.5} \geq 0$$

then this does constitute a function.

When domains are not specified then one should assume a sensible range for functions representing economic variables. For example, it is usually assumed $K^{0.5} > 0$ and $L^{0.5} > 0$ in a production function, as in Table 4.1 above.

Test Yourself, Exercise 4.1

1. An economist researching the market for tea assumes that

$$Q_t = f(P_t, Y, A, N, P_c)$$

where Q_t is the quantity of tea demanded, P_t is the price of tea, Y is average household income, A is advertising expenditure on tea, N is population and P_c is the price of coffee.

(a) What does $Q_t = f(P_t, Y, A, N, P_c)$ mean in words?
(b) Identify the dependent and independent variables.
(c) Make up a specific form for this function. (Use your knowledge of economics to deduce whether the coefficients of the different independent variables should be positive or negative.)

2. If a firm faces the total cost function

$$TC = 6 + x^2$$

where x is output, what is TC when x is (a) 14? (b) 1? (c) 0? What restrictions on the domain of this function would it be reasonable to make?

3. A firm's total expenditure E on inputs is determined by the formula

$$E = P_K K + P_L L$$

where K is the amount of input K used, L is the amount of input L used, P_K is the price per unit of K and P_L is the price per unit of L. Is one unique value for E determined by any given set of values for K, L, P_K and P_L? Does this mean that any one particular value for E must always correspond to the same set of values for K, L, P_K and P_L?

4.2 Inverse functions

An inverse function reverses the relationship in a function. If we confine the analysis to functions with only one independent variable, x, this means that if y is a function of x, i.e.

$$y = f(x)$$

then in the inverse function x will be a function of y, i.e.

$$x = g(y)$$

(The letter g is used to show that we are talking about a different function.)

Example 4.1

If the original function is

$$y = 4 + 5x$$

then $\qquad y - 4 = 5x$

$$0.2y - 0.8 = x$$

and so the inverse function is

$$x = 0.2y - 0.8$$

Not all functions have an inverse function. The mathematical condition necessary for a function to have a corresponding inverse function is that the original function must be 'monotonic'. This means that, as the value of the independent variable x is increased, the value of the dependent variable y must either always increase or always decrease. It cannot first increase and then decrease, or vice versa. This will ensure that, as well as there being one unique value of y for any given value of x, there will also be one unique value of x for any given value of y. This point will probably become clearer to you in the following sections on graphs of functions but it can be illustrated here with a simple example.

Example 4.2

Consider the function $y = 9x - x^2$ restricted to the domain $0 \leq x \leq 9$.
  Each value of x will determine a unique value of y. However, some values of y will correspond to two values of x, e.g.

when $x = 3$ then $y = 27 - 9 = 18$

when $x = 6$ then $y = 54 - 36 = 18$

This is because the function $y = 9x - x^2$ is not monotonic. This can be established by calculating y for a few selected values of x:

x	1	2	3	4	5	6	7
y	8	14	18	20	20	18	14

These figures show that y first increases and then decreases in value as x is increased and so there is no inverse for this non-monotonic function.

  Although mathematically it may be possible to derive an inverse function, it may not always make sense to derive the inverse of an economic function, or many other functions that are based on empirical data. For example, if we take the geometric function that the area A of a square is related to the length L of its sides by the function $A = L^2$, then we can also write the inverse function that relates the length of a square's side to its area: $L = A^{0.5}$ (assuming that L can only take non-negative values). Once one value is known then the other is determined by it. However, suppose that someone investigating expenditure on holidays abroad (H) finds that the level of average annual household income (M) is the main influence and the relationship can be explained by the function

$$H = 0.01M + 100 \quad \text{for } M \geq £10,000$$

This mathematical equation could be rearranged to give

$$M = 100H - 10,000?$$

but to say that H determines M obviously does not make sense. The amount of holidays taken abroad does not determine the level of average household income.

It is not always a clear-cut case though. The cause and effect relationship within an economic model is not always obviously in one direction only. Consider the relationship between price and quantity in a demand function. A monopoly may set a product's price and then see how much consumers are willing to buy, i.e. $Q = f(P)$. On the other hand, in a competitive industry firms may first decide how much they are going to produce and then see what price they can get for this output, i.e. $P = f(Q)$.

Example 4.3

Given the demand function $Q = 200 - 4P$, derive the inverse demand function.

Solution

$$Q = 200 - 4P$$
$$4P + Q = 200$$
$$4P = 200 - Q$$
$$P = 50 - 0.25Q$$

Test Yourself, Exercise 4.2

1. To convert temperature from degrees Fahrenheit to degrees Celsius one uses the formula

$$°C = \frac{5}{9}(°F - 32)$$

What is the inverse of this function?
2. What is the inverse of the demand function

$$Q = 1,200 - 0.5P?$$

3. The total revenue (TR) that a monopoly receives from selling different levels of output (q) is given by the function TR $= 60q - 4q^2$ for $0 \leq q \leq 15$. Explain why one cannot derive the inverse function $q = \mathrm{f}(\mathrm{TR})$.

4. An empirical study suggests that a brewery's weekly sales of beer are determined by the average air temperature given that the price of beer, income, adult population and most other variables are constant in the short run. This functional relationship is estimated as

$$X = 400 + 16T^{0.5} \quad \text{for } T^{0.5} > 0$$

where X is the number of barrels sold per week and T is the mean average air temperature, in °F. What is the mathematical inverse of this function? Does it make sense to specify such an inverse function in economics?

5. Make up your own examples for:

    (a) a function that has an inverse, and then derive the inverse function;
    (b) a function that does not have an inverse and then explain why this is so.

4.3   Graphs of linear functions

We are all familiar with graphs of the sort illustrated in Figure 4.1. This shows a firm's annual sales figures. To find what its sales were in 2002 you first find 2002 on the horizontal axis, move vertically up to the line marked 'sales' and read off the corresponding figure on the vertical axis, which in this case is £120,000. These graphs are often used as an alternative to tables of data as they make trends in the numbers easier to identify visually. These, however, are *not* graphs of functions. Sales are not determined by 'time'.

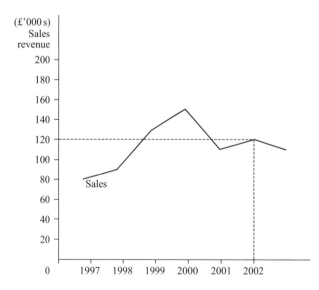

Figure 4.1

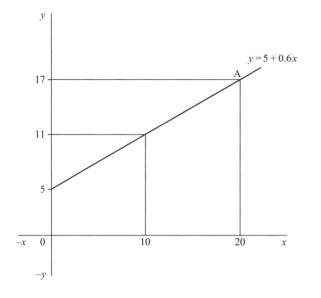

Figure 4.2

Mathematical functions are mapped out on what is known as a set of 'Cartesian axes', as shown in Figure 4.2. Variable x is measured by equal increments on the horizontal axis and variable y by equal increments on the vertical axis. Both x and y can be measured in positive or negative directions. Although obviously only a limited range of values can be shown on the page of a book, the Cartesian axes theoretically range from $+\infty$ to $-\infty$ (i.e. to plus or minus infinity).

Any point on the graph will have two 'coordinates', i.e. corresponding values on the x and y axes. For example, to find the coordinates of point A one needs to draw a vertical line down to the x axis and read off the value of 20 and draw a horizontal line across to the y axis and read off the value 17. The coordinates (20, 17) determine point A.

As only two variables can be measured on the two axes in Figure 4.2, this means that only functions with one independent variable can be illustrated by a graph on a two-dimensional sheet of paper. One axis measures the dependent variable and the other measures the independent variable. (However, in Section 4.9 a method of illustrating a two-independent-variable function is explained.)

Having set up the Cartesian axes in Figure 4.2, let us use it to determine the shape of the function

$$y = 5 + 0.6x$$

Calculating a few values of y for different values of x we get:

when $x = 0$ then $y = 5 + 0.6(0) = 5$

when $x = 10$ then $y = 5 + 0.6(10) = 5 + 6 = 11$

when $x = 20$ then $y = 5 + 0.6(20) = 5 + 12 = 17$

These points are plotted in Figure 4.2 and it is obvious that they lie along a straight line. The rest of the function can be shown by drawing a straight line through the points that have been plotted.

Any function that takes the format $y = a + bx$ will correspond to a straight line when represented by a graph (where a and b can be any positive or negative numbers). This is because the value of y will change by the same amount, b, for every one unit increment in x. For example, the value of y in the function $y = 5 + 0.6x$ increases by 0.6 every time x increases by one unit.

Usually the easiest way to *plot a linear function* is to find the points where it cuts the two axes and draw a straight line through them.

Example 4.4

Plot the graph of the function, $y = 6 + 2x$.

Solution

The y axis is a vertical line through the point where x is zero.
  When $x = 0$ then $y = 6$ and so this function must cut the y axis at $y = 6$.
  The x axis is a horizontal line through the point where y is zero.

$$\text{When } y = 0 \quad \text{then} \quad 0 = 6 + 2x$$
$$-6 = 2x$$
$$-3 = x$$

and so this function must cut the x axis at $x = -3$.

The function $y = 6 + 2x$ is linear. Therefore if we join up the points where it cuts the x and y axes by a straight line we get the graph as shown in Figure 4.3.

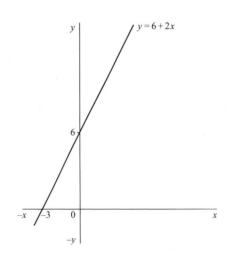

Figure 4.3

If no restrictions are placed on the domain of the independent variable in a function then the range of values of the dependent variable could possibly take any positive or negative value, depending on the nature of the function. However, in economics some variables may only take on positive values. A linear function that applies only to positive values of all the variables concerned may sometimes only intercept with one axis. In such cases, all one has to do is simply plot another point and draw a line through the two points obtained.

Example 4.5

Draw the graph of the function, $C = 200 + 0.6Y$, where C is consumer spending and Y is income, which cannot be negative.

Solution

Before plotting the shape of this function you need to note that the notation is different from the previous examples and this time C is the dependent variable, measured in the vertical axis, and Y is the independent variable, measured on the horizontal axis.

When $Y = 0$, then $C = 200$, and so the line cuts the vertical axis at 200.

However, when $C = 0$, then

$$0 = 200 + 0.6Y$$

$$-0.6Y = 200$$

$$Y = -\frac{200}{0.6}$$

As negative values of Y are unacceptable, just choose another pair of values, e.g. when $Y = 500$ then $C = 200 + 0.6(500) = 200 + 300 = 500$. This graph is shown in Figure 4.4.

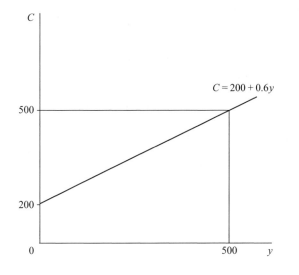

Figure 4.4

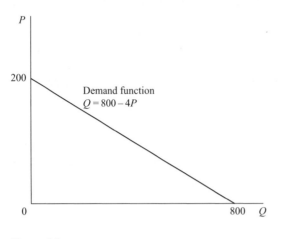

Figure 4.5

In mathematics the usual convention when drawing graphs is to measure the independent variable x along the horizontal axis and the dependent variable y along the vertical axis. However, in economic supply and demand analysis the usual convention is to measure price P on the vertical axis and quantity Q along the horizontal axis. This sometimes confuses students when a function in economics is specified with Q as the dependent variable, such as the demand function

$$Q = 800 - 4P$$

but then illustrated by a graph such as that in Figure 4.5. (Before you proceed, check that you understand why the intercepts on the two axes are as shown.)

Theoretically, it does not matter which axis is used to measure which variable. However, one of the main reasons for using graphs is to make analysis clearer to understand. Therefore, if one always has to keep checking which axis measures which variable this defeats the objective of the exercise. Thus, even though it may upset some mathematical purists, in this text we shall stick to the economist's convention of measuring quantity on the horizontal axis and price on the vertical axis, even if price is the independent variable in a function.

This means that care has to be taken when performing certain operations on functions. If necessary, one can transform monotonic functions to obtain the inverse function (as already explained) if this helps the analysis. For example, the demand function $Q = 800 - 4P$ has the inverse function

$$P = \frac{800 - Q}{4} = 200 - 0.25Q$$

Check again in Figure 4.5 for the intercepts of the graph of this function.

Test Yourself, Exercise 4.3

Sketch the graphs of the linear functions 1 to 8 below, identifying the relevant intercepts on the axes. Assume that variables represented by letters that suggest they are economic variables (i.e. all variables except x and y) are restricted to non-negative values.

1. $y = 6 + 0.5x$
2. $y = 12x - 40$
3. $P = 60 - 0.2Q$
4. $Q = 750 - 5P$
5. $1,200 = 50K + 30L$

    (Note that this budget constraint for a firm is an accounting identity rather than a function although a given value of K will still determine a unique value of L, and vice versa.)

6. $TR = 8Q$
7. $TC = 200 + 5Q$
8. $TFC = 75$
9. Make up your own example of a linear function and then sketch its graph.
10. Which of the following functions do you think realistically represents the supply schedule of a competitive industry? Why?

    (a) $P = 0.6Q + 2$ (b) $P = 0.5Q - 10$

    (c) $P = 4Q$ (d) $Q = -24 + 0.2P$

    Assume $P \geq 0$, $Q \geq 0$ in all cases.

4.4   Fitting linear functions

If you know that two points lie on a straight line then you can draw the rest of the line. You simply put your ruler on the page, join the two points and then extend the line in either direction as far as you need to go. For example, suppose that a firm faces a linear demand schedule and that 400 units of output Q are sold when price is £40 and 500 units are sold when price is £20. Once these two price and quantity combinations have been marked as points A and B in Figure 4.6 then the rest of the demand schedule can be drawn in.

One can then use this graph to predict the amounts sold at other prices. For example, when price is £29.50, the corresponding quantity can be read off as approximately 450. However, more accurate predictions of quantities demanded at different prices can be made if the information that is initially given is used to determine the algebraic format of the function.

A linear demand function must be in the format $P = a - bQ$, where a and b are parameters that we wish to determine the value of. From Figure 4.6 we can see that

when $P = 40$	then $Q = 400$	and so $40 = a - 400b$	(1)
when $P = 20$	then $Q = 500$	and so $20 = a - 500b$	(2)

Equations (1) and (2) are what is known as simultaneous linear equations. Various methods of solving such sets of simultaneous equations (i.e. finding the values of a and b) are explained

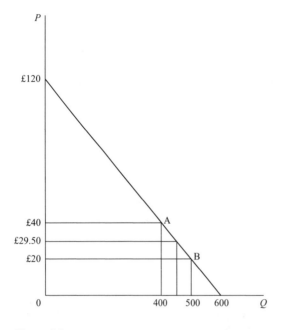

Figure 4.6

later in Chapter 5. Here we shall just use an intuitively obvious method of deducing the values of *a* and *b* from the graph in Figure 4.6.

Between points B and A we can see that a £20 rise in price causes a 100 unit decrease in quantity demanded. As this is a linear function then we know that further price rises of £20 will also cause quantity demanded to fall by 100 units. At A, quantity is 400 units. Therefore a rise in price of £80 is required to reduce quantity demanded from 400 to zero, i.e. a rise in price of $4 \times £20 = £80$ will reduce quantity demanded by $4 \times 100 = 400$ units. This means that the intercept of this function on the price axis is £80 plus £40 (the price at A), which is £120. This is the value of the parameter *a*.

To find the value of the parameter *b* we need to ask 'what will be the fall in price necessary to cause quantity demanded to increase by one unit?' Given that a £20 price fall causes quantity to rise by 100 units then it must be the case that a price fall of $£20/100 = £0.2$ will cause quantity to rise by one unit. This also means that a price rise of £0.2 will cause quantity demanded to fall by one unit. Therefore, $b = 0.2$. As we have already worked out that *a* is 120, our function can now be written as

$$P = 120 - 0.2Q$$

We can check that this is correct by substituting the original values of *Q* into the function.

If $Q = 400$ then $P = 120 - 0.2(400) = 120 - 80 = 40$

If $Q = 500$ then $P = 120 - 0.2(500) = 120 - 100 = 20$

These are the values of *P* originally specified and so we are satisfied that the line that passes through points A and B in Figure 4.6 is the linear function $P = 120 - 0.2Q$.

The inverse of this function will be $Q = 600 - 5P$. Precise values of Q can now be derived for given values of P. For example,

when $P = £29.50$ then $Q = 600 - 5(29.50) = 452.5$

This is a more accurate figure than the one read off the graph as approximately 450.

Having learned how to deduce the parameters of a linear downward-sloping demand function, let us now try to fit an upward-sloping linear function.

Example 4.6

It is assumed that consumption C depends on income Y and that this relationship takes the form of the linear function $C = a + bY$. When Y is £600, C is observed to be £660. When Y is £1,000, C is observed to be £900. What are the values of a and b in this function?

Solution

We expect b to be positive, i.e. consumption increases with income, and so our function will slope upwards, as shown in Figure 4.7. As this is a linear function then equal changes in Y will cause the same changes in C.

A decrease in Y of £400, from £1,000 to £600, causes C to fall by £240, from £900 to £660.

If Y is decreased by a further £600 (i.e. to zero) then the corresponding fall in C will be 1.5 times the fall caused by an income decrease of £400, since £600 = 1.5 × £400. Therefore the fall in C is 1.5 × £240 = £360.

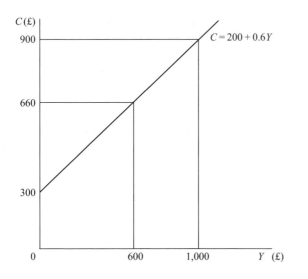

Figure 4.7

This means that the value of C when Y is zero is £660 – £360 = £300. Thus $a = 300$. A rise in Y of £400 causes C to rise by £240. Therefore a rise in Y of £1 will cause C to rise by £240/400 = £0.6. Thus $b = 0.6$.

The function can therefore be specified as

$$C = 300 + 0.6Y$$

Checking against original values:

When $Y = 600$ then predicted $C = 300 + 0.6(600)$

$$= 300 + 360 = 660. \text{ Correct.}$$

When $Y = 1{,}000$ then predicted $C = 300 + 0.6(1{,}000)$

$$= 300 + 600 = 900. \text{ Correct.}$$

Test Yourself, Exercise 4.4

1. A monopoly sells 30 units of output when price is £12 and 40 units when price is £10. If its demand schedule is linear, what is the specific form of the actual demand function? Use this function to predict quantity sold when price is £8. What domain restrictions would you put on this demand function?
2. Assume that consumption C depends on income Y according to the function $C = a + bY$, where a and b are parameters. If C is £60 when Y is £40 and C is £90 when Y is £80, what are the values of the parameters a and b?
3. On a linear demand schedule quantity sold falls from 90 to 30 when price rises from £40 to £80. How much further will price have to rise for quantity sold to fall to zero?
4. A firm knows that its demand schedule takes the form $P = a - bQ$. If 200 units are sold when price is £9 and 400 units are sold when price is £6, what are the values of the parameters a and b?
5. A firm notices that its total production costs are £3,200 when output is 85 and £4,820 when output is 130. If total cost is assumed to be a linear function of output what expenditure will be necessary to manufacture 175 units?

4.5   Slope

British road signs used to give warning of steep hills by specifying their slope in a format such as '1 in 10', meaning that the road rose vertically by 1 foot for every 10 feet travelled in a horizontal direction. Now the European format is used and so instead of '1 in 10' a road sign will say 10%. In mathematics the same concept of slope is used but it is expressed as a decimal fraction rather than in percentage terms.

The graph in Figure 4.8 shows the function $y = 2 + 0.1x$. The slope is obviously the same along the whole length of this straight line and so it does not matter where the slope is measured. To measure the slope along the stretch AB, draw a horizontal line across from A and drop a vertical line down from B. These intersect at C, forming the triangle ABC with a right angle at C. The horizontal distance AC is 20 and the vertical distance BC is 2, and so if this was a cross-section of a hill you would clearly say that the slope is 1 in 10, or 10%.

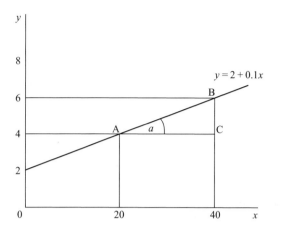

Figure 4.8

In mathematics the *slope of a line* is defined as

$$\text{slope} = \frac{\text{height}}{\text{base}}$$

where height and base measure the sides of a right-angled triangle drawn as above. (Note that this only applies to lines that slope upwards from left to right.) Thus in this example

$$\text{slope} = \frac{2}{20} = 0.1$$

This is also known as the tangent of the angle a.

One can see that the slope of this function (0.1) is the same as the coefficient of x. This is a general rule. For any linear function in the format $y = a + bx$, then b will always represent its slope.

Example 4.7

Find the slope of the function $y = -2 + 3x$.

Solution

The value of y increases by 3 for every 1 unit increase in x and so the slope of this linear function is 3.

When a line slopes downwards from left to right it has a negative slope. Thus the b in the function $y = a + bx$ will take a negative value.

Consider the function $P = 60 - 0.2Q$ where P is price and Q is quantity demanded. This is illustrated in Figure 4.9. As P and Q can be assumed not to take negative values, the whole

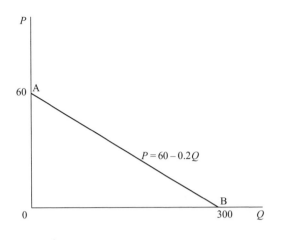

Figure 4.9

function can be drawn by joining the intercepts on the two axes which are found as follows.

$$\text{When } Q = 0 \quad \text{then} \quad P = 60$$

$$\text{When } P = 0 \quad \text{then} \quad 0 = 60 - 0.2Q$$

$$0.2Q = 60$$

$$Q = \frac{60}{0.2} = 300$$

The slope of a function which slopes down from left to right is found by applying the formula

$$\text{slope} = (-1)\frac{\text{height}}{\text{base}}$$

to the relevant right-angled triangle. Thus, using the triangle 0AB, the slope of the function in Figure 4.9 is

$$(-1)\frac{60}{300} = (-1)0.2 = -0.2$$

This, of course, is the same as the coefficient of Q in the function $P = 60 - 0.2Q$.

Remember that in economics the usual convention is to measure P on the vertical axis of a graph. If you are given a function in the format $Q = f(P)$ then you would need to derive the inverse function to read off the slope.

Example 4.8

What is the slope of the demand function $Q = 830 - 2.5P$ when P is measured on the vertical axis of a graph?

Solution

If $\quad Q = 830 - 2.5P$

then $2.5P = 830 - Q$

$$P = 332 - 0.4Q$$

Therefore the slope is the coefficient of Q, which is -0.4.

If the coefficient of x in a linear function is zero then the slope is also zero, i.e. the line is horizontal. For example, the function $y = 20$ means that y takes a value of 20 for every value of x.

Conversely, a vertical line will have an infinitely large slope. (Note, though, that a vertical line would not represent y as a function of x as no unique value of y is determined by a given value of x.)

Slope of a demand schedule and elasticity of demand   In Chapter 2, the calculation of arc elasticity was explained. Because elasticity of demand can alter along the length of a demand schedule the arc elasticity measure is used as a sort of 'average'. However, now that you understand how the slope of a line is derived we can examine how elasticity can be calculated at a specific point on a demand schedule. This is called '*point elasticity of demand*' and is defined as

$$e = (-1)\frac{P}{Q}\left(\frac{1}{\text{slope}}\right)$$

where P and Q are the price and quantity at the point in question. The slope refers to the slope of the demand schedule at this point although, of course, for a linear demand schedule the slope will be the same at all points. The derivation of this formula and its application to non-linear demand schedules is explained later in Chapter 8. Here we shall just consider its application to linear demand schedules.

Example 4.9

Calculate the point elasticity of demand for the demand schedule

$$P = 60 - 0.2Q$$

where price is (i) zero, (ii) £20, (iii) £40, (iv) £60.

Solution

This is the demand schedule referred to earlier and illustrated in Figure 4.9. Its slope must be -0.2 at all points as it is a linear function and this is the coefficient of Q.

To find the values of Q corresponding to the given prices we need to derive the inverse function. Given that

$$P = 60 - 0.2Q$$

then $0.2Q = 60 - P$

$$Q = 300 - 5P$$

(i) When P is zero, at point B, then $Q = 300 - 5(0) = 300$.
The point elasticity will therefore be

$$e = (-1)\frac{P}{Q}\left(\frac{1}{\text{slope}}\right) = (-1)\frac{0}{300}\left(\frac{1}{-0.2}\right) = 0$$

(ii) When $P = 20$ then $Q = 300 - 5(20) = 200$.

$$e = (-1)\frac{20}{200}\left(\frac{1}{-0.2}\right) = \frac{1}{10}\left(\frac{1}{0.2}\right) = \frac{1}{2} = 0.5$$

(iii) When $P = 40$ then $Q = 300 - 5(40) = 100$.

$$e = (-1)\frac{40}{100}\left(\frac{1}{-0.2}\right) = \frac{2}{5}\left(\frac{1}{0.2}\right) = \frac{2}{1} = 2$$

(iv) When $P = 60$ then $Q = 300 - 5(60) = 0$.
If $Q = 0$, then $P/Q \to \infty$.

$$\text{Therefore } e = (-1)\frac{P}{Q}\left(\frac{1}{\text{slope}}\right) = (-1)\frac{60}{0}\left(\frac{1}{-0.2}\right) \to \infty$$

Test Yourself, Exercise 4.5

1.   In Figure 4.10, what are the slopes of the lines 0A, 0B, 0C and EF?

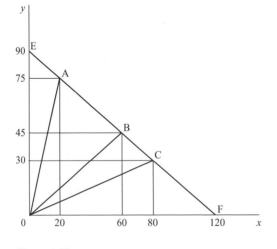

Figure 4.10

2.   A market has a linear demand schedule with a slope of -0.3. When price is £3, quantity sold is 30 units. Where does this demand schedule hit the price and

quantity axes? What is price if quantity sold is 25 units? How much would be sold at a price of £9?

3. For the demand schedule $P = 60 - 0.2Q$ illustrated in Figure 4.9, calculate point elasticity of demand when price is (a) £24 and (b) £45.

4. Consider the three demand functions

    (a) $P = 8 - 0.75Q$
    (b) $P = 8 - 1.25Q$
    (c) $Q = 12 - 2P$

    Which has the flattest demand schedule, assuming that P is measured on the vertical axis? In which case is quantity sold the greatest when price is (i) £1 and (ii) £5?

5. For positive values of x which, if any, of the functions below will intersect with the function $y = 1 + 0.5x$?

    (a) $y = 2 + 0.4x$ (c) $y = 4 + 0.5x$
    (b) $y = 2 + 1.5x$ (d) $y = 4$

6. In macroeconomics the average propensity to consume (APC) and the marginal propensity to consume (MPC) are defined as follows:

    APC $= C/Y$ where $C =$ consumption, $Y =$ income

    MPC $=$ increase in C from a 1 unit increase in Y

    Explain why APC will always be greater than MPC if $C = 400 + 0.5Y$.

7. For the demand schedule $P = 24 - 0.125Q$, calculate point elasticity of demand when price is

    (a) £5   (b) £10   (c) £15

8. Make up your own examples of linear functions that will

    (a) slope upwards and go through the origin;
    (b) slope downwards and cut the price axis at a positive value;
    (c) be horizontal.

4.6   Budget constraints

A frequently used application of the concept of slope in economics is the relationship between prices and the slope of a budget constraint. A budget constraint shows the combinations of two goods (or inputs) that it is possible to buy with a given budget and a given set of prices.

Assume that a firm has a budget of £3,000 to spend on the two inputs K and L and that input K costs £50 and input L costs £30 a unit. If it spends the whole £3,000 on K then it can buy

$$\frac{3,000}{50} = 60 \text{ units of K}$$

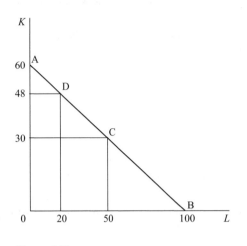

Figure 4.11

and if it spends all its budget on L then it can buy

$$\frac{3,000}{30} = 100 \text{ units of L}$$

These two quantities are marked on the axes of the graph in Figure 4.11. The firm could also split the budget between K and L. Many different combinations are possible, e.g.

30 of K and 50 of L

or

48 of K and 20 of L

If K and L are divisible into fractions of a unit then all the combinations of K and L that can be bought with the given budget of £3,000 can be shown by the line AB which is known as the 'budget constraint' or 'budget line'. The firm could in fact also purchase any of the combinations of K and L within the triangle OAB but only combinations along the budget constraint AB would entail it spending its entire budget.

Along the budget constraint any pairs of values of K and L must satisfy the equation

$$50K + 30L = 3,000$$

where K is the number of units of K bought and L is the number of units of L bought. All this equation says is that total expenditure on K (price of K × amount bought) plus total expenditure on L (price of L × amount bought) must sum to the total budget available.

We can check that this holds for the combinations of K and L shown in Figure 4.11.

At A   £50 × 60 + £30 × 0 = 3,000 + 0 = £3,000

At B   £50 × 0 + £30 × 100 = 0 + 3,000 = £3,000

At C   £50 × 30 + £30 × 50 = 1,500 + 1,500 = £3,000

At D   £50 × 48 + £30 × 20 = 2,400 + 600 = £3,000

As budget lines usually slope down from left to right they have a negative slope. From the graph one can see that this budget constraint has a slope of

$$\frac{-60}{100} = -0.6$$

The slope of a budget constraint can be deduced from the values of the prices of the two goods or inputs concerned. Consider the general case where the budget is M and the prices of the two goods X and Y are P_X and P_Y respectively. The maximum amount of X that can be bought will be M/P_X. This will be the intercept on the horizontal axis. Similarly the maximum amount of Y that can be purchased will be M/P_Y, which will be the intercept on the vertical axis. Therefore

$$\text{slope of budget constraint} = (-)\frac{\left(\dfrac{M}{P_Y}\right)}{\left(\dfrac{M}{P_X}\right)} = (-)\frac{M}{P_Y}\frac{P_X}{M} = (-)\frac{P_X}{P_Y}$$

Thus for any budget constraint the slope will be the negative of the price ratio. However, you should note that it is the price of the good measured on the *horizontal* axis that is at the top in this formula.

From this result we can also see that

- if the price ratio changes, the slope of the budget line changes
- if the budget alters, the slope of the budget line does not alter.

Example 4.10

A consumer has an income of £160 to spend on the two goods X and Y whose prices are £20 and £5 each, respectively.

 (i)   What is the slope of the budget constraint?
 (ii) What happens to this slope if P_Y rises to £10?
 (iii) What happens if income then falls to £100?

Solution

$$\text{(i) slope} = -\frac{P_X}{P_Y} = -\frac{20}{5} = -4$$

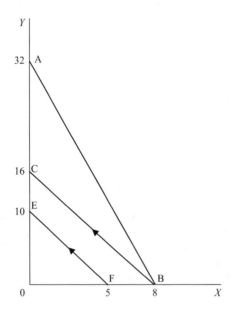

Figure 4.12

This can be checked by considering the intercepts on the X and Y axes shown in Figure 4.12 by points B and A.

If the total budget of £160 is spent on X then $160/20 = 8$ units are bought. If the total budget is spent on Y then $160/5 = 32$ units are bought. Therefore

$$\text{slope} = (-)\frac{\text{intercept on Y axis}}{\text{intercept on X axis}} = (-)\frac{32}{8} = -4$$

(ii) When P_Y rises to £10 the new slope of the budget constraint (shown by BC in Figure 4.12) becomes

$$-\frac{P_X}{P_Y} = -\frac{20}{10} = -2$$

(iii) The price ratio remains unchanged if income then falls to £100. There is a parallel shift inwards of the budget constraint to EF. The new intercepts are

$$\frac{M}{P_Y} = \frac{100}{10} = 10 \text{ on the } Y \text{ axis}$$

and

$$\frac{M}{P_X} = \frac{100}{20} = 5 \text{ on the } X \text{ axis}$$

The slope is thus $-10/5 = -2$, as before.

Example 4.11

A consumer can buy the two goods A and B at prices per unit of £6 and £4 respectively, and initially has an income of £120.

 (i) Show that a 25% rise in all prices will have a lesser effect on the consumer's purchasing possibilities than would a 25% reduction in money income with prices unchanged.
(ii) What is the opportunity cost of buying an extra unit of A? (Assume units of A and B are divisible.)

Solution

(i) The original intercept on the A axis $= \dfrac{120}{6} = 20$

The original intercept on the B axis $= \dfrac{120}{4} = 30$

If price of A rises by 25% to £7.50 the new intercept on the A axis $= \dfrac{120}{7.50} = 16$

If price of B rises by 25% to £5 the new intercept on the B axis $= \dfrac{120}{5} = 24$

Reducing income by 25% to 90 changes intercept on the A axis to $\dfrac{90}{6} = 15$

and intercept on the B axis to $\dfrac{90}{4} = 22.5$

Thus the 25% fall in income shifts the budget constraint towards the origin slightly more than does the 25% rise in prices, i.e. it reduces the consumer's purchasing possibilities by a greater amount.

(Note that the slope of the budget constraint always remains the same at $-6/4 = -1.5$.)
(ii) The opportunity cost of something is the next best alternative that one has to forgo in order to obtain it. In this context, the opportunity cost of an extra unit of A will be the amount of B the consumer has to forgo.

One unit of A costs £6 and one unit of B costs £4. Therefore, the opportunity cost of A in terms of B is 1.5, which is the negative of the slope of the budget line.

Test Yourself, Exercise 4.6

 1. A consumer can buy good A at £3 a unit and good B at £2 a unit and has a budget of £60. What is the slope of the budget constraint if quantity of A is measured on the horizontal axis?
      What happens to this slope if

   (a) the price of A falls to £2?
   (b) with A at its original price the price of B rises to £3?

    (c) both prices double?

    (d) the budget is cut by 25%?

2.   A firm has a budget of £800 per week to spend on the two inputs K and L. One week it is observed to buy 120 units of L and 25 of K. Another week it is observed to buy 80 units of L and 50 of K. Find out what the intercepts of its budget line on the K and L axes will be and use this information to deduce the prices of K and L, which are assumed to be unchanged from one week to the next.

3.   A firm can buy the two inputs K and L at £60 and £40 per unit respectively, and has a budget of £480. Explain why it would not be able to purchase 6 units of K plus 4 units of L and then calculate what price reduction in L would make this input combination a feasible purchase.

4.   If a firm buys the two inputs X and Y, what would the slope of its budget constraint be if the price of Y was £10 and

    (a) the price of X was £100?   (b) the price of X was £10?

    (c) the price of X was £1?   (d) the price of X was 25p?

    (e) X was free?

5.   If a consumer's income doubles and the prices of the two goods that she spends her entire income on also double, what happens to her budget constraint?

6.   An hourly paid worker can choose the number of hours per day worked, up to a maximum of 12, and gets paid £10 an hour. Leisure hours are assumed to be any hours not worked out of this 12. On a graph with leisure hours on the horizontal axis and total pay on the vertical axis draw in the budget constraint showing the feasible combinations of leisure and pay that this worker might choose from. Show that the slope of this budget constraint equals -1 multiplied by the hourly wage rate.

7.   A firm has a limited budget to spend on inputs K and L. Make up your own values for the budget and the prices of K and L and then say what the slope of the budget constraint and its intersection points on the K and L axes will be.

4.7   Non-linear functions

If the function $y = f(x)$ has a term with x to the power of anything other than 1, then it will be non-linear. For example,

$y = x^2$ is a non-linear function

$y = 6 + x^{0.5}$ is a non-linear function

but

$y = 5 + 0.2x$ is a linear function

Non-linear functions can take a variety of shapes. We shall only consider a few possibilities that will be useful at a later stage when looking at functions of economic variables.

If the function $y = f(x)$ has one term in x with x to the power of something greater than 1 then, as long as x takes positive values, it will rise at an increasing rate as x is increased. This is obvious from Table 4.2. The graphs of the functions $y = x^2$ and $y = x^3$ will curve upwards since y increases at a faster rate than x. These functions all go through the origin, as y is zero when x is zero. The table shows that the greater the power of x then the more quickly y rises.

Table 4.2

x	0	1	2	3	4	5	6
$y = x^2$	0	1	4	9	16	25	36
$y = x^3$	0	1	8	27	64	125	216

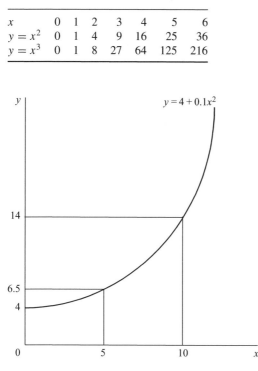

Figure 4.13

Although the intercept may vary if there is a constant term in a function, and the rate of change of y may be modified if the term in x has a coefficient other than 1, the general shape of an upward-sloping curve will still be retained. For example, Figure 4.13 illustrates the function

$$y = 4 + 0.1x^2$$

In economics the quantities one is working with are frequently constrained to positive values, e.g. price and quantity. However, if variables are allowed to take negative values then the functions $y = x^2$ and $y = x^3$ will take the shapes shown in Figure 4.14. Note that, when $x < 0$, $x^2 > 0$ but $x^3 < 0$.

If the power of x in a function lies between 0 and 1 then, as long as x is positive, the value of the function increases as x gets larger, but its rate of increase gets smaller and smaller. The values in Table 4.3 and Figure 4.15 illustrate this for the function $y = x^{0.5}$ (where only the positive square root is considered).

If the power of x in a function is negative then, as long as x is positive, the graph of the function will slope downwards and take the shape of a curve convex to the origin. The examples in Table 4.4 are illustrated in Figure 4.16 for positive values of x. Note that the value of y in these functions gets larger as x approaches zero.

A firm's average fixed cost (AFC) schedule typically takes a shape similar to the functions illustrated in Figure 4.16.

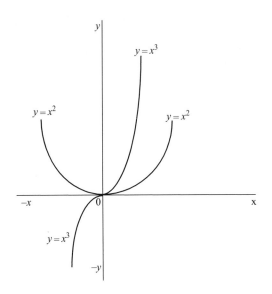

Figure 4.14

Table 4.3

x	0	1	2	3	4	5	6	7	8	9
$y = x^{0.5}$	0	1	1.414	1.732	2	2.236	2.449	2.646	2.828	3

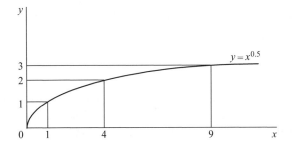

Figure 4.15

Table 4.4

x	0	0.1	1	2	3	4	5
$y = x^{-1}$	∞	10	1	0.5	0.33	0.25	0.2
$y = x^{-2}$	∞	100	1	0.25	0.11	0.0625	0.04

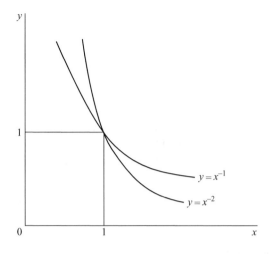

Figure 4.16

Example 4.12

A firm has to pay a fixed annual cost of £90,000 for leasing its premises. Derive its average fixed cost function (AFC).

Solution

$$\text{AFC} = \frac{\text{total fixed cost}}{Q} = \frac{90,000}{Q} = 90,000Q^{-1}$$

Although all values of Q^{-1} will be multiplied by 90,000, this will not alter the general shape of the function which will be similar to the graph of $y = x^{-1}$ illustrated in Figure 4.16 above.

Test Yourself, Exercise 4.7

1. Sketch the approximate shapes of the following functions for positive values of x and y.

   (a) $y = -8 + 0.2x^3$ (b) $y = 250 - 0.01x^2$

   (c) $y = x^{-1.5}$ (d) $y = x^{-0.5}$

   (e) $y = 20 - 0.2x^{-1}$

2. Sketch the approximate shapes of the following functions when x and y are allowed to take both positive and negative values.

   (a) $y = 4 + 0.1x^2$ (b) $y = 0.01x^3$

   (c) $y = 10 - x^{-1}$

3. Will the non-linear demand schedule $p = 570 - 0.4q^2$ get flatter or steeper as q rises?
4. A firm has to pay fixed costs of £65,000 before it starts production. What will its average fixed cost function look like? What will AFC be when output is 250?
5. Make up your own example of a non-linear function and sketch its approximate shape.

4.8   Composite functions

When a function has more than one term then one can build up the shape of the overall function from its different components. We have already done this when showing how a constant term determines the starting point of a function on the vertical axis of a graph. Now some more complex functions are considered.

Example 4.13

A firm faces the average fixed cost function

$$AFC = 200x^{-1}$$

where x is output, and the average variable cost (AVC) function

$$AVC = 0.2x^2$$

What shape will its average total cost function (AC) take?

Solution

The graphs of AFC and AVC are illustrated in Figure 4.17. By definition,

$$AC = AFC + AVC$$

Therefore, substituting the given AFC and AVC functions, average total cost is

$$AC = 200x^{-1} + 0.2x^2$$

For any given value of x, this means that the position of the AC function can be found by vertically summing the corresponding values on the AFC and AVC schedules.

As x gets larger then the value of AFC gets closer and closer to zero and so the value of AC gets closer and closer to AVC. Therefore the AC function will take the U-shape shown.

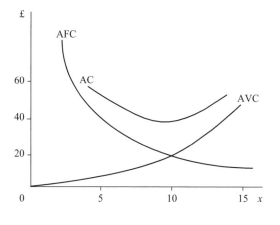

Figure 4.17

A composite function that takes the form

$$y = a_0 + a_1x + a_2x^2 + \cdots + a_nx^n$$

where $a_0, a_1, \ldots, a_n$ are constants and n is a non-negative integer is what is called a 'polynomial'. The 'degree' of the polynomial is the highest power value of x. For example, the total cost (TC) function

$$TC = 4 + 6x - 0.2x^2 + 0.1x^3$$

is a polynomial of the third degree. (See if you can sketch the shape of this function. Don't worry of you can't. In Section 6.6 we will return to polynomials and Section 4.9 explains how computer spreadsheets can help to plot functions.)

To see how the graph of a composite function is constructed when one term has a negative value, an example of a total revenue function is worked through below.

Example 4.14

If a demand schedule is represented by the function $P = 80 - 0.2Q$, what shape will the corresponding total revenue function take?

Solution

Total revenue (TR) is simply the total amount of money raised by selling a good and so

$$TR = PQ$$

If we substitute the demand function $P = 80 - 0.2Q$ for P in this TR function then

$$TR = (80 - 0.2Q)Q = 80Q - 0.2Q^2$$

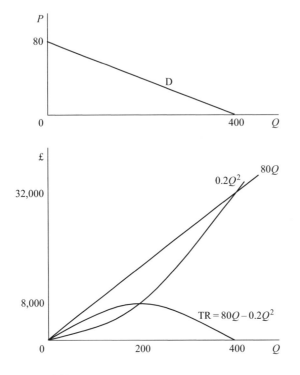

Figure 4.18

Now that we have derived the function for TR in terms of the single variable Q, its shape can be built up as shown in Figure 4.18. The component $80Q$ is clearly a straight line from the origin. The component $0.2Q^2$ is a curve rising at an increasing rate. One can easily see that, for low values of Q, $80Q > 0.2Q^2$. However, as Q becomes larger, the value of Q^2, and hence $0.2Q^2$, rapidly increases and eventually exceeds $80Q$.

Given that TR is the difference between $80Q$ and $0.2Q^2$, its value is the vertical distance between these two functions. This gets larger as Q increases to 200 and then decreases. It is zero when Q is 400 (when $80Q = 0.2Q^2$) and then becomes negative. Thus we get the inverted U-shape shown.

We can check that this shape makes sense by referring to the demand schedule $P = 80 - 0.2Q$ illustrated at the top of Figure 4.18. When Q is zero, nothing is sold and so TR must be zero. To sell 400, price must fall to zero and so again TR will be zero. Between these two output levels, TR will rise and then fall.

Slope of non-linear functions

We have seen how the slopes of non-linear composite functions can change along their length, but how can the slope of non-linear functions be measured from a graph? In Chapter 8 a mathematical method for finding the precise value of the slope of a function at any point using calculus is explained. Here we shall just consider an approximate geometrical method assuming that the graph of the function has already been drawn.

Example 4.15

The graph of the composite function

$$y = 40x - 2x^2$$

is illustrated in Figure 4.19. Find its slope at point A where $x = 5$.

Solution

First find the y coordinate of point A which will be

$$y = 40x - 2x^2 = 40(5) - 2(5)^2 = 200 - 50 = 150$$

Draw a straight line that just touches the curve at point A. This line is known as a 'tangent' and is shown by TT′ in Figure 4.19. The slope of the line is the same as the slope of the function at A.

To understand why this is so, first consider point B which is slightly to the left of A. The function is steeper at B than at A and also has a greater slope than the tangent at A. On the other hand, at point C (slightly to the right of A) the function is flatter than at A and has a slope less than that of the tangent TT′. If the slope of the tangent TT′ is less than the slope of the function for points slightly to the left of A and greater than the slope of the function for points slightly to the right of A, then it will be equal to the slope of the function at point A itself.

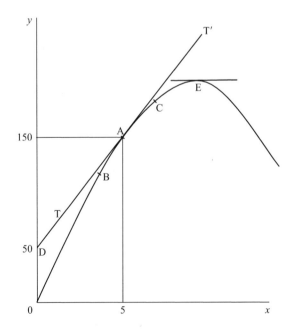

Figure 4.19

To determine the actual value of the slope of the tangent TT′ and hence the value of the slope of the function at A, it is necessary to find two sets of coordinates for the line TT′. We already know that it goes through A where $x = 5$ and $y = 150$. If TT′ is extended leftwards, it cuts the y axis at D where y is 50.

Using the method explained in Section 4.5, we can now fit an equation in the format $y = a + bx$ to this straight line.

We have already worked out that the y intercept a is 50 when $x = 0$. Between points D and A the value of x increases from 0 to 5 and the value of y increases by 100 from 50 to 150. Therefore, for every one unit increment in x the increase in y must be

$$\frac{100}{5} = 20 = b = \text{the slope of the tangent}$$

Therefore the equation $y = 50 + 20x$ can then be fitted to TT′.

The slope of this tangent is 20 and so the slope of the function $y = 40 - 2x^2$ will also be 20 at point A.

The slope of the function at other points can be determined in the same way by drawing other tangents. For example, the slope at the highest point of the function E will be the same as the slope of the tangent at E. This tangent is a horizontal line. A horizontal line always has a slope of zero and so *at its maximum point the slope of this function will also be zero.*

Sometimes you may encounter composite functions with similar terms. The summed function can then usually be simplified so that it does not remain a composite function.

Example 4.16

A firm's manufacturing system requires two processes for each unit produced. Process A involves a fixed cost of £650 plus £15 for each unit produced and process B involves a fixed cost of £220 plus £45 for each unit. What is the composite total cost function?

Solution

For process A

$$TC_A = 650 + 15Q$$

For process B

$$TC_B = 220 + 45Q$$

The overall total cost is therefore

$$TC = TC_A + TC_B$$
$$= (650 + 15Q) + (220 + 45Q)$$

The summed function is thus

$$TC = 870 + 60Q$$

Test Yourself, Exercise 4.8

1. Sketch the approximate shape of the following composite functions for positive values of all independent variables.

    (a) $TR = 40q - 4q^2$

    (b) $TC = 12 + 4q + 0.2q^2$

    (c) $\pi = -12 + 36q - 3.8q^2$

    (d) $y = 15 - 2x^{-1}$

    (e) $AVC = 8 - 3q + 0.5q^2$

2. Make up your own example of a composite function and sketch its approximate shape.
3. A firm is able to sell all its output at a fixed price of £50 per unit. If its average cost of production is given by the function

    $$AC = 100x^{-1} + 0.4x^2$$

    where x is output, derive a function for profit (π) in terms of x. What approximate shape will this profit function take?
4. A small group of companies operate in an industry where all firms face the average cost function $AC = 40 + 1,250q^{-1}$ where q is output per week. This function refers only to production costs. They then decide to launch an advertising campaign, not just to try to increase sales but also to try to raise the total average cost of low output levels and deter potential smaller-scale rival firms from competing in the same market. The cost of the advertising campaign is £2,000 per week per firm and any competitor would have to spend the same sum on advertising if it wished to compete in this market.

    (a) Derive a function for the new total average cost function including advertising, and sketch its approximate shape.
    (b) Explain why this advertising campaign will deter competition if the original companies sell a 100 units a week at a price of £100 each and new competitors cannot produce more than 25 units a week.

4.9 Using Excel to plot functions

It may not immediately be obvious what shape some composite functions take. If this is the case then it may help to set up the function as a formula on a spreadsheet and then see how the value of function changes over a range of values for the independent variable. Learning how to set up your own formulae on a spreadsheet can help you to in a number of ways. In particular, spreadsheets can be very useful and save you a lot of time and effort when tackling problems that entail very complex and time-consuming numerical calculations. They can also be used to plot graphs to get a picture of how functions behave and to check that answers to mathematical problems derived from manual calculations are correct. This book will not teach you how to use Excel, or any other computer spreadsheet package, from scratch. It is assumed that most students will already know the basics of creating files and spreadsheets, or will learn about them as part of their course. What we will do here is run through some methods of using spreadsheets to help solve, or illustrate and make clearer, certain aspects of economic analysis. In particular, spreadsheet applications will be explained when manual calculation would be very time-consuming. The detailed instructions for constructing spreadsheets are given in Excel format, as this is now the most commonly used spreadsheet package. However, the basic principles for constructing the formulae relevant to economic analysis can also be applied to other spreadsheet programmes.

Although Excel offers a range of in-built formulae for commonly used functions, such as square root, for many functions you will encounter in economics you will need to create your own formulae. A few reminders on *how to enter a formula* in an Excel spreadsheet cell:

- Start with the sign $=$
- Use the usual arithmetic $+$ and $-$ signs on your keyboard, with $*$ for multiplication and $/$ for division.
- Do not leave any spaces between characters and make sure you use brackets properly.
- For powers use the sign $^\wedge$ and also for roots which must be specified as powers, e.g. use $^\wedge 0.5$ to denote square root.
- Arithmetic operations can be performed on numbers typed into a formula or on cell references that contain a number.
- When you copy a formula to another cell all the references to other cells change unless you anchor their row or column by typing the $ sign in front of it in the formula.
- The quickest way to copy cell contents in Excel is to

    (a) highlight the cells to be copied
    (b) hold the cursor over the bottom right corner of the cell (or block of cells) to be copied until the $+$ sign appears
    (c) drag highlighted block over the cells where copy is to go.

Example 4.17

Use an Excel spreadsheet to calculate values for TR for the function $TR = 80Q - 0.2Q^2$ from *Example 4.14* above for range the range where both Q and TR take positive values. and then plot these values on a graph.

Solution

To answer this question, the essential features of the required spreadsheet are:

- A column of values for Q.
- Another column that calculates the value of TR corresponding to the value in the Q column.

Table 4.5 shows what to enter in the different cells of a spreadsheet to generate the relevant ranges of values and also gives a brief explanation of what each entry means. Once a formula has been entered only the calculated value appears in the cell where the formula is. However, when you put the cursor on a cell containing a formula, the full formula should always appear in the formula bar just above the spreadsheet.

When a formula is copied down a column any cell's numbers that the formula contains should also change. As the main formulae in this example are entered initially in row 4 and contain reference to cell A4, when they are copied to row 5 the reference should change to cell A5.

Table 4.5

CELL	Enter	*Explanation*
A1	Ex. 4.17	Label to remind you what example this is
B1	TR= 80Q – 0.2Q^2	Label to remind you what the demand schedule is. NB This is NOT an actual Excel formula because it does not start with the sign =
A3	Q	Column heading label
B3	TR	Column heading label
A4	0	Initial value for Q
B4	=80*A4– 0.2*A4^2 *(The value 0 should appear)*	This formula calculates the value for TR that corresponds to the value of Q in cell A4.
A5	=A4+20	Calculates a 20 unit increase in Q.
A6 to A25	*Copy cell A5 formula down column A*	Calculates a series of values of Q in 20 unit increments (so we will only need 25 rows in the spreadsheet rather than 400 plus.)
B5 to B25	*Copy cell B4 formula down column B*	Calculates values for TR in each row corresponding to the values of Q in column A.

If you follow these instructions you should end up with a spreadsheet that looks like Table 4.6. This clearly shows that TR increases as Q increases from 0 to 200 and then starts to decrease.

We can also use this spreadsheet to read off the value of TR for any given quantity. This can save you entering the whole formula in a calculator every time you have to find a value of the function. (Although we have only used increments of 20 units for Q to keep down the number of rows, the same formula can be used to calculate TR for any value of Q.)

Plotting a graph using Excel

Although it is obvious just by looking at the values of TR that this function rises and then falls, it is not quite so easy to get an idea of the exact shape of the function. It is easy, though, to use Excel to plot a graph for the columns of data for Q and TR generated in the spreadsheet.

1. Put the cursor on a cell in the region of the spreadsheet where you want the chart to go. You can adjust the position and size of the chart afterwards so don't worry too much about this, but try to choose a cell, such as F5, that is well away from the data columns so that you will still be able to see the data when the chart instructions appear.
2. Click on the Chart Wizard button at the top of your screen (the one with coloured columns) so that you enter Step 1 Chart Type.
3. Select 'Line' for the Chart Type and click on the first box in the Chart Sub-type examples. (This will give a plain line graph.) Then hit the Next button.

4. The cursor should now be flashing in the Data Range box. Use the mouse to take the cursor to cell A3, where the data start, then drag so that the dotted lines enclose the whole range A3 to B25, including the column headings. Once you let go of the left side of the mouse these cells should appear in the Data Range box.
5. Now click on the Series tag at the top of the grey instruction box.
6. At the bottom where it says 'Category (X) axis label' click on the white box and then use the mouse to take the cursor to cell A3 in the data and then drag down the Q column so that the dotted lines enclose the Q range A3 to A25. (This is to put Q on the horizontal axis.)
7. In the other box that says 'Series', make sure that Q is highlighted then click the 'Remove' button. (Otherwise the chart would draw a graph of Q.)
8. Click Next to go to Step 3.
9. You can choose your own labels, but probably best to enter 'TR $= 80Q - 0.2Q^2$' in the Chart title box and 'Q' in the Category (X) axis label box.
10. Click Next to go to Step 4.
11. Make sure 'Sheet 1' is shown in the bottom box and the 'As object in' button is clicked and has a black dot in the circle.
12. Click the Finish button, and your chart should appear.

If you want to enlarge or reposition the chart just click on it and then click on a corner or edge and drag. Clicking on the chart itself will allow you to change colours, which may be helpful if pale colours on graphs don't come out clearly on your black-and-white printer. You

Table 4.6

	A	B	C	D	E
1	Ex 4.17	TR = 80 - 0.2Q^2			
2					
3	Q	TR			
4	0	0			
5	20	1520			
6	40	2880			
7	60	4080			
8	80	5120			
9	100	6000			
10	120	6720			
11	140	7280			
12	160	7680			
13	180	7920			
14	200	8000			
15	220	7920			
16	240	7680			
17	260	7280			
18	280	6720			
19	300	6000			
20	320	5120			
21	340	4080			
22	360	2880			
23	380	1520			
24	400	0			
25	420	-1680			

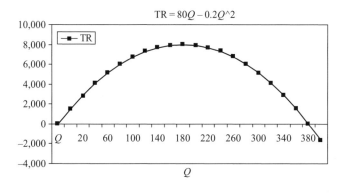

Figure 4.20

can also click on Chart in the toolbar at the top of the screen to go back and alter any of the formatting details, e.g. print font size. (The Data button in the toolbar only changes to Chart when the chart itself is clicked on.) Try experimenting to learn how to get the chart format that suits you best.

Your finished graph should look similar to Figure 4.20. This confirms that this function takes a smooth inverted U-shape. It has zero value when Q is 0 and 400 and has its maximum value of 8,000 when Q is 200. We will use this tool again in Section 6.6 to help find solutions to polynomial equations.

Test Yourself, Exercise 4.9

Use an Excel spreadsheet to plot values and draw graphs of the following functions:

1. $TR = 40q - 4q^2$
2. $TC = 12 + 4q + 0.2q^2$
3. $\pi = -12 + 36q - 3.8q^2$
4. $AC = 24q^{-1} + 8 - 3q + 0.5q^2$

4.10 Functions with two independent variables

On a two-dimensional sheet of paper you cannot sketch a function with more than one independent variable as this would require more than two axes (one for the dependent variable and one each for the independent variables). However, in economics we often need to analyse functions that have two or more independent variables, e.g. production functions. When there are more than two independent variables then a function cannot really be visually represented (and mathematical analysis has to be employed), but when there are only two independent variables a 'contour line' graphing method can be used.

Consider the production function

$$Q = f(K, L)$$

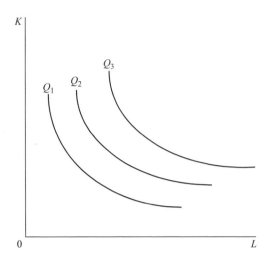

Figure 4.21

Table 4.7

K	L	$K^{0.5}$	$L^{0.5}$	Q
64	4	8	2	320
16	16	4	4	320
4	64	2	8	320
256	1	16	1	320
1	256	1	16	320

Assume that the way in which Q depends on K and L is represented by the height above the two-dimensional surface on which K and L are measured. To show this production 'height' economics borrows the idea of contour lines from geography. On a map, contour lines join points of equal height and so, for example, a steep hill will be represented by closely spaced contour lines. In production theory a line that joins combinations of inputs K and L that will give the same production level (when used efficiently) is known as an 'isoquant'. An 'isoquant map' is shown in Figure 4.21. Isoquants normally show equal increments in output level which enables one to get an idea of how quickly output responds to changes in the inputs. If isoquants are spaced far apart then output increases relatively slowly, and if they are spaced closely together then output increases relatively quickly.

One can plot the position of an isoquant map from a production function although this is a rather tedious, long-winded business. As we shall see later, it is not usually necessary to draw in all the isoquants in order to tackle some of the resource allocation problems that this concept can be used to illustrate. Examples of some of the different combinations of K and L that would produce an output of 320 with the production function

$$Q = 20K^{0.5}L^{0.5}$$

are shown in Table 4.7. In this particular case there is a symmetrical curve known as a 'rectangular hyperbola' for the isoquant $Q = 320$.

A quicker way of finding out the shape of an isoquant is to transform it into a function with only two variables.

Example 4.18

For the production function $Q = 20K^{0.5}L^{0.5}$ derive a two-variable function in the form $K = f(L)$ for the isoquant $Q = 100$.

Solution

$$20K^{0.5}L^{0.5} = Q = 100$$

Thus $K^{0.5}L^{0.5} = 5$.

$$K^{0.5} = \frac{5}{L^{0.5}}.$$

Squaring both sides gives the required function

$$K = \frac{25}{L} = 25L^{-1}$$

From Section 4.7 we know that this form of function will give a curve convex to the origin since the value of K gets closer to zero as L increases in value.

Example 4.19

For the production function $Q = 4.5K^{0.4}L^{0.7}$ derive a function in the form $K = f(L)$ for the isoquant representing an output of 54.

Solution

$$Q = 54 = 4.5K^{0.4}L^{0.7}$$
$$12 = K^{0.4}L^{0.7}$$
$$12L^{-0.7} = K^{0.4}$$

Taking both sides to the power 2.5

$$12^{2.5}L^{-1.75} = K$$
$$K = 498.83L^{-1.75}$$

This function will also give a curve convex to the origin since the value of $L^{-1.75}$ (and hence K) gets closer to zero as L increases in value.

The Cobb–Douglas production function

The production functions given in this section are examples of what are known as 'Cobb–Douglas' production functions. The general format of a Cobb–Douglas production function with two inputs K and L is

$$Q = AK^{\alpha}L^{\beta}$$

where A, α and β are parameters. (The Greek letter α is pronounced 'alpha' and β is 'beta'.) Many years ago, the two economists Cobb and Douglas found this form of function to be a good match to the statistical evidence on input and output levels that they studied. Although economists have since developed more sophisticated forms of production functions, this basic Cobb–Douglas production function is a good starting point for students to examine the relationship between a firm's output level and the inputs required, and hence costs.

Cobb–Douglas production functions fall into the mathematical category of *homogeneous* functions. In general terms, a function is said to be homogeneous of degree m if, when all inputs are multiplied by any given positive constant λ, the value of y increases by the proportion λ^{m}. (λ is the Greek letter 'lambda'.) Thus if

$$y = f(x_1, x_2, \ldots, x_n)$$
$$\text{then} \quad y\lambda^{m} = f(\lambda x_1, \lambda x_2, \ldots, \lambda x_n)$$

An example of a function that is homogeneous of degree 1 is the production function

$$Q = 20K^{0.5}L^{0.5}.$$

The powers in a Cobb–Douglas production function determine the degree of returns to scale present.

Assume that initially the input amounts are K_1 and L_1, giving production level

$$Q_1 = 20K_1^{0.5}L_1^{0.5}$$

If input amounts are doubled (i.e. $\lambda = 2$) then the new input amounts are

$$K_2 = 2K_1 \quad \text{and} \quad L_2 = 2L_1$$

giving the new output level

$$Q_2 = 20K_2^{0.5}L_2^{0.5} \tag{1}$$

This can be compared with the original output level by substituting $2K_1$ for K_2 and $2L_1$ for L_2. Thus

$$Q_2 = 20(2K_1)^{0.5}(2L_1)^{0.5} = 20(2^{0.5}K_1^{0.5}2^{0.5}L_1^{0.5}) = 2(20K_1^{0.5}L_1^{0.5}) = 2Q_1$$

Therefore, when inputs are doubled, output doubles, and so this production function exhibits *constant returns to scale*.

The degree of homogeneity of a Cobb–Douglas production function can easily be determined by adding up the indices of the input variables. This can be demonstrated for the two-input function

$$Q = AK^{\alpha}L^{\beta}$$

If we let initial input amounts be K_1 and L_1, then

$$Q_1 = AK_1^\alpha L_1^\beta$$

If all inputs are multiplied by the constant λ then new input amounts will be

$$K_2 = \lambda K_1 \quad \text{and} \quad L_2 = \lambda L_1$$

The new output level will then be

$$Q_2 = AK_2^\alpha L_2^\beta = A(\lambda K_1)^\alpha (\lambda L_1)^\beta = \lambda^{\alpha+\beta} AK_1^\alpha L_1^\beta = \lambda^{\alpha+\beta} Q_1$$

Given that λ, α and β are all assumed to be positive numbers, this result tells us the relationship between α and β and the three possible categories of returns to scale.

1. If $\alpha + \beta = 1$ then $\lambda^{\alpha+\beta} = \lambda$ and so $Q_2 = \lambda Q_1$, i.e. constant returns to scale.
2. If $\alpha + \beta > 1$ then $\lambda^{\alpha+\beta} > \lambda$ and so $Q_2 > \lambda Q_1$, i.e. increasing returns to scale.
3. If $\alpha + \beta < 1$ then $\lambda^{\alpha+\beta} < \lambda$ and so $Q_2 < \lambda Q_1$, i.e. decreasing returns to scale.

Example 4.20

What type of returns to scale does the production function $Q = 45K^{0.4}L^{0.4}$ exhibit?

Solution

Indices sum to $0.4 + 0.4 = 0.8$. Thus the degree of homogeneity is less than 1 and so there are decreasing returns to scale.

To estimate the parameters of Cobb–Douglas production functions requires the use of logarithms. The standard linear regression analysis method (that you should cover in your statistics module) allows you to use data on p and q to estimate the parameters a and b in linear functions such as the supply schedule

$$p = a + bq$$

If you have a non-linear function, logarithms can be used to transform it into a linear form so that linear regression analysis method can be used to estimate the parameters. For example, the Cobb–Douglas production function

$$Q = AK^a L^b$$

can be put into log form as

$$\log Q = \log A + a \log K + b \log L$$

so that a and b can be estimated by linear regression analysis.

In your economics course you should learn how the optimum input combination for a firm can be discovered using budget constraints, production functions and isoquant maps. We shall return to these concepts in Chapters 8 and 11, when mathematical solutions to optimization problems using calculus are explained.

Test Yourself, Exercise 4.10

For the production functions below, assume fractions of a unit of K and L can be used, and

(a) derive a function for the isoquant representing the specified output level in the form $K = f(L)$

(b) find the level of K required to achieve the given output level if $L = 100$, and

(c) say what type of returns to scale are present.

1. $Q = 9K^{0.5}L^{0.5}$, $Q = 36$
2. $Q = 0.3K^{0.4}L^{0.6}$, $Q = 24$
3. $Q = 25K^{0.6}L^{0.6}$, $Q = 800$
4. $Q = 42K^{0.6}L^{0.75}$, $Q = 5{,}250$
5. $Q = 0.4K^{0.3}L^{0.5}$, $Q = 65$
6. $Q = 2.83K^{0.35}L^{0.62}$, $Q = 52$
7. Use logs to put the production function $Q = AK^{\alpha}L^{\beta}R^{\gamma}$ into a linear format.

4.11   Summing functions horizontally

In economics, there are several occasions when theory requires one to sum certain functions 'horizontally'. Students are most likely to encounter this concept when studying the theory of third-degree price discrimination and the theory of multiplant monopoly and/or cartels. By 'horizontally' summing a function we mean summing it along the horizontal axis. This idea is best explained with an example.

Example 4.21

A price-discriminating monopolist sells in two separate markets at prices P_1 and P_2 (measured in £). The relevant demand and marginal revenue schedules are (for positive values of Q)

$$P_1 = 12 - 0.15Q_1 \qquad P_2 = 9 - 0.075Q_2$$
$$MR_1 = 12 - 0.3Q_1 \qquad MR_2 = 9 - 0.15Q_2$$

It is assumed that output is allocated between the two markets according to the price-discrimination revenue-maximizing criterion that $MR_1 = MR_2$. Derive a formula for the aggregate marginal revenue schedule which is the horizontal sum of MR_1 and MR_2.

(Note: In Chapter 5, we shall return to this example to find out how this summed MR schedule can help determine the profit-maximizing prices P_1 and P_2 when marginal cost is known.)

Solution

The two schedules MR_1 and MR_2 are illustrated in Figure 4.22. What we are required to do is find a formula for the summed schedule MR. This tells us what aggregate output will correspond to a given level of marginal revenue and vice versa, assuming that output is adjusted so that the marginal revenue from the last unit sold in each market is the same.

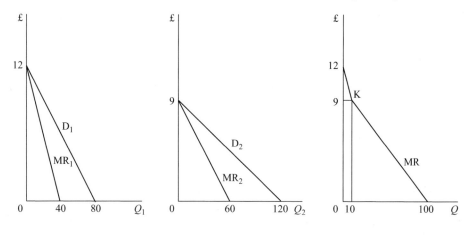

Figure 4.22

As you can see in Figure 4.22, the summed MR schedule is in fact kinked at point K. This is because the MR schedule sums the horizontal distances of MR_1 and MR_2 from the price axis. Given that MR_2 starts from a price of £9, then above £9 the only distance being summed is the distance between MR_1 and the price axis. Thus between £12 and £9 MR is the same as MR_1, i.e.

$$MR = 12 - 0.3Q$$

where Q is aggregate output. If MR = £9 then

$$9 = 12 - 0.3Q$$
$$0.3Q = 3$$
$$Q = 10$$

Thus the coordinates of the kink K are £9 and 10 units of output.
The proper summation occurs below £9. We are given the schedules

$$MR_1 = 12 - 0.3Q_1 \quad \text{and} \quad MR_2 = 9 - 0.15Q_2$$

but if we simply added MR_1 and MR_2 we would be summing vertically instead of horizontally. To be summed horizontally, these marginal revenue functions first have to be transposed to obtain their inverse functions as follows:

$$MR_1 = 12 - 0.3Q_1 \qquad\qquad MR_2 = 9 - 0.15Q_2$$
$$0.3Q_1 = 12 - MR_1 \qquad\qquad 0.15Q_2 = 9 - MR_2$$
$$Q_1 = 40 - 3\tfrac{1}{3}MR_1 \quad (1) \qquad Q_2 = 60 - 6\tfrac{2}{3}MR_2 \quad (2)$$

Given that the theory of price discrimination assumes that a firm will adjust the amount sold in each market until $MR_1 = MR_2 = MR$, then

$$Q = Q_1 + Q_2$$

$$= \left(40 - 3\tfrac{1}{3}MR\right) + \left(60 - 6\tfrac{2}{3}MR\right) \qquad \text{by substituting (1) and (2)}$$

$$= 100 - 10MR$$

$$10MR = 100 - Q$$

$$MR = 10 - 0.1Q$$

This summed MR function will apply above an aggregate output of 10.

From the above example it can be seen that the basic procedure for summing functions horizontally is as follows:

1. transform the functions so that quantity is the dependent variable;
2. sum the functions representing quantities;
3. transform the function back so that quantity is the independent variable again;
4. note the quantity range that this summed function applies to, given the intersection points of the functions to be summed on the price axis.

This procedure can also be applied to multiplant monopoly examples where it is necessary to find the horizontally summed marginal cost schedule.

Example 4.22

A monopoly operates two plants whose marginal cost schedules are

$$MC_1 = 2 + 0.2Q_1 \quad \text{and} \quad MC_2 = 6 + 0.04Q_2$$

Find the function which describes the horizontal summation of these two functions.
(As with the previous example, we shall return to the use of the summed function in determining profit-maximizing price and output levels in Chapter 5.)

Solution

The relevant schedules are illustrated in Figure 4.23. The horizontal sum of MC_1 and MC_2 will be the function MC which is kinked at K. Below £6 only MC_1 is relevant. Therefore, MC is the same as MC_1 from £2 to £6. The corresponding output range can be found by substituting £6 for MC_1. Thus

$$MC_1 = 6 = 2 + 0.2Q_1$$

$$4 = 0.2Q_1$$

$$20 = Q_1$$

Therefore $MC = 2 + 0.2Q$ between $Q = 0$ and $Q = 20$.

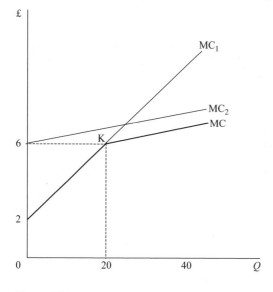

Figure 4.23

Above this output we need to derive the proper sum of the two functions. Given

$$MC_1 = 2 + 0.2Q_1 \qquad \text{and} \qquad MC_2 = 6 + 0.04Q_2$$

then
$$MC_1 - 2 = 0.2Q_1 \qquad \text{and} \qquad MC_2 - 6 = 0.04Q_2$$

$$5MC_1 - 10 = Q_1 \qquad (1) \qquad 25MC_2 - 150 = Q_2 \qquad (2)$$

Summing the functions (1) and (2) gives

$$Q = Q_1 + Q_2 = (5MC_1 - 10) + (25MC_2 - 150) \qquad (3)$$

A profit-maximizing monopoly will adjust output between two plants until

$$MC_1 = MC_2 = MC$$

Therefore, substituting MC into (3) gives

$$Q = 5MC - 10 + 25MC - 150$$
$$Q = 30MC - 160$$
$$160 + Q = 30MC$$
$$5\tfrac{1}{3} + \tfrac{1}{30}Q = MC$$

This summed MC function applies above an output level of 20.

In the examples above the summation of only two linear functions was considered. The method can easily be adapted to situations when three or more linear functions are to be summed. However, the inverses of some non-linear functions are not in forms that can easily be summed and so this method is best confined to applications involving linear functions.

Test Yourself, Exercise 4.11

Sum the following sets of marginal revenue and marginal cost schedules horizontally to derive functions in the form $MR = f(Q)$ or $MC = f(Q)$ and define the output ranges over which the summed function applies.

1. $MR_1 = 30 - 0.01Q_1$ and $MR_2 = 40 - 0.02Q_2$
2. $MR_1 = 80 - 0.4Q_1$ and $MR_2 = 71 - 0.5Q_2$
3. $MR_1 = 48.75 - 0.125Q_1$ and $MR_2 = 75 - 0.3Q_2$ and $MR_3 = 120 - 0.15Q_3$
4. $MC_1 = 20 + 0.25Q_1$ and $MC_2 = 34 + 0.1Q_2$
5. $MC_1 = 60 + 0.2Q_1$ and $MC_2 = 48 + 0.4Q_2$
6. $MC_1 = 3 + 0.2Q_1$ and $MC_2 = 1.75 + 0.25Q_2$ and $MC_3 = 4 + 0.2Q_3$

5 Linear equations

Learning objectives

After completing this chapter students should be able to:

- Solve sets of simultaneous linear equations with two or more variables using the substitution and row operations methods.
- Relate mathematical solutions to simultaneous linear equations to economic analysis.
- Recognize when a linear equations system cannot be solved.
- Derive the reduced-form equations for the equilibrium values of dependent variables in basic linear economic models and interpret their meaning.
- Derive the profit-maximizing solutions to price discrimination and multiplant monopoly problems involving linear functions
- Set up linear programming constrained maximization and minimization problems and solve them using the graphical method.

5.1 Simultaneous linear equation systems

The way to solve single linear equations with one unknown was explained in Chapter 3. We now turn to sets of linear equations with more than one unknown. A simultaneous linear equation system exists when:

1. there is more than one functional relationship between a set of specified variables, and
2. all the functional relationships are in linear form.

The solution to a set of simultaneous equations involves finding values for all the unknown variables.

Where only two variables and equations are involved, a simultaneous equation system can be related to familiar graphical solutions, such as supply and demand analysis. For example, assume that in a competitive market the demand schedule is

$$p = 420 - 0.2q \tag{1}$$

and the supply schedule is

$$p = 60 + 0.4q \tag{2}$$

If this market is in equilibrium then the equilibrium price and quantity will be where the demand and supply schedules intersect. As this will correspond to a point which is on both the demand schedule and the supply schedule then the equilibrium values of p and q will be such that both equations (1) and (2) hold. In other words, when the market is in equilibrium (1) and (2) above form a set of simultaneous linear equations.

Note that in most of the examples in this chapter the 'inverse' demand and supply functions are used, i.e. $p = f(q)$ rather than $q = f(p)$. This is because price is normally measured on the vertical axis and we wish to relate the mathematical solutions to graphical analysis. However, simultaneous linear equations systems often involve more than two unknown variables in which case no graphical illustration of the problem will be possible. It is also possible that a set of simultaneous equations may contain non-linear functions, but these are left until the next chapter.

5.2   Solving simultaneous linear equations

The basic idea involved in all the different methods of algebraically solving simultaneous linear equation systems is to manipulate the equations until there is a single linear equation with one unknown. This can then be solved using the methods explained in Chapter 3. The value of the variable that has been found can then be substituted back into the other equations to solve for the other unknown values.

It is important to realize that not all sets of simultaneous linear equations have solutions. The general rule is that the number of unknowns must be equal to the number of equations for there to be a unique solution. However, even if this condition is met, one may still come across systems that cannot be solved, e.g. functions which are geometrically parallel and therefore never intersect (see Example 5.2 below).

We shall first consider four different methods of solving a 2×2 set of simultaneous linear equations, i.e. one in which there are two unknowns and two equations, and then look at how some of these methods can be employed to solve simultaneous linear equation systems with more than two unknowns.

5.3   Graphical solution

The graphical solution method can be used when there are only two unknown variables. It will not always give 100% accuracy, but it can be useful for checking that algebraic solutions are not widely inaccurate owing to analytical or computational errors.

Example 5.1

Solve for p and q in the set of simultaneous equations given previously in Section 5.1:

$$p = 420 - 0.2q \tag{1}$$
$$p = 60 + 0.4q \tag{2}$$

Solution

These two functional relationships are plotted in Figure 5.1. Both hold at the intersection point X. At this point the solution values

$$p = 300 \quad \text{and} \quad q = 600$$

can be read off the graph.

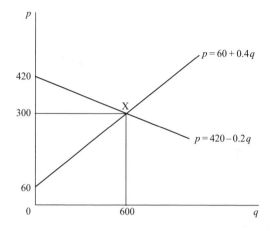

Figure 5.1

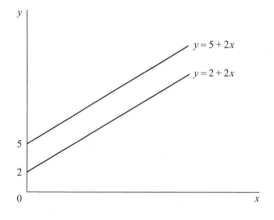

Figure 5.2

A graph can also illustrate why some simultaneous linear equation systems cannot be solved.

Example 5.2

Attempt to use graphical analysis to solve for y and x if

$$y = 2 + 2x \quad \text{and} \quad y = 5 + 2x$$

Solution

These two functions are plotted in Figure 5.2. They are obviously parallel lines which never intersect. This problem therefore does not have a solution.

Test Yourself, Exercise 5.1

Solve the following (if a solution exists) using graph paper.
1. In a competitive market, the demand and supply schedules are respectively

$$p = 9 - 0.075q \quad \text{and} \quad p = 2 + 0.1q$$

Find the equilibrium values of p and q.

2. Find x and y when

$$x = 80 - 0.8y \quad \text{and} \quad y = 10 + 0.1x$$

3. Find x and y when

$$y = -2 + 0.5x \quad \text{and} \quad x = 2y - 9$$

5.4   Equating to same variable

The method of equating to the same variable involves rearranging both equations so that the same unknown variable appears by itself on one side of the equality sign. This variable can then be eliminated by setting the other two sides of the equality sign in the two equations equal to each other. The resulting equation in one unknown can then be solved.

Example 5.3

Solve the set of simultaneous equations in Example 5.1 above by the equating method.

Solution

In this example no preliminary rearranging of the equations is necessary because a single term in p appears on the left-hand side of both. As

$$p = 420 - 0.2q \tag{1}$$

and

$$p = 60 + 0.4q \tag{2}$$

then it must be true that

$$420 - 0.2q = 60 + 0.4q$$

Therefore

$$360 = 0.6q$$
$$600 = q$$

The value of p can be found by substituting this value of 600 for q back into either of the two original equations. Thus

from (1) $p = 420 - 0.2q = 420 - 0.2(600) = 420 - 120 = 300$

or

from (2) $p = 60 + 04q = 60 + 0.4(600) = 60 + 240 = 300$

Example 5.4

Assume that a firm can sell as many units of its product as it can manufacture in a month at £18 each. It has to pay out £240 fixed costs plus a marginal cost of £14 for each unit produced. How much does it need to produce to break even?

Solution

From the information in the question we can work out that this firm faces the total revenue function TR $= 18q$ and the total cost function TC $= 240 + 14q$, where q is output. These functions are plotted in Figure 5.3, which is an example of what is known as a *break-even chart*. This is a rough guide to the profit that can be expected for any given production level.

The break-even point is clearly at B, where the TR and TC schedules intersect. Since

TC $= 240 + 14q$ and TR $= 18q$

and the break-even point is where TR $=$ TC, then

$$18q = 240 + 14q$$
$$4q = 240$$
$$q = 60$$

Therefore the output required to break even is 60 units.

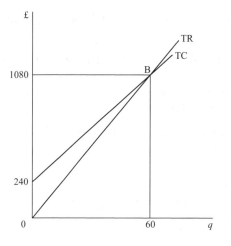

Figure 5.3

Note that in reality at some point the TR schedule will start to flatten out when the firm has to reduce price to sell more, and TC will get steeper when diminishing marginal productivity causes marginal cost to rise. If this did not happen, then the firm could make infinite profits by indefinitely expanding output. Break-even charts can therefore only be used for the range of output where the specified linear functional relationships hold.

What happens if you try to use this algebraic method when no solution exists, as in Example 5.2 above?

Example 5.5

Attempt to use the equating to same variable method to solve for y and x if

$$y = 2 + 2x \quad \text{and} \quad y = 5 + 2x$$

Solution

Eliminating y from the system and equating the other two sides of the equations, we get

$$2 + 2x = 5 + 2x$$

Subtracting $2x$ from both sides gives $2 = 5$. This is clearly impossible, and hence no solution can be found.

Test Yourself, Exercise 5.2

1. A competitive market has the demand schedule $p = 610 - 3q$ and the supply schedule $p = 20 + 2q$. Calculate equilibrium price and quantity.
2. A competitive market has the demand schedule $p = 610 - 3q$ and the supply schedule $p = 50 + 4q$ where p is measured in pounds.

    (a) Find the equilibrium values of p and q.
    (b) What will happen to these values if the government imposes a tax of £14 per unit on q?

3. Make up your own linear functions for a supply schedule and a demand schedule and then:

    (a) plot them on graph paper and read off the values of price and quantity where they intersect, and
    (b) algebraically solve your set of linear simultaneous equations and compare your answer with the values you got for (a).

4. A firm manufactures product x and can sell any amount at a price of £25 a unit. The firm has to pay fixed costs of £200 plus a marginal cost of £20 for each unit produced.

    (a) How much of x must be produced to make a profit?
    (b) If price is cut to £24 what happens to the break-even output?

5. If $y = 16 + 22x$ and $y = -2.5 + 30.8x$, solve for x and y.

5.5 Substitution

The substitution method involves rearranging one equation so that one of the unknown variables appears by itself on one side. The other side of the equation can then be substituted into the second equation to eliminate the other unknown.

Example 5.6

Solve the linear simultaneous equation system

$$20x + 6y = 500 \tag{1}$$
$$10x - 2y = 200 \tag{2}$$

Solution

Equation (2) can be rearranged to give

$$10x - 200 = 2y$$
$$5x - 100 = y \tag{3}$$

If we substitute the left-hand side of equation (3) for y in equation (1) we get

$$20x + 6y = 500$$
$$20x + 6(5x - 100) = 500$$
$$20x + 30x - 600 = 500$$
$$50x = 1,100$$
$$x = 22$$

To find the value of y we now substitute this value of x into (1) or (2). Thus, in (1)

$$20x + 6y = 500$$
$$20(22) + 6y = 500$$
$$440 + 6y = 500$$
$$6y = 60$$
$$y = 10$$

Example 5.7

Find the equilibrium level of national income in the basic Keynesian macroeconomic model

$$Y = C + I \tag{1}$$
$$C = 40 + 0.5Y \tag{2}$$
$$I = 200 \tag{3}$$

Solution

Substituting the consumption function (2) and given I value (3) into (1) we get

$$Y = 40 + 0.5Y + 200$$

Therefore

$$0.5Y = 240$$
$$Y = 480$$

Test Yourself, Exercise 5.3

1. A consumer has a budget of £240 and spends it all on the two goods A and B whose prices are initially £5 and £10 per unit respectively. The price of A then rises to £6 and the price of B falls to £8. What combination of A and B that uses up all the budget is it possible to purchase at both sets of prices?

2. Find the equilibrium value of Y in a basic Keynesian macroeconomic model where

    $Y = C + I$ the accounting identity
    $C = 20 + 0.6Y$ the consumption function
    $I = 60$ exogenously determined

3. Solve for x and y when

    $$600 = 3x + 0.5y$$
    $$52 = 1.5y - 0.2x$$

5.6   Row operations

Row operations entail multiplying or dividing all the terms in one equation by whatever number is necessary to get the coefficient of one of the unknowns equal to the coefficient of that same unknown in another equation. Then, by subtraction of one equation from the other, this unknown can be eliminated.

Alternatively, if two rows have the same absolute value for the coefficient of an unknown but one coefficient is positive and the other is negative, then this unknown can be eliminated by adding the two rows.

Example 5.8

Given the equations below, use row operations to solve for x and y.

$$10x + 3y = 250 \tag{1}$$
$$5x + y = 100 \tag{2}$$

Solution

Multiplying (2) by 3	$15x + 3y = 300$
Subtracting (1)	$10x + 3y = 250$
Gives	$5x = 50$
	$x = 10$

Substituting this value of x back into (1),

$$10(10) + 3y = 250$$
$$100 + 3y = 250$$
$$3y = 150$$
$$y = 50$$

Example 5.9

A firm makes two goods A and B which require two inputs K and L. One unit of A requires 6 units of K plus 3 units of L and one unit of B requires 4 units of K plus 5 units of L. The firm has 420 units of K and 300 units of L at its disposal. How much of A and B should it produce if it wishes to exhaust its supplies of K and L totally?

(NB. This question requires you to use the economic information given to set up a mathematical problem in a format that can be used to derive the desired solution. Learning how to set up a problem is just as important as learning how to solve it.)

Solution

The total requirements of input K are 6 for every unit of A and 4 for each unit of B, which can be written as

$$K = 6A + 4B$$

Similarly, the total requirements of input L can be specified as

$$L = 3A + 5B$$

As we know that $K = 420$ and $L = 300$ because all resources are used up, then

$$420 = 6A + 4B \qquad (1)$$

and

$$300 = 3A + 5B \qquad (2)$$

Multiplying (2) by 2 $600 = 6A + 10B$

Subtracting (1) $\underline{420 = 6A + 4B}$

gives $180 = 6B$

$$30 = B$$

Substituting this value for B into (1) gives

$$420 = 6A + 4(30)$$
$$420 = 6A + 120$$
$$300 = 6A$$
$$50 = A$$

The firm should therefore produce 50 units of A and 30 units of B.

(Note that the method of setting up this problem will be used again when we get to linear programming in the Appendix to this chapter.)

Test Yourself, Exercise 5.4

1. Solve for x and y if

   $$420 = 4x + 5y \quad \text{and} \quad 600 = 2x + 9y$$

2. A firm produces the two goods A and B using inputs K and L. Each unit of A requires 2 units of K plus 6 units of L. Each unit of B requires 3 units of K plus 4 units of L. The amounts of K and L available are 120 and 180, respectively. What output levels of A and B will use up all the available K and L?
3. Solve for x and y when

   $$160 = 8x - 2y \quad \text{and} \quad 295 = 11x + y$$

5.7   More than two unknowns

With more than two unknowns it is usually best to use the row operations method. The basic idea is to use one pair of equations to eliminate one unknown and then bring in another equation to eliminate the same variable, repeating the process until a single equation in one unknown is obtained. The exact operations necessary will depend on the format of the particular problem. There are several ways in which row operations can be used to solve most problems and you will only learn which is the quickest method to use through practising examples yourself.

Example 5.10

Solve for x, y and z, given that

$$x + 12y + 3z = 120 \tag{1}$$
$$2x + y + 2z = 80 \tag{2}$$
$$4x + 3y + 6z = 219 \tag{3}$$

Solution

Multiplying (2) by 2	$4x + 2y + 4z = 160$	(4)
Subtracting (4) from (3)	$y + 2z = 59$	(5)

We have now eliminated x from equations (2) and (3) and so the next step is to eliminate x from equation (1) by row operations with one of the other two equations. In this example the easiest way is

Multiplying (1) by 2	$2x + 24y + 6z = 240$
Subtracting (2)	$2x + y + 2z = 80$

$$23y + 4z = 160 \tag{6}$$

We now have the set of two simultaneous equations (5) and (6) involving two unknowns to solve. Writing these out again, we can now use row operations to solve for y and z.

$$y + 2z = 59 \tag{5}$$
$$23y + 4z = 160 \tag{6}$$

Multiplying (5) by 2	$2y + 4z = 118$
Subtracting (6)	$23y + 4z = 160$
Gives	$-21y = -42$
	$y = 2$

Substituting this value for y into (5) gives

$$2 + 2z = 59$$
$$2z = 57$$
$$z = 28.5$$

These values for y and z can now be substituted into any of the original equations. Thus using (1) we get

$$x + 12(2) + 3(28.5) = 120$$
$$x + 24 + 85.5 = 120$$
$$x = 120 - 109.5$$
$$x = 10.5$$

Therefore, the solutions are $x = 10.5$, $y = 2$, $z = 28.5$.

Example 5.11

Solve for x, y and z in the following set of simultaneous equations:

$$14.5x + 3y + 45z = 340 \tag{1}$$
$$25x - 6y - 32z = 82 \tag{2}$$
$$9x + 2y - 3z = 16 \tag{3}$$

Solution

Multiplying (1) by 2	$29x + 6y + 90z = 680$	
Adding (2)	$25x - 6y - 32z = 82$	(2)
Gives	$54x + 58z = 762$	(4)

Having used equations (1) and (2) to eliminate y we now need to bring in equation (3) to derive a second equation containing only x and z.

Multiplying (3) by 3	$27x + 6y - 9z = 48$	
Adding (2)	$25x - 6y - 32z = 82$	(2)
gives	$52x - 41z = 130$	(5)

Multiplying (5) by 27	$1,404x - 1,107z = 3,510$
Multiplying (4) by 26	$1,404x + 1,508z = 19,812$
Subtracting gives	$-2,615z = -16,302$
	$z = 6.234$

(Note that although final answers are more neatly specified to one or two decimal places, more accuracy will be maintained if the full value of z above is entered when substituting to calculate remaining values of unknown variables.)

Substituting the above value of z into (5) gives

$$52x - 41(6.234) = 130$$
$$52x = 130 + 255.594$$
$$x = 7.415$$

Substituting for both x and z in (1) gives

$$14.5(7.415) + 3y + 45(6.234) = 340$$
$$3y = -48.05$$
$$y = -16.02$$

Thus, solutions to 2 decimal places are

$$x = 7.42 \quad y = -16.02 \quad z = 6.23$$

The above examples show how the solution to a 3×3 set of simultaneous equations can be solved by row operations. The same method can be used for larger sets but obviously more stages will be required to eliminate the unknown variables one by one until a single equation with one unknown is arrived at.

It must be stressed that it is only practical to use the methods of solution for linear equation systems explained here where there are a relatively small number of equations and unknowns. For large systems of equations with more than a handful of unknowns it is more appropriate to use matrix algebra methods and an Excel spreadsheet (see Chapter 15).

Test Yourself, Exercise 5.5

1. Solve for x, y and z when

    $2x + 4y + 2z = 144$ (1)

    $4x + y + 0.5z = 120$ (2)

    $x + 3y + 4z = 144$ (3)

2. Solve for x, y and z when

    $12x + 15y + 5z = 158$ (1)

    $4x + 3y + 4z = 50$ (2)

    $5x + 20y + 2z = 148$ (3)

3. Solve for A, B and C when

    $32A + 14B + 82C = 664$ (1)

    $11.5A + 8B + 52C = 349$ (2)

    $18A + 26.2B - 62C = 560.4$ (3)

4. Find the values of x, y and z when

    $4.5x + 7y + 3z = 128.5$

    $6x + 18.2y + 12z = 270.8$

    $3x + 8y + 7z = 139$

5. Solve for A, B, C and D when

    $A + 6B + 25C + 17D = 843$

    $3A + 14B + 60C + 21D = 1{,}286.5$

    $10A + 3B + 4C + 28D = 1{,}206$

    $6A + 2B + 12C + 51D = 1{,}096$

5.8 Which method?

There is no hard and fast rule regarding which of the different methods for solving simultaneous equations should be used in different circumstances. The row operations method can be used for most problems but sometimes it will be quicker to use one of the other methods, particularly in 2×2 systems. It may also be quicker to change methods midway. For example, one may find that in a 3×3 problem it may be quicker to revert to the substitution method after one of the unknowns has been eliminated by row operations. Only by practising solving problems will you learn how to spot the quickest methods of solving them.

Not all economic problems are immediately recognizable as linear simultaneous equation systems and one first has to apply economic analysis to set up a problem. Try solving Test Yourself, Exercise 5.6 below when you have covered the relevant topics in your economics course.

Example 5.12

A firm uses the three inputs K, L and R to manufacture its final product. The prices per unit of these inputs are £20, £4 and £2 respectively. If the other two inputs are held fixed then the marginal product functions are

$$MP_K = 200 - 5K$$
$$MP_L = 60 - 2L$$
$$MP_R = 80 - R$$

What combination of inputs should the firm use to maximize output if it has a fixed budget of £390?

Solution

The basic rule for optimal input determination is that the last £1 spent on each input should add the same amount to output, i.e.

$$\frac{MP_K}{P_K} = \frac{MP_L}{P_L} = \frac{MP_R}{P_R}$$

Therefore, substituting the given marginal product functions, we get

$$\frac{200 - 5K}{20} = \frac{60 - 2L}{4} = \frac{80 - R}{2}$$

Multiplying out two of the three pairwise combinations of equations to get K and R in terms of L gives

$$4(200 - 5K) = 20(60 - 2L) \qquad\qquad 2(60 - 2L) = 4(80 - R)$$
$$800 - 20K = 1,200 - 40L \qquad\qquad 120 - 4L = 320 - 4R$$
$$40L - 400 = 20K \qquad\qquad\qquad 4R = 4L + 200$$
$$2L - 20 = K \quad (1) \qquad\qquad\qquad R = L + 50 \quad (2)$$

The third pairwise combination will not add any new information. Instead we use the budget constraint

$$20K + 4L + 2R = 390 \tag{3}$$

Substituting (1) and (2) into (3),

$$20(2L - 20) + 4L + 2(L + 50) = 390$$
$$40L - 400 + 4L + 2L + 100 = 390$$
$$46L = 690$$
$$L = 15$$

Substituting this value for L into (1)

$$K = 2(15) - 20 = 10$$

and into (2)

$$R = 15 + 50 = 65$$

Therefore the optimal input combination is

$$K = 10 \quad L = 15 \quad R = 65$$

Example 5.13

In a closed economy where the usual assumptions of the basic Keynesian macroeconomic model apply,

$$C = £60m + 0.7Y_t$$
$$Y = C + I + G$$
$$Y_t = 0.6Y$$

where C is consumption, Y is national income, Y_t is disposable income, I is investment and G is government expenditure. If the values of I and G are exogenously determined as £90 million and £140 million respectively, what is the equilibrium level of national income?

Solution

Once the given values of I and G are substituted, we have a 3×3 set of simultaneous equations with three unknowns:

$$C = 60 + 0.7Y_t \tag{1}$$
$$Y = C + 90 + 140 = C + 230 \tag{2}$$
$$Y_t = 0.6Y \tag{3}$$

This sort of problem is most easily solved by substitution. Substituting (3) into (1) gives

$$C = 60 + 0.7(0.6Y)$$
$$C = 60 + 0.42Y \tag{4}$$

Substituting (4) into (2) gives

$$Y = (60 + 0.42Y) + 230$$
$$0.58Y = 290$$
$$Y = 500$$

Therefore the equilibrium value of the national income is £500 million.

Example 5.14

In a competitive market where the supply price (in £) is $p = 3 + 0.25q$

and demand price (in £) is $p = 15 - 0.75q$

the government imposes a per-unit tax of £4. How much of a price rise will this tax mean to consumers? What will be the tax revenue raised?

Solution

The original equilibrium price and quantity can be found by equating demand and supply price. Hence

$$15 - 0.75q = 3 + 0.25q$$
$$12 = q$$

Substituting this value of q into the supply schedule gives

$$p = 3 + 0.25(12) = 3 + 3 = 6$$

If a per-unit tax is imposed each quantity would be offered for sale by suppliers at the old price plus the amount of the tax. In this case the tax is £4 and so the supply schedule shifts upwards by £4. Thus the new supply schedule becomes

$$p = 3 + 0.25q + 4 = 7 + 0.25q$$

Again equating demand and supply price

$$15 - 0.75q = 7 + 0.25q$$
$$8 = q$$

Substituting this value of q into the demand schedule

$$p = 15 - 0.75(8) = 15 - 6 = 9$$

Therefore, consumers see a price rise of £3 from £6 to £9 (and producers will incur a £1 price reduction and receive a net price of £5).

$$\text{Total tax revenue} = \text{quantity sold} \times \text{tax per unit} = 8 \times 4 = £32$$

Test Yourself, Exercise 5.6

1. A firm faces the demand schedule $p = 400 - 0.25q$
    the marginal revenue schedule $MR = 400 - 0.5q$
    and the marginal cost schedule $MC = 0.3q$
    What price will maximize profit?

2. A firm buys the three inputs K, L and R at prices per unit of £10, £5 and £3 respectively. The marginal product functions of these three inputs are

    $MP_K = 150 - 4K$

    $MP_L = 72 - 2L$

    $MP_R = 34 - R$

    What input combination will maximize output if the firm's budget is fixed at £285?

3. In a competitive market, the supply schedules is $p = 4 + 0.25q$
    and the demand schedule is $p = 16 - 0.5q$

    What would happen to the price paid by consumers and the quantity sold if

    (a) a per-unit tax of £3 was imposed, and
    (b) a proportional sales tax of 20% was imposed?

4. In a Keynesian macroeconomic model of an economy with no foreign trade it is assumed that

    $Y = C + I + G$

    $C = 0.75Y_t$

    $Y_t = (1 - t)Y$

    where the usual notation applies and the following are exogenously fixed: $I = £600\,m$, $G = £900\,m$, $t = 0.2$ is the tax rate. Find the equilibrium value of Y and say whether or not the government's budget is balanced at this value.

5. In an economy which engages in foreign trade, it is assumed that

    $Y = C + I + G + X - M$

    $C = 0.9Y_t$

    $Y_t = (1 - t)Y$

    and imports

    $M = 0.15Y_t$

    The usual notation applies and the following values are given:

    $I = £200m$ $G = £270m$ $X = £180m$ $t = 0.2$

What is the equilibrium value of Y? What is the balance of payments surplus/deficit at this value?

6. (Leave this question if you have not yet covered factor supply theory.) In a factor market for labour, a monopsonistic buyer faces

the marginal revenue product schedule $MRP_L = 244 - 2L$
the supply of labour schedule $w = 20 + 0.4L$
and the marginal cost of labour schedule $MC_L = 20 + 0.8L$

How much labour should it employ, and at what wage, if MRP_L must equal MC_L in order to maximize profit?

5.9   Comparative statics and the reduced form of an economic model

Now that you are familiar with the basic methods for solving simultaneous linear equations, this section will explain how these methods can help you to derive predictions from some economic models. Although no new mathematical methods will be introduced in this section it is important that you work through the examples in order to *learn how to set up economic problems in a mathematical format* that can be solved. This is particularly relevant for those students who can master mathematical methods without too many problems but find it difficult to set up the problem that they need to solve. It is important that you understand the *application of mathematical techniques* to economics, which is the reason why you are studying mathematics as part of your economics course.

Equilibrium and comparative statics

In Section 5.1 we saw how two simultaneous equations representing the supply and demand functions in a competitive market could be solved to determine equilibrium price and quantity. Markets need not always be in equilibrium, however. For example, if

$$\text{Quantity demanded} = q_d = 90 - 0.05p$$

and

$$\text{Quantity supplied} = q_s = -12 + 0.8p$$

then if price is £100

$$q_d = 90 - 0.05(100) = 90 - 5 = 85$$
$$q_s = -12 + 0.8(100) = -12 + 80 = 68$$

and so there would be excess demand equal to

$$q_d - q_s = 85 - 68 = 17$$

In a freely competitive market this situation of excess demand would result in price rising. As price rises the quantity demanded will fall and the quantity supplied will increase until quantity demanded equals quantity supplied and the market is in equilibrium.

The time it takes for adjustment to equilibrium to take place will vary from market to market and the analysis of this dynamic adjustment process between equilibrium situations is considered later in Chapters 13 and 14. Here we shall just examine how the equilibrium values in an economic model change when certain variables alter. This is known as *comparative static analysis*.

If a market is in equilibrium it means that quantity supplied equals quantity demanded and so there are no market forces pushing price up or pulling it down. Therefore price and quantity will remain stable unless something disturbs the equilibrium. One factor that might cause this to happen is a change in the value of an independent variable. In the simple supply and demand model above both quantity demanded and quantity supplied are determined within the model and so there are no independent variables, but consider the following market model

$$\text{Quantity supplied} = q_s = -20 + 0.4p$$

and

$$\text{Quantity demanded} = q_d = 160 - 0.5p + 0.1m$$

where m is average income.

The value of m cannot be worked out from the model. Its value will just be given as it will be determined by factors outside this model. It is therefore an independent variable, sometimes known as an exogenous variable. Without knowing the value of m we cannot work out the values for the dependent variables determined within the model (also known as endogenous variables) which are the equilibrium values of p and q.

Once the value of m is known then equilibrium price and quantity can easily be found. For example, if m is £270 then

$$q_d = 160 - 0.5p + 0.1m = 160 - 0.5p + 0.1(270) = 187 - 0.5p$$

In equilibrium

$$q_s = q_d$$

and so

$$-20 + 0.4p = 187 - 0.5p$$
$$0.9p = 207$$
$$p = 230$$

Substituting this value for p into the supply function to get equilibrium quantity gives

$$q = -20 + 0.4p = -20 + 0.4(230) = -20 + 92 = 72$$

If factors outside this model cause the value of m to alter, then the equilibrium price and quantity will also change. For example, if income rises to £360 then quantity demanded becomes

$$q_d = 160 - 0.5p + 0.1m = 160 - 0.5p + 0.1(360) = 196 - 0.5p$$

and so equating supply and demand quantities to find equilibrium price and quantity

$$-20 + 0.4p = 196 - 0.5p$$
$$0.9p = 216$$
$$p = 240$$

and so
$$q = -20 + 0.4(240) = 76$$

To save having to work out the new equilibrium values in an economic model from first principles every time an exogenous variable changes it can be useful to derive the reduced form of an economic model.

Reduced form

The reduced form specifies each of the dependent variables in an economic model as a function of the independent variable(s). This reduced form can then be used to:

- Predict what happens to the dependent variables when an independent variable changes.
- Estimate the parameters of the model from data using regression analysis (which you should learn about in your statistics or econometrics module).

It is usually possible to derive a reduced form equation for every dependent variable in an economic model.

Example 5.15

A per unit tax t is imposed by the government in a competitive market with the

demand function $q = 20 - 1\frac{1}{3}p$

and

supply function $q = -12 + 4p$

Derive reduced form equations for the equilibrium values of p and q in terms of the tax t.

Solution

Firms have to pay the government a per unit tax of t on each unit they sell. This means that to supply any given quantity firms will require an additional amount t on top of the supply price without the tax, i.e. the supply schedule will shift up vertically by the amount of the tax. To show the effect of this it is easier to work with the inverse demand and supply functions, where price is a function of quantity.

Thus the demand function $q = 20 - 1\frac{1}{3}p$ becomes $p = 15 - 0.75q$ (1)

and the supply function $q = -120 + 4p$ becomes $p = 3 + 0.25q$

After the tax is imposed the inverse supply function becomes

$$p = 3 + 0.25q + t \tag{2}$$

In equilibrium the supply price equals the demand price and so equating (1) and (2)

$$3 + 0.25q + t = 15 - 0.75q$$
$$q = 12 - t \tag{3}$$

This is the reduced form equation for equilibrium quantity. From this reduced form we can easily work out that

when $t = 0$ then $q = 12$

when $t = 4$ then $q = 8$

(You can check these solutions are the same as those in Example 5.14 which had the same supply and demand functions.)

In a model with two dependent variables, like this supply and demand model, once the reduced form equation for one dependent variable has been derived then the reduced form equation for the other dependent variable can be found. This is done by substituting the reduced form for the first variable into one of the functions that make up the model. Thus, in this example, if the reduced form equation for equilibrium quantity (3) is substituted into the demand function $p = 15 - 0.75q$

it becomes $p = 15 - 0.75(12 - t)$

giving $p = 6 + 0.75t$ \hfill (4)

which is the reduced form equation for equilibrium price.

The reduced form equations can also be used to work out the *comparative static effect of a change in t on equilibrium quantity or price*, i.e. what happens to these equilibrium values when tax is increased by one unit.

In this example the reduced form equation for price (4) tells us that for every one unit increase in t the equilibrium price p increases by 0.75. This is illustrated below for a few values of t:

> when $t = 4$ then $p = 6 + 0.75(4) = 6 + 3 = 9$
>
> when $t = 5$ then $p = 6 + 0.75(5) = 6 + 3.75 = 9.75$
>
> when $t = 6$ then $p = 6 + 0.75(6) = 6 + 4.5 = 10.5$

Note that this method can only be used with linear functions. If a dependent variable is a non-linear function of an independent variable then calculus must be used (see Chapter 9).

Before proceeding any further, students should make sure that they understand an important difference between the supply and demand functions and the reduced form of an economic model. The supply and demand functions give the quantities supplied and demanded for *any* price, which includes prices out of equilibrium. The *reduced form only includes the equilibrium values* of p and q.

Reduced form and comparative static analysis of monopoly

The basic principles for deriving reduced form equations for dependent variables can be applied in various types of economic models, and are not confined to supply and demand analysis. The example below shows how the comparative static effect of a per unit tax on a monopoly can be derived from the reduced form equations.

Example 5.16

A monopoly operates with the marginal cost function $MC = 20 + 4q$

and faces the demand function $p = 400 - 8q$

If a per unit tax t is imposed on its output derive reduced form equations for the profit maximizing values of p and q in terms of the tax t and use them to predict the effect of a one unit increase in the tax on price and quantity. Assume that fixed costs are low enough to allow positive profits to be made.

Solution

The per unit tax will cause the cost of supplying each unit to rise by amount t and so the monopoly's marginal cost function will change to

$$MC = 20 + 4q + t$$

For any linear demand function the corresponding marginal revenue function will have the same intercept on the price axis but twice the slope. (See Section 8.3 for a proof of this result.) Therefore, if

$$p = 400 - 8q \quad \text{then} \quad MR = 400 - 16q$$

If the monopoly is maximizing profit then

$$MC = MR$$
$$20 + 4q + t = 400 - 16q$$
$$20q + t = 380$$
$$q = 19 - 0.05t \tag{1}$$

From this reduced form equation for equilibrium q we can see that for every one unit increase in the sales tax the monopoly's output will fall by 0.05 units.

To find the reduced form equation for equilibrium p we can substitute (1), the reduced form for q, into the demand function. Thus

$$p = 400 - 8q = 400 - 8(19 - 0.05t) = 400 - 152 + 0.4t = 248 + 0.4t$$

Thus the reduced form equation for equilibrium p is

$$p = 248 + 0.4t$$

This tells us that for every one unit increase in t the monopoly's price will rise by 0.4. So, for example, a £1 tax increase will cause price to rise by 40p.

The effect of a proportional sales tax

In practice sales taxes are often specified as a percentage of the pre-tax price rather than being set at a fixed amount per unit. For example, in the UK, VAT (value added tax) is levied at a rate of 17.5% on most goods and services at the point of sale. To work out the reduced form equations, a proportional tax needs to be specified in decimal format. Thus a sales tax of 17.5% becomes 0.175 in decimal format.

Example 5.17

A proportional sales tax t is imposed in a competitive market where

$$\text{demand price} = p_d = 375 - 2.5q$$

and

$$\text{supply price} = p_s = 55 + 4q$$

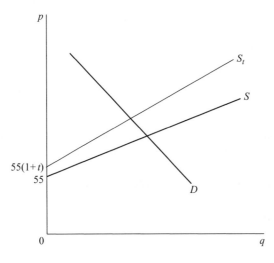

Figure 5.4

Derive reduced form equations for the equilibrium values of p and q in terms of the tax rate t and use them to predict the effect of an increase in the tax rate on the equilibrium values of p and q.

Solution

To supply any given quantity firms will require the original pre-tax supply price p_s plus the proportional tax that is levied at that price. Therefore the total new price p_s^* that firms will require to supply any given quantity will be

$$p_s^* = p_s(1+t) = (55 + 4q)(1+t) \tag{1}$$

The supply function therefore swings up as shown in Figure 5.4. (Instead of the parallel shift caused by a per unit tax.)

Setting this new supply price function (1) equal to demand price

$$p_s^* = p_d$$
$$(55 + 4q)(1+t) = 375 - 2.5q$$
$$55 + 55t + 4q + 4qt = 375 - 2.5q$$
$$6.5q + 4qt = 320 - 55t$$
$$q(6.5 + 4t) = 320 - 55t$$
$$q = \frac{320 - 55t}{6.5 + 4t}$$

This reduced form equation for equilibrium q is a bit more complicated than the one we derived for the per unit sales tax case. However, we can still use it to work out the predicted value of q for a few values of t. Normally we would expect sales taxes to lie between 0% and 100%, giving a value of t in decimal format between 0 and 1.

$$\text{If } t = 10\% = 0.1 \quad \text{then} \quad q = \frac{320 - 55(0.1)}{6.5 + 4(0.1)} = \frac{320 - 5.5}{6.5 + 0.4} = \frac{314.5}{6.9} = 45.58$$

$$\text{If } t = 20\% = 0.2 \quad \text{then} \quad q = \frac{320 - 55(0.2)}{6.5 + 4(0.2)} = \frac{320 - 11}{6.5 + 0.8} = \frac{309}{7.3} = 42.33$$

$$\text{If } t = 30\% = 0.3 \quad \text{then} \quad q = \frac{320 - 55(0.3)}{6.5 + 4(0.3)} = \frac{320 - 16.5}{6.5 + 1.2} = \frac{303.5}{7.7} = 39.42$$

These examples show that as the tax rate increases the value of q falls, as one would expect. However, these equal increments in the tax rate do not bring about equal changes in q because the reduced form equation for equilibrium q is not a simple linear function of t.

Lastly, we can derive the reduced form equation for equilibrium p by substituting the reduced form for q that we have already found into the demand schedule. Thus

$$p = 375 - 2.5q = 375 - 2.5 \left(\frac{320 - 55t}{6.5 + 4t} \right)$$

$$= \frac{2{,}437.5 + 1{,}500t - 800 + 137.5t}{6.5 + 4t} = \frac{1{,}637.5 + 1{,}637.5t}{6.5 + 4t}$$

$$= \frac{1{,}637.5(1 + t)}{6.5 + 4t}$$

To check this reduced form equation, we can calculate p for some extreme values of t to see if the prices calculated lie in a reasonable range for this demand schedule.

$$\text{If } t = 0 \text{ (i.e. no tax)} \quad \text{then} \quad p = \frac{1{,}637.5 + 1{,}637.5(0)}{6.5 + 4(0)} = \frac{1{,}637.5}{6.5} = 251.92$$

$$\text{If } t = 100\% = 1 \quad \text{then} \quad p = \frac{1{,}637.5 + 1{,}637.5}{6.5 + 4} = \frac{3{,}275}{10.5} = 311.81$$

These values lie in a range that one would expect for this demand schedule.

The reduced form of a Keynesian macroeconomic model

Consider the basic Keynesian macroeconomic model used in *Example 5.7* earlier where

$$Y = C + I \tag{1}$$

$$C = 40 + 0.5Y \tag{2}$$

As the value of investment is exogenously determined we can derive a reduced form equation for the equilibrium value of the dependent variable Y in terms of this independent variable I. Substituting the consumption function (2) into the accounting identity (1) gives

$$Y = 40 + 0.5Y + I$$

$$0.5Y = 40 + I$$

$$Y = 80 + 2I \tag{3}$$

From this reduced form we can directly predict the equilibrium value of Y for any given level of I. For example

when $I = 200$ then $Y = 80 + 2(200) = 80 + 400 = 480$ (check with Example 5.7)

when $I = 300$ then $Y = 80 + 2(300) = 80 + 600 = 680$

From the reduced form equation (3) we can also see that for every £1 increase in I the value of Y will increase by £2. This ratio of 2 to 1 is the investment multiplier.

Reduced forms in models with more than one independent variable

Equilibrium values of dependent variables in an economic model may be determined by more than one independent variable. If this is the case then all the independent variables will appear in the reduced form equations for these dependent variables.

Consider the Keynesian macroeconomic model

$$Y = C + I + G \tag{1}$$

$$C = 50 + 0.8Y_d \tag{2}$$

and disposable income $Y_d = (1 - t)Y \tag{3}$

The values of investment, government expenditure and the tax rate (I, G and t) are exogenously determined. Substituting the function for disposable income (3) into the consumption function (2) gives

$$C = 50 + 0.8Y_d = 50 + 0.8(1 - t)Y \tag{4}$$

Substituting (4) into (1) gives

$$Y = 50 + 0.8(1 - t)Y + I + G$$

$$Y(1 - 0.8 + 0.8t) = 50 + I + G$$

$$Y = \frac{50 + I + G}{0.2 + 0.8t}$$

This reduced form equation tells us that the equilibrium value of Y will be determined by the values of the three exogenous variables I, G and t. For example

when $I = 180$, $G = 150$ and $t = 0.375$ then

$$Y = \frac{50 + I + G}{0.2 + 0.8t} = \frac{50 + 180 + 150}{0.2 + 0.8(0.375)} = \frac{380}{0.5} = 760$$

The comparative static effect of an increase in one of the three independent variables can only be worked out if the values of the other two are held constant. For example,

if $I = 180$ and $t = 0.375$

then

$$Y = \frac{50 + 180 + G}{0.2 + 0.3} = \frac{230 + G}{0.5} = 460 + 2G$$

From this new reduced form equation we can see that (when I is 180 and t is 0.375) for every £1 increase in G there will be a £2 increase in Y, i.e. the government expenditure multiplier is 2.

In Chapter 9 we will return to this form of analysis when we have shown how calculus can be used to derive comparative static effects for economic models with non-linear functions.

Test Yourself, Exercise 5.7

1. In a competitive market

    $$q_s = -12 + 0.3p \quad \text{and} \quad q_d = 80 - 0.2p + 0.1a$$

    where a is the price of an alternative substitute good.
    Derive reduced form equations for equilibrium price and quantity and use them to predict the values of p and q when a is 160.
2. A per unit tax t is imposed on all items sold in a competitive market where

    $$q_s = -10 + 0.5p \quad \text{and} \quad q_d = 200 - 2p$$

    Derive reduced form equations for equilibrium price and quantity and use them to predict the values of p and q when t is 5.
3. A monopoly faces the marginal cost function $MC = 12 + 6q$

    and the demand function $p = 150 - 2q$

    If a per unit tax t is imposed on its output derive reduced form equations for the profit maximizing values of p and q in terms of the tax t and use them to predict these values when t is 5.
4. In a Keynesian macroeconomic model $Y = C + I + G$

    $$C = 20 + 0.75Y_d$$

    and disposable income $Y_d = (1 - t)Y$

    (a) If the values of investment and government expenditure (I and G) are exogenously fixed at 50 and 30, respectively, derive a reduced form equation for equilibrium Y in terms of t and use it to predict Y when the tax rate t is 20%.
    (b) Explain what will happen to this reduced form equation and the equilibrium level of Y if G changes to 40.

5. A proportional sales tax v is imposed in a competitive market where

    $$p_d = 800 - 4q \quad \text{and} \quad p_s = 50 + 5q$$

    Derive reduced form equations for the equilibrium values of p and q in terms of the tax rate v and use them to predict p and when v is 15%.

5.10 Price discrimination

In Section 4.10 we examined how linear functions could be summed 'horizontally'. We shall now use this method to help tackle some problems involving price discrimination and, in the following section, multiplant firm/cartel pricing. It is assumed that the main economic principles underpinning these models will be explained in your economics course and only the methods of calculating prices and output are explained here.

In *third-degree price discrimination*, firms charge different prices in separate markets. To maximize profits the theory of price discrimination says that firms should

1. split total sales between the different markets so that the marginal revenue from the last unit sold in each market is the same, and
2. decide on the total sales level by finding the output level where the aggregate marginal revenue function (derived by horizontally summing the marginal revenue schedules from each individual market) intersects the firm's marginal cost function.

It is usually assumed that the firm practising price discrimination is a monopoly. All the examples in this section assume that the firm faces linear demand schedules in each of the separate markets. We shall also make use of the rule that the marginal revenue schedule corresponding to a linear demand schedule will have the same intercept on the price axis but twice the slope. The method of solution is best explained with some examples.

Example 5.18

A monopoly can sell in two separate markets at different prices (in £) and faces the marginal cost schedule

$$MC = 1.75 + 0.05q$$

The two demand schedules are

$$p_1 = 12 - 0.15q_1 \quad \text{and} \quad p_2 = 9 - 0.075q_2$$

What price should it charge and how much should it sell in each market to maximize profit?

Solution

It helps to draw a sketch diagram when tackling this type of problem so that you can relate the different quantities to the economic model. Note that the demand schedules in this example, illustrated in Figure 5.5, are the same as those in Example 4.20 in the last chapter when the marginal revenue summation process was explained in more detail. You can refer back if you do not follow the steps below.

First, the relevant MR schedules and their inverse functions are derived from the demand schedules. Given

$$p_1 = 12 - 0.15q_1 \qquad\qquad p_2 = 9 - 0.075q_2$$

$$\text{then} \quad MR_1 = 12 - 0.3q_1 \qquad\qquad MR_2 = 9 - 0.15q_2$$

$$\text{and so} \quad q_1 = 40 - \frac{MR_1}{0.3} \quad (1) \qquad\qquad q_2 = 60 - \frac{MR_2}{0.15} \quad (2)$$

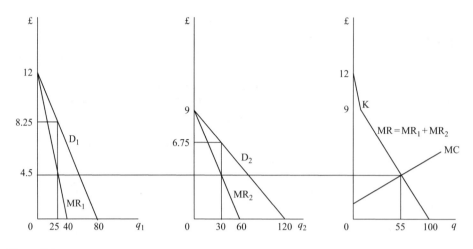

Figure 5.5

For profit maximization

$$MR_1 = MR_2 = MR \tag{3}$$

and by definition

$$q = q_1 + q_2 \tag{4}$$

Therefore, substituting (1), (2) and (3) into (4)

$$q = \left(40 - \frac{MR}{0.3}\right) + \left(60 - \frac{MR}{0.15}\right)$$

$$= \frac{12 - MR + 18 - 2MR}{0.3}$$

$$= \frac{30 - 3MR}{0.3} = 100 - 10MR$$

and so $MR = 10 - 0.1q \tag{5}$

This function does not apply above £9 as only MR_1 applies above this price. In this example Figure 5.5 shows that MC will cut MR in the section below the kink K.

The aggregate profit-maximizing output is found where

$$MR = MC$$

Thus using the MC function given in the question and the aggregated marginal revenue function (5) derived above we get

$$10 - 0.1q = 1.75 + 0.05q$$

$$8.25 = 0.15q$$

$$55 = q$$

Therefore

$$MR = 10 - 0.1(55) = 10 - 5.5 = 4.5$$

and so

$$MR_1 = 4.5 \qquad MR_2 = 4.5$$

To determine the prices and output levels in each market we now just substitute these MR values into the inverse marginal revenue functions (1) and (2) derived above. Thus

$$q_1 = 40 - \frac{MR_1}{0.3} = 40 - \frac{4.5}{0.3} = 40 - 15 = 25$$

$$q_2 = 60 - \frac{MR_2}{0.15} = 60 - \frac{4.5}{0.15} = 60 - 30 = 30$$

You can check these output figures to ensure that $q_1 + q_2 = q$.

Relating these calculations to Figure 5.5, what we have done is found the intersection point of MR and MC to determine the profit-maximizing levels of q and MR. Then a horizontal line is drawn across to see where this level of marginal revenue cuts MR_1 and MR_2. This enables us to read off q_1 and q_2 and the corresponding prices p_1 and p_2 These prices can be determined by simply substituting the above values of q_1 and q_2 into the two demand schedules specified in the question. Thus

$$p_1 = 12 - 0.15q_1 = 12 - 0.15(25) = 12 - 3.75 = £8.25$$

$$p_2 = 9 - 0.075q_2 = 9 - 0.075(30) = 9 - 2.25 = £6.75$$

Finally, refer back to the sketch diagram to ensure that the relative magnitudes of the answer correspond to those read off the graph. In this type of problem it is easy to get mixed up in the various stages of the calculation. From Figure 5.5, we can see that p_1 should be greater than p_2 which checks out with the above answers.

Not all price discrimination models involve the horizontal summation of demand schedules.

In **first-degree (perfect) price discrimination** each individual unit is sold at a different price. Because the prices of other units do not have to be reduced for a firm to increase sales, the marginal revenue from each unit is the price it sells for. Therefore the marginal revenue schedule is the same as the demand schedule, instead of lying below it.

In **second-degree price discrimination** a firm breaks the market up into a series of price bands. In a two-part pricing scheme this might mean that the first few units are sold at a previously determined price and then a price is chosen for the remaining units that will maximize profits, given the first price and the marginal cost schedule.

The example below explains how the relevant prices and quantities can be calculated under these different forms of price discrimination.

Example 5.19

A monopoly faces the demand schedule $\qquad p = 16 - 0.064q$

and the marginal cost schedule $\qquad MC = 2.2 + 0.019q$

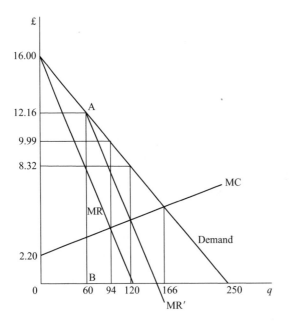

Figure 5.6

It has already been decided that the first 60 units will be sold at a price of £12.16. Given this constraint, what price for the remaining units will maximize profits? How will total output compare with output when

(i) the firm can only set a single price?
(ii) perfect price discrimination takes place?

Solution

We can check that the price of £12.16 for the first 60 units corresponds to point A on the demand schedule in Figure 5.6 since

$$p = 16 - 0.064q = 16 - 0.064(60) = 16 - 3.84 = £12.16$$

If the firm wishes to sell more output it will not have to reduce the price of these first 60 units. It therefore effectively faces the marginal revenue schedule MR'. This is constructed by assuming that the zero on the quantity axis is moved 60 units to the right to point B. MR' is then drawn in the usual way with the same 'intercept' on the price axis (effectively point A) but twice the slope of the demand schedule. The firm should then employ the usual rule for profit maximization, which is to produce the output level at which marginal revenue MR' equals marginal cost.

To derive a function for MR', define $q' = q - 60$, i.e. q' measures output from point B on the quantity axis. The demand schedule over this output range has the same slope as the original demand schedule (-0.064) but the new 'intercept' value of £12.16. It is therefore described by the function

$$p = 12.16 - 0.064q'$$

and therefore

$$MR' = 12.16 - 0.128q' \tag{1}$$

using the rule that a marginal revenue function has twice the slope of a linear demand schedule. Substituting the original definition of output for q' into (1)

$$MR' = 12.16 - 0.128(q - 60)$$
$$= 12.16 - 0.128q + 7.68$$
$$= 19.84 - 0.128q \tag{2}$$

Profit maximization, subject to the given price constraint on the first 60 units, requires

$$MR' = MC$$

Therefore, equating (2) and the given MC function

$$19.84 - 0.128q = 2.2 + 0.019q$$
$$17.64 = 0.147q$$
$$120 = q$$

This is total output. The amount sold at the second (lower) price will be

$$q' = q - 60 = 120 - 60 = 60$$

The price for these units will be

$$p = 12.16 - 0.064q' = 12.16 - 0.064(60) = 12.16 - 3.84 = £8.32$$

The total output level of 120 units under this second-degree price discrimination policy can be compared with

(a) single-price profit maximization:

given the demand schedule $p = 16 - 0.064q$

then $MR = 16 - 0.128q$

Single-price profit maximization occurs when

$$MR = MC$$
$$16 - 0.128q = 2.2 + 0.019q$$
$$13.8 = 0.147q$$
$$93.877 = q \quad \text{(marked as 94 on graph)}$$

Output is therefore lower than the 120 units produced under a two-part pricing scheme. This is what one would expect given that price discrimination allows a firm to sell extra output without reducing the price of all previously sold units and hence shifts the relevant marginal revenue schedule to the right.

The profit-maximizing single price can be found from the demand schedule as

$$p = 16 - 0.064q = 16 - 0.064(93.877) = £9.99$$

This is higher than the £8.32 price for the second batch of output in the second-degree price discrimination example above.

(b) Perfect price discrimination:
  If all units are sold at different prices then marginal revenue is the same as the demand schedule, i.e.

$$MR = 16 - 0.064q$$

The profit-maximizing output is determined where

$$MR = MC$$
$$16 - 0.064q = 2.2 + 0.019q$$
$$13.8 = 0.083q$$
$$166.265 = q \quad \text{(166 on graph)}$$

This is greater than the two-part pricing discrimination output, which is what is expected. The greater the number of different segments a market can be broken up into the higher will be the profits that can be extracted and the output that can be sold.

Note that in the above example, and in all the others in this section, we shall assume that total costs, which are not actually specified, are low enough to allow an overall profit to be made.

Test Yourself, Exercise 5.8

1. A price-discriminating monopoly sells in two markets whose demand functions are

   $$q_1 = 160 - 10p_1 \quad \text{and} \quad q_2 = 240 - 20p_2$$

   and it faces the marginal cost schedule $MC = 1.2 + 0.02q$, where $q = q_1 + q_2$. How much should it sell in each market, and at what prices, in order to maximize profits?

2. A monopoly faces the marginal cost schedule $MC = 1.1 + 0.01q$ and can price-discriminate between the two markets where

   $$p_1 = 10 - 0.1q_1 \quad \text{and} \quad p_2 = 6 - 0.04q_2$$

   How much should it sell in each market to maximize profit, and at what prices?

3. A price-discriminating monopoly sells in two markets whose demand schedules are

   $$p_1 = 12.5 - 0.0625q_1 \qquad p_2 = 7.2 - 0.002q_2$$

   and faces the horizontal marginal cost schedule $MC = 5$.
     What price and output should it choose for each market?

4. A monopoly faces the horizontal marginal cost schedule MC $= 42$ and can operate a two-part pricing scheme in the market with the demand schedule

$$p = 180 - 0.6q$$

If the first 100 units are sold at a price of £120 each, what price should be charged for the remaining units in order to maximize profit?

5. A monopoly sells in a market where $\quad p = 12 - 0.06q$

and has the marginal cost schedule $\quad$ MC $= 3 + 0.04q$

If it can operate second-degree price discrimination, what price should it sell the remaining units for if it has already been decided to sell the first 50 units for a price of £9?

6. A price-discriminating monopoly sells in two markets whose demand schedules are

$$q_1 = 120 - 6p_1 \qquad q_2 = 110 - 8p_2$$

If its marginal cost function is MC $= 2.26 + 0.02q$ calculate the profit-maximizing price and sales levels for each market.

7. A monopoly has the demand schedule $\quad p = 210 - 0.2q$

and the marginal cost schedule $\quad$ MC $= 20 + 0.8q$

(a) If it can practise first-degree price discrimination how much should it sell?
(b) If it can practise second-degree price discrimination and it has already made the decision to sell the first 100 units at a price of £190, what price should it charge for the rest of the units it sells?

5.11 Multiplant monopoly

The theory of multiplant monopoly is analogous to the model of third-degree price discrimination explained above except that it is marginal cost schedules that are summed rather than marginal revenue schedules. The basic principles of the multiplant model are:

1. The firm should adjust production so that the marginal cost of the last unit produced in each plant is equal to the marginal cost of the last unit produced by the other plant(s).
2. Total output is determined where the aggregate marginal cost schedule (derived by horizontally summing the marginal cost schedules in each individual plant) intersects the firm's marginal revenue schedule.

The firm is usually assumed to be a monopoly so that the demand and marginal revenue schedules can be clearly defined. The multiplant monopoly model can also be used to determine price and output levels for the different (single-plant) firms in a cartel where perfect collusion takes place. This is a less likely scenario, however, as perfect collusion within cartels is beset with many problems, as you will know from your economics course.

Example 5.20

A firm operates two plants whose marginal cost schedules are

$$MC_1 = 2 + 0.2q_1 \qquad MC_2 = 6 + 0.04q_2$$

It is a monopoly seller in a market where the demand schedule is

$$p = 66 - 0.1q$$

where q is aggregate output and all costs and prices are measured in £.

How much should the firm produce in each plant, and at what price should total output be sold, if it wishes to maximize profits?

Solution

We need to derive the horizontally summed marginal cost schedule MC, find where it intersects MR, and then see which output levels this marginal cost value corresponds to in each plant. Price is read off the demand schedule at the aggregate output level. (You will note that the marginal cost schedules to be summed are the same as those in Example 4.21 which was illustrated in Figure 4.23.)

Given the demand schedule

$$p = 66 - 0.1q$$

we know that the marginal revenue schedule will have the same intercept and twice the slope. Thus

$$MR = 66 - 0.2q \tag{1}$$

To be able to set MC = MR and solve for q we need to derive MC as a function of q. To do this, we first derive the inverse functions of the individual plant MC schedules, as shown below.

$$MC_1 = 2 + 0.2q_1 \qquad\qquad MC_2 = 6 + 0.04q_2$$
$$MC_1 - 2 = 0.2q_1 \qquad\qquad MC_2 - 6 = 0.04q_2$$
$$5MC_1 - 10 = q_1 \qquad\qquad 25MC_2 - 150 = q_2$$

Given that $q = q_1 + q_2$ by definition and MC = MC_1 = MC_2 for profit maximization then by substituting the above inverse functions for q_1 and q_2 we get

$$q = (5MC - 10) + (25MC - 150)$$
$$q = 30MC - 160$$
$$q + 160 = 30MC$$
$$\frac{q + 160}{30} = MC \tag{2}$$

Setting MC = MR, from (1) and (2) we now get

$$\frac{q + 160}{30} = 66 - 0.2q$$

$$q + 160 = 1{,}980 - 6q$$

$$7q = 1{,}820$$

$$q = 260 \tag{9}$$

Substituting this aggregate output level into (2) gives

$$MC = \frac{q + 160}{30} = \frac{260 + 160}{30} = \frac{420}{30} = 14$$

Therefore $MC_1 = MC_2 = MC = 14$

and so $q_1 = 5(14) - 10 = 70 - 10 = 60$

and $q_2 = 25(14) - 150 = 350 - 150 = 200$

We can easily check that these output levels for the individual plants correspond to the aggregate output of 260 calculated above since

$$q_1 + q_2 = 60 + 200 = 260 = q$$

To find the price at which this aggregate output is sold, simply substitute this value of q into the demand schedule. Therefore

$$p = 66 - 0.1q = 66 - 0.1(260) = 66 - 26 = £40$$

The basic principles explained above can also be applied to more complex problems where there are more than two plants.

Example 5.21

A firm operates four plants whose marginal cost schedules are

$$MC_1 = 20 + q_1 \qquad MC_3 = 40 + q_3$$
$$MC_2 = 40 + 0.5q_2 \qquad MC_4 = 60 + 0.5q_4$$

and it is a monopoly seller in a market where

$$p = 580 - 0.3q$$

How much should it produce in each plant and at what price should its output be sold if it wishes to maximize profit?

Solution

First we find the inverses of the marginal cost functions. Thus

$$\text{MC}_1 = 20 + q_1 \qquad \text{MC}_2 = 40 + 0.5q_2 \qquad \text{MC}_3 = 40 + q_3 \qquad \text{MC}_4 = 60 + 0.5q_4$$

$$q_1 = \text{MC}_1 - 20 \qquad q_2 = 2\text{MC}_2 - 80 \qquad q_3 = \text{MC}_3 - 40 \qquad q_4 = 2\text{MC}_4 - 120$$

Given that

$$q = q_1 + q_2 + q_3 + q_4$$

and for profit maximization

$$\text{MC} = \text{MC}_1 = \text{MC}_2 = \text{MC}_3 = \text{MC}_4$$

then, by summing all the inverses of the individual MC functions and substituting MC, we can write

$$q = (\text{MC} - 20) + (2\text{MC} - 80) + (\text{MC} - 40) + (2\text{MC} - 120)$$

$$q = 6\text{MC} - 260$$

$$\frac{q + 260}{6} = \text{MC} \tag{1}$$

Since

$$p = 580 - 0.3q$$

$$\text{MR} = 580 - 0.6q \tag{2}$$

To maximize profits MC = MR and so equating (1) and (2)

$$\frac{q + 260}{6} = 580 - 0.6q$$

$$q + 260 = 3,480 - 3.6q$$

$$4.6q = 3,220$$

$$q = 700$$

For this aggregate output level the marginal cost is

$$\text{MC} = \frac{q + 260}{6} = \frac{700 + 260}{6} = \frac{960}{6} = 160$$

Substituting this value of MC into the individual inverse marginal cost functions to find plant output levels gives

$$q_1 = \text{MC}_1 - 20 = 160 - 20 = 140$$

$$q_2 = 2\text{MC}_2 - 80 = 320 - 80 = 240$$

$$q_3 = \text{MC}_3 - 40 = 160 - 40 = 120$$

$$q_4 = 2\text{MC}_4 - 120 = 320 - 120 = 200$$

These total to 700, which checks out with the answer for q above.

The price to sell at is found by substituting the total output of 700 units into the demand schedule given in the question. Thus

$$p = 580 - 0.3q = 580 - 0.3(700) = 580 - 210 = £370$$

Note that we did not draw a sketch diagram for the above example to check whether or not the MR schedule cuts the aggregated MC schedule at a level where output by all four plants is positive, i.e. where the value of MC is above the intercept on the vertical axis for each individual MC schedule. However, as all four output levels were calculated as positive numbers we know that this must be the case.

In this type of question, if the usual mathematical method throws up a negative quantity for output by one or more plants (or a negative sales figure in a price discrimination model), then this means that output in this plant (or plants) should be zero. The question should then be reworked with the marginal cost schedule for any such plants excluded from the aggregated MC schedule.

Price discrimination with multiplant monopoly

It is possible to apply the principles of both price discrimination and multiplant monopoly at the same time, if all the necessary conditions hold.

Example 5.22

A multiplant monopoly operates two plants whose marginal cost schedules are

$$MC_1 = 42.5 + 0.5q_1 \qquad MC_2 = 130 + 2q_2$$

It also sells its product in two separable markets whose demand schedules are

$$p_A = 360 - q_A \qquad p_B = 280 - 0.4q_B$$

(Note that the subscripts A and B are used to distinguish quantities sold in the two markets from the quantities q_1 and q_2 produced in the two plants.)

Calculate how much it should produce in each plant, how much it should sell in each market, and how much it should charge in each market.

Solution

First derive the aggregate MC function by the usual method. Given

$$MC_1 = 42.5 + 0.5q_1 \qquad MC_2 = 130 + 2q_2$$

then $\qquad q_1 = 2MC - 85 \qquad\qquad q_2 = 0.5MC_2 - 65$

To maximize profits, output is adjusted so that $MC_1 = MC_2 = MC$

Therefore $\qquad\qquad q = q_1 + q_2 = (2MC - 85) + (0.5MC - 65)$

$$q = 2.5MC - 150$$

$$60 + 0.4q = MC \qquad\qquad (1)$$

Next, derive the aggregate MR function. Given

$$p_A = 360 - q_A \qquad p_B = 280 - 0.4q_B$$

then

$$MR_A = 360 - 2q_A \qquad MR_B = 280 - 0.8q_B$$
$$q_A = 180 - 0.5MR_A \qquad q_B = 350 - 1.25MR_B$$

To maximize profits, sales are adjusted so that $MR_A = MR_B = MR$

Therefore
$$q = q_A + q_B = (180 - 0.5MR) + (350 - 1.25MR)$$
$$q = 530 - 1.75MR$$
$$MR = \frac{530 - q}{1.75} \qquad\qquad (2)$$

To maximize profits $MC = MR$. Therefore, equating (1) and (2)

$$60 + 0.4q = \frac{530 - q}{1.75}$$
$$105 + 0.7q = 530 - q$$
$$1.7q = 425$$
$$q = 250$$

Thus

$$MC = 60 + 0.4q = 60 + 0.4(250) = 60 + 100 = 160$$

and also

$$MR = MC = 160$$

To find production levels in the two plants, substitute this value of MC into the inverse MC functions above. Thus

$$q_1 = 2MC - 85 = 2(160) - 85 = 320 - 85 = 235$$
$$q_2 = 0.5MC - 65 = 0.5(160) - 65 = 80 - 65 = 15$$

To find sales levels in each market, substitute this value of MR into the inverse MR functions above. Thus

$$q_A = 180 - 0.5MR = 180 - 0.5(160) = 180 - 80 = 100$$
$$q_B = 350 - 1.25MR = 350 - 1.25(160) = 350 - 200 = 150$$

A quick check shows that the two production levels and the two sales levels both add to 250, which is what is expected.

Finally, the prices charged in the two markets A and B are found by substituting the above values of q_A and q_B into the demand schedules. Thus

$$p_A = 360 - q_A = 360 - 100 = £260$$
$$p_B = 280 - 0.4q_B = 280 - 0.4(150) = 280 - 60 = £220$$

Test Yourself, Exercise 5.9

1. A monopoly operates two plants whose marginal cost schedules are

   $$MC_1 = 2 + 0.1q_1 \qquad MC_2 = 4 + 0.08q_2$$

   and sells in a market where the demand function is $q = 1160 - 20p$.
   How much should it produce in each plant and at what price should its product be sold?

2. A multiplant monopoly sells in a market where the demand schedule is

   $$p = 253.4 - 0.025q$$

   and produces in two plants whose marginal cost schedules are

   $$MC_1 = 20 + 0.0625q_1 \qquad MC_2 = 50 + 0.1q_2$$

   How should it split output between the two plants in order to maximize profit? What price should it sell at?

3. A firm operates two plants whose marginal cost schedules are

   $$MC_1 = 22.5 + 0.25q_1 \qquad MC_2 = 15 + 0.25q_2$$

   It is also a monopoly which can price-discriminate between two markets, A and B, whose demand schedules are

   $$p_A = 600 - 0.125q_A \qquad p_B = 850 - 0.1q_B$$

   If it wishes to maximize profits, how much should it produce in each plant, how much should it sell in each market, and what prices should it sell at?

4. A multiplant monopoly produces using two plants with the marginal cost schedules

   $$MC_1 = 8 + 0.2q_1 \qquad MC_2 = 10 + 0.05q_2$$

   It can also price-discriminate between three markets whose demand schedules are

   $$p_A = 150 - 0.1875q_A \qquad p_B = 80 - 0.15q_B \qquad p_C = 80 - 0.1q_C$$

   In order to maximize profits, how much should it produce in each plant, how much should it sell in each market, and what prices should it sell at?

5. A monopoly operates three plants with marginal cost schedules

   $$MC_1 = 0.1 + 0.02q_1 \qquad MC_2 = 0.3 + 0.004q_2 \qquad MC_3 = 0.2 + 0.008q_3$$

   How much should it make in each plant to maximize profit if its market demand schedule is

   $$p = 28 - 0.2q$$

   and what price will the total output be sold at?

Appendix: linear programming

Although basically an extension of the linear algebra covered in the main body of this chapter, the technique of linear programming involves special features which distinguish it from other linear algebra applications. When all the relevant functions are linear, this technique enables one to:

- calculate the profit-maximizing output mix of a multi-product firm subject to restrictions on input availability, or
- calculate the input mix that will minimize costs subject to minimum quality standards being met.

This makes it an extremely useful tool for managerial decision-making.

However, it should be noted that, from a pure economic theory viewpoint, linear programming cannot make any general predictions about price or output for a large number of firms. Its usefulness lies in the realm of managerial (or business) economics where economic techniques can help an individual firm to make efficient decisions.

Constrained maximization

A resource allocation problem that a firm may encounter is how to decide on the product mix which will maximize profits when it has limited amounts of the various inputs required for the different products that it makes. The firm's objective is to maximize profit and so profit is what is known as the 'objective function'. It tries to optimize this function subject to the constraint of limited input availability. This is why it is known as a 'constrained optimization' problem.

When both the objective function and the constraints can be expressed in a linear form then the technique of linear programming can be used to try to find a solution. (Constrained optimization of non-linear functions is explained in Chapter 11.) We shall restrict the analysis here to objective functions which have only two variables, e.g. when only two goods contribute to a firm's profit. This enables us to use graphical analysis to help find a solution, as explained in the example below.

Example 5.A1

A firm manufactures two goods A and B using three inputs K, L and R. The firm has at its disposal 150 units of K, 120 units of L and 40 units of R. The net profit contributed by each unit sold is £4 for A and £1 for B. Each unit of A produced requires 3 units of K, 4 units of L plus 2 units of R. Each unit of B produced requires 5 units of K, 3 units of L and none of R. What combination of A and B should the firm manufacture to maximize profits given these constraints on input availability?

Solution

From the per-unit profit figures in the question we can see that the linear objective function for profit which the firm wishes to maximize will be

$$\pi = 4A + B$$

where A and B represent the quantities of goods A and B that are produced.

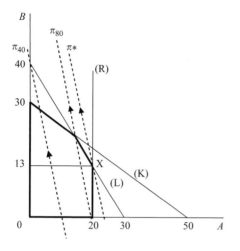

Figure 5.A1

The total amount of input K required will be 3 for each unit of A plus 5 for each unit of B and we know that only 150 units of K are available. The constraint on input K is thus

$$3A + 5B \leq 150 \tag{1}$$

Similarly, for L

$$4A + 3B \leq 120 \tag{2}$$

and for R

$$2A \leq 40 \tag{3}$$

As the firm cannot produce negative quantities of the two goods, we can also add the two non-negativity constraints on the solutions for the optimum values of A and B, i.e.

$$A \geq 0 \tag{4}$$

and

$$B \geq 0 \tag{5}$$

Now turn to the graph in Figure 5.A1 which measures A and B on its axes. The first step in the graphical solution of a linear programming problem is to mark out what is known as the *'feasible area'*. This will contain all the values of A and B that satisfy all the above constraints (1) to (5). This is done *by eliminating the areas which could not possibly contain the solution.*

We can easily see that the non-negativity constraints (4) and (5) mean that the solution must lie on, or above, the A axis and on, or to the right of, the B axis.

To mark out the other constraints we consider in turn what would happen if the firm entirely used up its quota of each of the inputs K, L and R. If all the available K was used up, then in constraint (1) an equality sign would replace the $\leq$ sign and it would become the function

$$3A + 5B = 150 \tag{6}$$

This linear constraint can easily be marked out by joining its intercepts on the two axes. When $A = 0$ then $B = 30$ and when $B = 0$ then $A = 50$. Thus the constraint will be the straight line marked (K). This is rather like a budget constraint. If all the available K is used then the firm's production mix will correspond to a point somewhere on the constraint line (K). It is also possible to use less than the total amount available, in which case the firm would produce a combination of A and B below this constraint. Points above this constraint are not feasible, though, as they correspond to more than 150 units of K.

In a similar fashion we can deduce that all points above the constraint line (L) are not feasible because when all the available L is used up then

$$4A + 3B = 120 \tag{7}$$

The constraint on R is shown by the vertical line (R) since when all available R is used up then

$$2A = 40 \tag{8}$$

Points to the right of this line will not be feasible.

Having marked out the individual constraints, we can now delineate the area which contains combinations of A and B which satisfy all five constraints. This is shown by the heavier black lines in Figure 5.A1.

We know that the firm's objective function is $\pi = 4A + B$. But as we do not yet know what the profit is, how can we draw in this function? To overcome this problem, first make up a figure for profit, which when divided by the two per-unit profit figures (£4 and £1) will give numbers within the range shown on the graph. For example, if we suppose profit is £40, then we can draw in the broken line π_{40} corresponding to the function

$$40 = 4A + B$$

If we had chosen a figure for profit of more than £40 then we would have obtained a line parallel to this one, but further away from the origin; e.g. the line π_{80} corresponds to the function $80 = 4A + B$.

If the firm is seeking to maximize profit then it needs to *find the furthest profit line from the origin that passes through the feasible area.* All profit lines will have the same slope and so, using π_{40} as a guideline, we can see that the highest feasible profit line is π_* which just touches the edge of the feasible area at X. The optimum values of A and B can then simply be read off the graph, giving $A = 20$ and $B = 13$ (approximately).

A more accurate answer may be obtained algebraically, once the graph has been used to determine which is the optimum point, since the solution to a linear programming problem will nearly always be at the intersection of two or more constraints. (Exceptionally the objective function may be parallel to a constraint – see Example 5.A3.)

The graph in Figure 5A.1 tells us that the solution to this problem is where the constraints (L) and (R) intersect. Thus we have the two simultaneous equations

$$4A + 3B = 120 \tag{7}$$
$$2A = 40 \tag{8}$$

which can easily be solved to find the optimum values of A and B.

From (8) $\qquad\qquad\qquad\qquad A = 20$

Substituting in (7) $\quad 4(20) + 3B = 120$

$$3B = 40$$

$$B = 13.33 \quad \text{(to 2 dp)}$$

Thus maximum profit is

$$\pi = 4A + B = 4(20) + 13.33 = 80 + 13.33 = £93.33$$

The optimum combination X is on the constraints for L and R, but below the constraint for K. Thus, as the K constraint does not 'bite', there must be some spare capacity, or what is often called 'slack', for K. When the firm produces 20 of A and 13.33 of B, then its usage of K is

$$3A + 5B = 3(20) + 5(13.33) = 60 + 66.67 = 126.67$$

The amount of K available is 150 units; therefore slack is

$$150 - 126.67 = 23.33 \text{ units of K}$$

Now that the different steps involved in solving a linear programming problem have been explained let us work through another problem.

Example 5.A2

A firm produces two goods A and B, which each contribute a net profit of £1 per unit sold. It uses two inputs K and L. The input requirements are:

> 3 units of K plus 2 units of L for each unit of A
>
> 2 units of K plus 3 units of L for each unit of B

If the firm has 600 units of K and 600 units of L at its disposal, how much of A and B should it produce to maximize profit?

Solution

Using the same method as in the previous example we can see that the constraints are:

for input K	$3A + 2B \leq 600$	(1)
for input L	$2A + 3B \leq 600$	(2)
non-negativity	$A \geq 0 \quad B \geq 0$	

The feasible area is therefore as marked out by the heavy black lines in Figure 5A.2.
  As profit is £1 per unit for both A and B, the objective function is

$$\pi = A + B$$

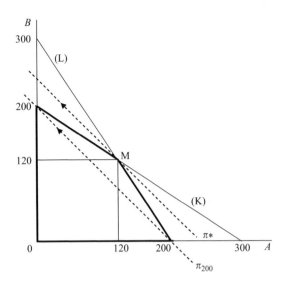

Figure 5.A2

If we suppose profit is £200, then

$$200 = A + B$$

This function corresponds to the line π_{200} which can be used as a guideline for the slope of the objective function. The line parallel to π_{200} that is furthest away from the origin but still within the feasible area will represent the maximum profit. This is the line $\pi*$ through point M. The optimum values of A and B can thus be read off the graph as 120 of each.

Alternatively, once we know that the optimum combination of A and B is at the intersection of the constraints (K) and (L), the values of A and B can be found from the simultaneous equations

$$3A + 2B = 600 \tag{1}$$
$$2A + 3B = 600 \tag{2}$$

From (1) $2B = 600 - 3A$

$$B = 300 - 1.5A \tag{3}$$

Substituting (3) into (2)

$$2A + 3(300 - 1.5A) = 600$$
$$2A + 900 - 4.5A = 600$$
$$300 = 2.5A$$
$$120 = A$$

Substituting this value of A into (3)

$$B = 300 - 1.5(120) = 120$$

As both A and B equal 120 then

$$\pi* = 120 + 120 = £240$$

The optimum combination at M is where both constraints (K) and (L) bite. There is therefore no slack for either K or L.

It is possible that the objective function will have the same slope as one of the constraints. In this case there will not be one optimum combination of the inputs as all points along the section of this constraint that forms part of the boundary of the feasible area will correspond to the same value of the objective function.

Example 5.A3

A firm produces two goods x and y which require inputs of raw material (R), labour (L) and components (K) in the following quantities:

1 unit of x requires 12 kg of R, 10 hours of L and 15 units of K

1 unit of y requires 21 kg of R, 10 hours of L and 6 units of K

Both x and y add £200 per unit sold to the firm's profits. The firm can use up to a total of 252 kg of R, 150 hours of L and 180 units of K. What production mix of x and y will maximize profits?

Solution

The constraints can be written as

$$12x + 21y \leq 252 \qquad \text{(R)}$$
$$10x + 10y \leq 150 \qquad \text{(L)}$$
$$15x + 6y \leq 180 \qquad \text{(K)}$$
$$x \geq 0, \quad y \geq 0$$

These are shown in Figure 5.A3 where the feasible area is marked out by the shape ABCD0. The objective function is

$$\pi = 200x + 200y$$

To find the slope of this objective function, assume profit is £2,000. This could be achieved by producing 10 of x and none of y, or 10 of y and no x, and is therefore shown by the broken line π_{2000}. This line is parallel to the constraint (L). Therefore if we slide out the objective function π to find the maximum value of profit within the feasible area we can see that it coincides with the boundary of the feasible area along the stretch BC.

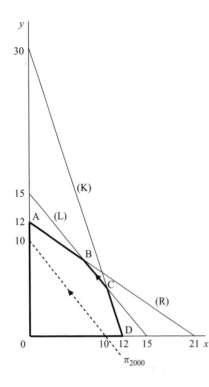

Figure 5.A3

What this means is that both points B and C, and anywhere along the portion of the constraint line (L) between these points, will give the same (maximum) profit figure.

At B the constraints (R) and (L) intersect. Therefore these two resources are used up completely and so

$$12x + 21y = 252 \tag{1}$$

$$10x + 10y = 150 \tag{2}$$

From (2) $x = 15 - y$ (3)

Substituting (3) into (1)

$$12(15 - y) + 21y = 252$$

$$180 - 12y + 21y = 252$$

$$9y = 72$$

$$y = 8$$

Substituting this value of y into (3)

$$x = 15 - 8 = 7$$

Thus profit at B is

$$\pi = 200x + 200y = 200(7) + 200(8) = £1,400 + £1,600 = £3,000$$

At C the constraints (L) and (K) intersect, giving the simultaneous equations

$$10x + 10y = 150 \tag{2}$$

$$15x + 6y = 180 \tag{4}$$

Using (3) again to substitute for x in (4),

$$15(15 - y) + 6y = 180$$

$$225 - 15y + 6y = 180$$

$$45 = 9y$$

$$5 = y$$

Substituting this value of y into (3)

$$x = 15 - 5 = 10$$

Thus, profit at C is

$$\pi = 200x + 200y = 200(10) + 200(5) = 2,000 + 1,000 = £3,000$$

which is the same as the profit achieved at B, as expected. This example therefore illustrates how a linear programming problem may not have a unique solution if the objective function has the same slope as one of the constraints that bounds the feasible area.

You should also note that the solution to a linear programming problem may be on one of the axes, where a non-negativity constraint operates. Some students who do not fully understand linear programming sometimes manage to draw in the constraints correctly, but then incorrectly assume that the solution must lie where the constraints they have drawn intersect. However, it is, of course, also necessary to draw in the objective function to find the solution. The example below illustrates such a case.

Example 5.A4

A company uses inputs K and L to manufacture goods A and B. It has available 200 units of K and 180 units of L and the input requirements are

10 units of K plus 30 units of L for each unit of A

25 units of K plus 15 units of L for each unit of B

If the per-unit profit is £80 for A and £30 for B, what combination of A and B should it produce to maximize profit and how much of K and L will be used in doing this?

Solution

The resource constraints are

$$10A + 25B \leq 200 \quad \text{(K)}$$

$$30A + 15B \leq 180 \quad \text{(L)}$$

$$A \geq 0 \quad B \geq 0$$

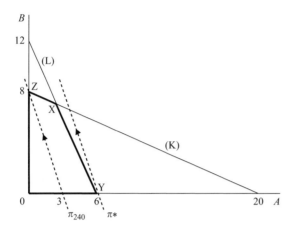

Figure 5.A4

The corresponding feasible area ZXY0 is marked out in Figure 5.A4. The objective function is

$$\pi = 80A + 30B$$

To find the slope of the objective function, assume total profit is £240. This could be obtained by selling 8 of B or 3 of A, and so the broken line π_{240} in Figure 5A.4 illustrates the combinations of A and B that would yield this level of profit. The maximum profit mix is obtained when a line parallel to π_{240} is drawn as far from the origin as possible but still within the feasible area. This will be line $\pi*$ through point Y.

Therefore, profit is maximized at Y, where no B is produced and 6 units of A are produced. Maximum profit = 6 × £80 = £480.

In this example only the constraint (L) bites and so there will be slack in the (K) constraint. The total requirement of K to produce 6 units of A will be 60. There are 200 units of K available and so 140 remain unused. All 180 units of L are used up.

Test Yourself, Exercise 5.A1

1. A firm manufactures products A and B using the two inputs X and Y in the following quantities:

    1 tonne of A requires 80 units of X plus 148 units of Y

    1 tonne of B requires 200 units of X plus 120 units of Y

    The profit per unit of A is £20, and the per-unit profit of B is £30. If the firm has at its disposal 1,600 units of X and 1,800 units of Y, what combination of A and B should it manufacture in order to maximize profit? (Fractions of a tonne may be produced.)
      Should the firm change its production mix if per-unit profits alter to

    (a) £25 each for both A and B, or (b) £30 for A and £20 for B?

2. A firm produces the goods A and B using the four inputs W, X, Y and Z in the following quantities:

    1 unit of A requires 9 units of W, 30 of X, 20 of Y and 20 of Z

    1 unit of B requires 13 units of W, 55 of X, 28 of Y and 20 of Z

    The firm has available 468 units of W, 1,980 units of X, 1,120 units of Y and 800 units of Z. What production mix will maximize its total profit if each unit of A adds £60 to profit and each unit of B adds £75?

3. A firm sells two versions of a device for cutting and drilling. Version A is sold direct to the public in DIY stores, yielding a profit per unit of £50, and version B is sold to other firms for industrial use, yielding a per-unit profit of £20. Each day the firm is able to use 400 hours of labour, 750 kg of raw material and 240 metres of packaging material. These inputs are required to produce A and B in the following quantities: one version A device requires 20 hours of labour, 50 kg of raw material and 20 metres of packaging, whilst one of version B only requires 20 hours of labour plus 30 kg of raw material. How many of each version should be produced each day in order to maximize profit?

4. A firm uses three inputs X, Y and Z to manufacture two goods A and B. The requirements per tonne are as follows.

    A: 5 loads of X, 4 containers of Y and 6 hours of Z

    B: 5 loads of X, 6 containers of Y and 2 hours of Z

    Each tonne of A brings in £400 profit and each tonne of B brings in £300. What combination of A and B should the firm produce to maximize profit if it has at its disposal 150 loads of X, 240 containers of Y and 150 hours of Z?

5. A firm makes the two food products A and B and the contribution to profit is £2 per unit of A and £3 per unit of B. There are three stages in the production process: cleaning, mixing and tinning. The number of hours of each process required for each product and the total number of hours available for each process are given in Table 5.A1. Given these constraints what combination of A and B should the firm produce to maximize profit?

Table 5.1

	Hours of		
	Cleaning	Mixing	Tinning
1 unit of A requires	3	6	2
1 unit of B requires	6	2	1.5
Total hours available	210	120	60

6. Make up your own values for the per-unit profit of A and B in the above question and then say what the optimum production combination is.

7. A firm manufactures two compounds A and B using two raw materials R and Q, in addition to labour and a mixing additive. Input requirements per tonne are:

   For A: 1 container of R, 3 sacks of Q, 4 hours labour and 2 tins of mixing additive
   For B: 2 containers of R, 5 sacks of Q and 3 hours labour, but no mixing additive

   Both A and B add £200 per tonne to the firm's profits and it has at its disposal 60 containers of R, 150 sacks of Q, 120 hours of labour and 50 tins of mixing additive.
   What combination of A and B should it produce to maximize profits, assuming that fractions of a tonne can be manufactured? What will these profits be? What surplus amounts of the inputs will there be?

8. A firm manufactures two products A and B which sell for respectively £900 and £2,000 each. It uses the four processes cutting, drilling, finishing and assembly and the requirements per unit of output are:

   A: 5 hours cutting, 18 hours drilling, 9 hours finishing and 10 hours assembly

   B: 15 hours cutting, 7 hours drilling, 15 hours finishing and 10 hours assembly

   How can this firm maximize its weekly sales revenue if the capacity of its factory is limited to 390 hours cutting, 630 hours drilling, 450 hours finishing and 400 hours assembly per week?

9. If a firm is faced with the constraints described in question 2 in Test Yourself, Exercise 5.4, what combination of A and B will maximize profit if A contributes £30 per unit to profit and B contributes £10?

10. Show that more than one solution exists if one tries to maximize the objective function

    $$\pi = 4A + 4B$$

    subject to the constraints

    $$20A + 20B \leq 60$$

    $$20A + 80B \leq 120$$

    $$A \geq 0 \quad B \geq 0$$

11. A firm has £120,000 to invest. It can buy shares in company X which cost £2 each and give an expected annual return of 6%, or shares in company Y which cost £4 each and give an expected annual return of 8%. It is advised not to put more than 60% of its total investments into any one type of share. What investment portfolio will maximize the expected return? (You may answer this question with or without a diagram.)

12. Make up your own linear programming problem involving the constrained maximization of an objective function with two variables and at least two constraints, and solve it.

Constrained minimization

Another problem a firm might be faced with is how to minimize the cost of producing a good subject to constraints regarding its quality. If the objective function and the constraints are both linear then the method used for constrained minimization is analogous to that used in the maximization problems. The main differences in constrained minimization problems are that:

- the feasible area is usually *above* the constraint lines,
- one needs to find the objective function line that is *nearest* to the origin within the feasible area.

The following examples show how this method operates.

Example 5.A5

A firm manufactures a medicinal product containing three ingredients X, Y and Z. Each unit produced must contain at least 100 g of X, 30 g of Y and 75 g of Z. The product is made by mixing the inputs A and B which come in containers costing respectively £3 and £6 each.
   These contain X, Y and Z in the following quantities:

   1 container of A contains 50 g of X, 10 g of Y and 15 g of Z

   1 container of B contains 20 g of X, 10 g of Y and 50 g of Z

What mix of A and B will minimize the cost per unit of the product subject to the above quality constraints? (It does not matter if these minimum requirements are exceeded and all other production costs can be ignored.)

Solution

Total usage of X will be 50 g for each container of A plus 20 g for each container of B. Total usage must be at least 100 g. This quality constraint for X can thus be written as

$$50A + 20B \geq 100$$

Note that the constraint has the $\geq$ sign instead of $\leq$. The quality constraints on Y and Z can also be written as

$$10A + 10B \geq 30$$
$$15A + 50B \geq 75$$

As negative amounts of the inputs A and B are not feasible there are also the two non-negativity constraints

$$A \geq 0 \qquad B \geq 0$$

If the quality constraint for X is only just met then

$$50A + 20B = 100 \quad (X)$$

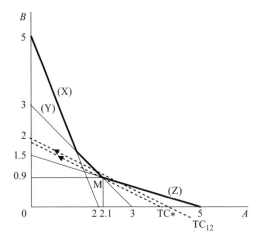

Figure 5.A5

The line representing this function is drawn as (X) in Figure 5.A5. Any combination of A and B above this line will more than satisfy the quality constraint for X. Any combination of A and B below this line will not satisfy this constraint and will therefore not be feasible.

In a similar fashion the constraints for Y and Z are shown by the lines representing the functions

$$10A + 10B = 30 \quad (Y)$$

and

$$15A + 50B = 75 \quad (Z)$$

Taking all the constraints into account, the feasible area is marked out by the heavy black lines in Figure 5A.5, or at least its lower bounds are. As these are minimum constraints then theoretically there are no upper limits to the amounts of A and B that could be used to make a unit of the final product.

The objective function is total cost (TC) which the firm is seeking to minimize. Given the prices of A and B of £3 and £6 respectively, then

$$TC = 3A + 6B$$

To obtain a guideline for the slope of the TC function assume any value for TC that is easily divisible by the two prices of £3 and £6. For example, if TC is assumed to be £12 then the line TC_{12} representing the function

$$12 = 3A + 6B$$

can be drawn, which has a slope of -0.5.

One now needs to ask the question 'can a line with this slope be drawn closer to the origin (thus representing a smaller value for TC) but still going through the feasible area?' In this case the answer is 'yes'. The line TC_* through M represents the lowest cost method of combining A and B that still satisfies the three quality constraints. The optimum amounts of A and B can now be read off the graph at M as approximately 2.1 and 0.9 respectively.

More accurate answers can be obtained algebraically. The optimum combination M is where the quality constraints for Y and Z intersect. These correspond to the linear equations

$$10A + 10B = 30 \tag{1}$$

$$15A + 50B = 75 \tag{2}$$

Dividing (2) by 5 we get $3A + 10B = 15$

Subtracting (1) $\underline{10A + 10B = 30}$

$$-7A = -15$$

$$A = \frac{15}{7} = 2\tfrac{1}{7}$$

Substituting this value for A into (1)

$$10\left(\frac{15}{7}\right) + 10B = \frac{150}{7} + 10B = 30 \tag{3}$$

Multiplying (3) by 7

$$150 + 70B = 210$$

$$70B = 60$$

$$B = \frac{6}{7}$$

Thus the firm should use $2\tfrac{1}{7}$ containers of A plus $\tfrac{6}{7}$ of a container of B for every unit of the final product it makes. As long as large quantities of the product are made, the firm does not have to worry about unused fractions of containers. It just needs to use containers A and B in the ratio $2\tfrac{1}{7}$ to $\tfrac{6}{7}$ which is the same as the ratio 2.5 to 1.

The constraint on X does not bite and so there is some slack. In a minimization problem slack means overabundance. The total amount of X contained in a unit of the final product will be

$$50A + 20B = 50\left(\frac{15}{7}\right) + 20\left(\frac{6}{7}\right) = \frac{750 + 120}{7} = \frac{970}{7} = 138.57\,\text{mg}$$

This exceeds the minimum requirement of 100 g of X by 38.57 g.

Example 5.A6

A firm makes a product that has minimum input requirements for the four ingredients W, X, Y and Z. These cannot be manufactured individually and can only be supplied as part of the composite inputs A and B.

1 litre of A includes 20 g of W, 5 g of X, 5 g of Y and 20 g of Z

1 litre of B includes 90 g of W, 7 g of X and 4 g of Y but no Z

One drum of the final product must contain at least 7,200 g of W, 1,400 g of X, 1,000 g of Y and 1,200 g of Z. (The volume of the drum is fixed and not related to the volume of inputs A and B as evaporation occurs during the production process.) If a litre of A costs £9 and a litre of B costs £16 how many litres of A and B should the firm use to minimize the cost of a drum of the final product? Assume that all other costs can be ignored.

Solution

The minimum input requirements can be written as

$$20A + 90B \geq 7,200 \quad \text{(W)}$$
$$5A + 7B \geq 1,400 \quad \text{(X)}$$
$$5A + 4B \geq 1,000 \quad \text{(Y)}$$
$$20A \geq 1,200 \quad \text{(Z)}$$

plus the non-negativity conditions $A \geq 0$, $B \geq 0$. These constraints are shown in Figure 5.A6.

If only the minimum 7,200 g of W is included in the final product then $20A + 90B = 7,200$. If no B was used then one would need $7,200/20 = 360$ litres of A to satisfy this constraint. If no A was used then $7,200/90 = 80$ litres of B would be needed. Thus the values where the linear constraint (W) hits the A and B axes are 360 and 80 respectively. Combinations of A and B below this line do not satisfy the minimum amount of W requirement. The other constraints, for X, Y and Z, are constructed in a similar fashion and the feasible area is marked out by the heavy black lines in Figure 5.A6.

To find a guideline for the slope of the objective function, assume that the total cost (TC) of A and B is £1,440, giving the budget constraint

$$1,440 = 9A + 16B$$

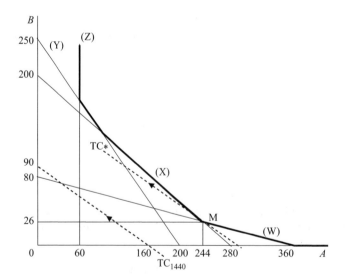

Figure 5.A6

This particular budget constraint is shown by the broken line TC_{1440} and does not go through the feasible area. Therefore total cost must be greater than £1,440. An increased budget will mean a budget line further from the origin but still with the same slope as TC_{1440}. The budget line with this slope that is closest to the origin and that also passes through the feasible area is TC_*.

Minimum TC is therefore achieved by using the combination of A and B corresponding to point M. Approximate values read off the graph at M are 244 litres of A and 26 litres of B.

More accurate answers can be obtained algebraically as we know that M is at the intersection of the constraints for W and X. This means that the minimum requirements for W and X are only just met and so

$$20A + 90B = 7{,}200 \qquad \text{for W} \qquad (1)$$
$$5A + 7B = 1{,}400 \qquad \text{for X} \qquad (2)$$

Multiplying (2) by 4 gives
$$20A + 28B = 5{,}600$$

Subtracting (1)
$$20A + 90B = 7{,}200$$
$$\overline{}$$
$$-62B = -1{,}600$$

$$B = \frac{1{,}600}{62} = 25.8 \quad \text{(to 1 dp)}$$

Substituting this value for B into (1) gives

$$20A + 90(25.8) = 7{,}200$$
$$20A + 2{,}322 = 7{,}200$$
$$A = \frac{4{,}878}{20} = 243.9$$

Therefore the firm should use 243.9 litres of A and 25.8 litres of B for each drum of the final product.

The total input cost will be

$$243.9 \times £9 + 25.8 \times £16 = £2{,}195.10 + £412.80 = £2{,}607.90$$

Test Yourself, Exercise 5.A2

1. Find the minimum value of the function $C = 40A + 20B$ subject to the constraints

   $$10A + 40B \geq 40 \quad \text{(x)}$$
   $$30A + 20B \geq 60 \quad \text{(y)}$$
   $$10A \geq 10 \quad\qquad \text{(z)}$$
   $$A \geq 0, B \geq 0$$

   Will there be slack in any of the constraints at the optimum combination of A and B? If so, what is the excess capacity?

2. A firm manufactures a product that, per litre, must contain at least 18 g of chemical X and 10 g of chemical Y. The rest of the product is water whose costs can be ignored. The two inputs A and B contain X and Y in the following quantities:

    1 unit of A contains 6 g of X and 5 g of Y

    1 unit of B contains 9 g of X and 2 g of Y

    The per-unit costs of A and B are £2 and £6 respectively. What combination of A and B will give the cheapest way of producing a litre of the final product?

3. A firm mixes the two inputs Q and R to make a vitamin supplement in liquid form. The inputs Q and R contain the four vitamins A, B, C and D in the following amounts:

    6 mg of A, 50 mg of B, 35 mg of C and 12 mg of D per unit of Q

    30 mg of A, 25 mg of B, 30 mg of C and 20 mg of D per unit of R

    The inputs Q and R cost respectively 5 p and 12 p per unit. Each centilitre of the final product must contain at least 60 mg of A, 100 mg of B, 105 mg of C and 60 mg of D. What is the cheapest way of making the final product? Which vitamins will exceed the minimum requirements per centilitre using this method?

4. A delivery firm has two types of van, A and B, and carries three types of load, X, Y and Z. Each van is capable of carrying a mixed load, but only in certain proportions, given the special size and weight of the different loads. When fully loaded,

    type A can carry 20 of X, 15 of Y and 15 of Z

    type B can carry 10 of X, 60 of Y and 15 of Z

    A typical daily delivery schedule requires the firm to carry 200 loads of X, 450 loads of Y and 225 loads of Z. Each van is only loaded for deliveries once a day. The smaller van, A, costs £50 a day to run and the larger van, B, costs £100 a day. How many of each type of van should the firm use to minimize total running costs? Will there be space in the vans for any more of any of the loads X, Y or Z should more orders be placed?

5. A firm uses two inputs R and T which cost £40 each per tonne. They both contain the chemical compounds G and H in the following quantities:

    1 tonne of R contains 6 kg of G and 3 kg of H

    1 tonne of T contains 15 kg of G and 4 kg of H

    The final product must contain at least 180 kg of G and 60 kg of H per batch. How many tonnes of R and T should the firm use to minimize the cost of a batch of the final product? Will the amount of G or H it contains exceed the minimum requirement?

6. An aircraft manufacturer fitting out the interior of a plane can use two fitments A and B, which contain components X, Y and Z in the following quantities:

    1 unit of A contains 3 units of X, 4 units of Y plus 2 units of Z

    1 unit of B contains 6 units of X, 5 units of Y plus 8 units of Z

The aircraft design is such that there must be at least 540 units of X, 600 units of Y and 480 units of Z in total in the plane. If each unit of A weighs 4 kg and each unit of B weighs 6 kg, what combination of A and B will minimize the total weight of these fitments in the plane?

7. Construct your own linear programming problem involving the minimization of an objective function and then solve it.

Mixed constraints

Some linear programming problems may contain both 'less than or equal to' and 'greater than or equal to' constraints. It is also possible to have equality constraints, i.e. where one variable must equal a specified quantity.

Example 5.A7

Minimize the objective function $C = 12A + 8B$ subject to the constraints

$$10A + 40B \geq 40 \tag{1}$$

$$12A + 16B \leq 48 \tag{2}$$

$$A = 1.5 \tag{3}$$

Solution

The constraints are marked out in Figure 5.A7. Constraint (1) means that the feasible area must be above the line

$$10A + 40B = 40$$

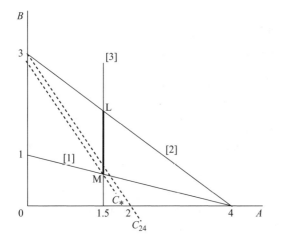

Figure 5.A7

Constraint (2) means that the feasible area must be below the line

$$12A + 16B = 48$$

Constraint (3) means that the feasible area must be along the vertical line through $A = 1.5$. The only section of the graph that satisfies all three of these constraints is the heavy black section LM of the vertical line through $A = 1.5$.

If C is assumed to be 24 then the line C_{24} representing the function

$$24 = 12A + 8B$$

can be drawn in and has a slope of -1.5. To minimize C, one needs to find the closest line to the origin that has this slope and also passes through the feasible area. This will be the line C_* through M.

The optimum value of A is therefore obviously 1.5.

The optimum value of B occurs at the intersection of the two lines

$$A = 1.5$$

and $\qquad 10A + 40B = 40$

Thus $\quad 10(1.5) + 40B = 40$

$$15 + 40B = 40$$

$$40B = 25$$

$$B = 0.625$$

Test Yourself, Exercise 5.A3

1. A firm makes two goods A and B using the three inputs X, Y and Z in the following quantities:

    20 units of X, 8 units of Y and 20 units of Z per unit of A

    20 units of X, 20 units of Y and 14 units of Z per unit of B

    The per-unit profit of A is £1,500, and for B the figure is £1,000. Input availability is restricted to 60 units of X, 40 units of Y and 70 units of Z. The firm has already committed itself to a contract to supply one customer with 1 unit of B. What combination of A and B should it produce to maximize total profit?

2. A company produces two industrial compounds X and Y that are mixed in a final product. They both contain one common input, R. The amount of R in one tonne of X is 8 litres and the amount of R in one tonne of Y is 12 litres. A load of the final product must contain at least 240 litres of R to ensure that its quality level is met. No R is lost in the production process of combining X and Y.

    The total cost of a tonne of X is £30 and the total cost of a tonne of Y is £15. If the firm has already signed a contract to buy 7.5 tonnes of X per week, what mix of X and Y should the firm use to minimize the cost of a load of the final product?

3. A firm manufactures two goods A and B which require the two inputs K and L in the following amounts:

   1 unit of A requires 6 units of K and 4 of L

   1 unit of B requires 8 units of K and 10 of L

   The firm has at its disposal 96 units of K and 100 of L. The per-unit profit of A is £600 and for B the figure is £300. The firm is under contract to produce a minimum of 6 units of B. How many units of A should it make to maximize profit?

4. Construct and solve your own linear programming problem that has two variables in the objective function and three constraints of at least two different types.

More than two variables

When the objective function in a linear programming problem contains more than two variables then it cannot be solved by graphical analysis. An advanced mathematical technique known as the *simplex method* can be used for these problems. This is based on the principle that the optimum value of the objective function will usually be at the intersection of two or more constraints.

It is an iterative method that can be very time-consuming to use manually and for most practical purposes it is best to use a computer program package to do the necessary calculations. If you have access to a linear programming computer package then you may try to use it now that you understand the basic principles of linear programming. The way that data are entered will depend on the computer package you use and you will need to consult the relevant handbook.

6 Quadratic equations

Learning objectives

After completing this chapter students should be able to:

- Use factorization to solve quadratic equations with one unknown variable.
- Use the quadratic equation solution formula.
- Identify quadratic equations that cannot be solved.
- Set up and solve economic problems that involve quadratic functions.
- Construct a spreadsheet to plot quadratic and higher order polynomial functions.

6.1 Solving quadratic equations

A quadratic equation is one that can be written in the form

$$ax^2 + bx + c = 0$$

where x is an unknown variable and a, b and c are constant parameters with $a \neq 0$. For example,

$$6x^2 + 2.5x + 7 = 0$$

A quadratic equation that includes terms in both x and x^2 cannot be rearranged to get a single term in x, so we cannot use the method used to solve linear equations.

There are three possible methods one might try to use to solve for the unknown in a quadratic equation:

  (i) by plotting a graph
 (ii) by factorization
(iii) using the quadratic 'formula'

In the next three sections we shall see how each can be used to tackle the following question.
    If a monopoly can face the linear demand schedule

$$p = 85 - 2q \tag{1}$$

at what output will total revenue be 200?

It is not immediately obvious that this question involves a quadratic equation. We first need to use economic analysis to set up the mathematical problem to be solved. By definition we know that total revenue will be

$$TR = pq \tag{2}$$

So, substituting the function for p from (1) into (2), we get

$$TR = (85 - 2q)q = 85q - 2q^2$$

This is a quadratic function that cannot be 'solved' as it stands. It just tells us the value of TR for any given output. What the question asks is 'at what value of q will this function be equal to 200'? The mathematical problem is therefore to solve the quadratic equation

$$200 = 85q - 2q^2 \tag{3}$$

All three solution methods require like terms to be brought together on one side of the equality sign, leaving a zero on the other side. It is also necessary to put the terms in the order given in the above definition of a quadratic equation, i.e.

unknown squared (q^2),   unknown (q),   constant

Thus (3) above can be rewritten as

$$2q^2 - 85q + 200 = 0$$

It is this quadratic equation that each of the three methods explained in the following sections will be used to solve.

Before we run through these methods, however, you should note that an equation involving terms in x^2 and a constant, but not x, can usually be solved by a simpler method. For example, suppose that

$$5x^2 - 80 = 0$$

this can be rearranged to give

$$5x^2 = 80$$
$$x^2 = 16$$
$$x = 4$$

6.2   Graphical solution

Drawing a graph of a quadratic function can be a long-winded and not very accurate process that involves separately plotting each individual value of the variable within the range that is being considered. It is therefore usually not a very practical method of solving a quadratic equation. The graphical method can be useful, however, not so much for finding an approximate value for the solution, but for explaining why certain quadratic equations do not have a solution whilst others have two solutions. Only a rough sketch diagram is necessary for this purpose.

Example 6.1

Show graphically that a solution does exist for the quadratic equation

$$2q^2 - 85q + 200 = 0$$

Solution

We first need to define a new function

$$y = 2q^2 - 85q + 200$$

If the graph of this function cuts the q axis then $y = 0$ and we have a solution to the quadratic equation specified in the question. Next, we calculate a few values of the function to get an approximate idea of its shape.

When $q = 0$,   then $y = 200$
When $q = 1$,   then $y = 2 - 85 + 200 = 117$

and so the graph initially falls.

When $q = 3$,   then $y = 18 - 255 + 200 = -37$

and so it must cut the q axis as y has gone from a positive to a negative value.

When $q = 50$ then $y = 5{,}000 - 4{,}250 + 200 = 950$

and so the value of y rises again and must cut the q axis a second time.

These values indicate that the graph is a U-shape, as shown in Figure 6.1. This cuts the horizontal axis twice and so there are two values of q for which y is zero, which means that there are two solutions to the question. The precise values of these solutions, 2.5 and 40, can be found by the other two methods explained in the following sections or by computation of y for different values of q. (See spreadsheet solution method below.)

If we slightly change the problem in Example 6.1 we can see why there may not always be a solution to a quadratic equation.

Example 6.2

Find out if there is an output level at which total revenue is 1,500 for the function

$$TR = 85q - 2q^2$$

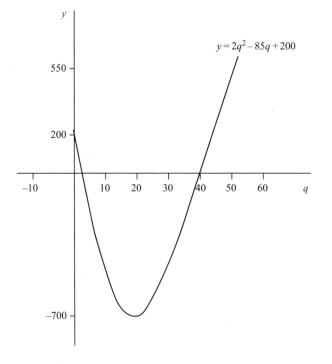

Figure 6.1

Solution

The quadratic equation to be solved is

$$1,500 = 85q - 2q^2$$

which can be rewritten as

$$2q^2 - 85q + 1,500 = 0$$

If we now specify the new function

$$y = 2q^2 - 85q + 1,500$$

and calculate a few values, we can see that it falls and then rises again but never cuts the q axis, as Figure 6.2 shows.

When $q = 0$,   then $y = 1,500$

When $q = 10$,   then $y = 850$

When $q = 20$,   then $y = 600$

When $q = 25$,   then $y = 625$

There are therefore no solutions to this quadratic equation, i.e. there is no output at which total revenue will be 1,500.

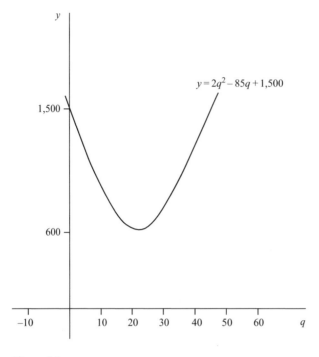

Figure 6.2

Although one would never try to plot the whole graph of a quadratic function manually, one may of course get a computer plot. The accuracy of the answer you obtain will depend on the graphics package that you use.

Plotting quadratic functions with Excel

An Excel spreadsheet for calculating different values of the function y in Example 6.1 above can be constructed by following the instructions in Table 6.1. Rather than building in formulae that are specific to this example, this spreadsheet is constructed in a format that can be used to plot any function in the form $y = aq^2 + bq + c$ once the parameters a, b and c are entered in the relevant cells. The range for q has been chosen to ensure that it includes the values when y is zero, which is what we are interested in finding.

If you construct this spreadsheet you should get the series of values shown in Table 6.2. The q values which correspond to a y value of zero can now be read off, giving the solutions 2.5 and 40.

You may also use the Excel spreadsheet you have created to plot a graph of the function $y = 2q^2 - 85q + 200$. Assuming that you have q and y in single columns, then you just use the Chart Wizard command to obtain a plot with q measured on the X axis and y as variable A on the vertical axis. (It you don't know how to use this chart command, refer back to Example 4.17.) To make the chart clearer to read, enlarge it a bit by dragging the corner. The legend box for y can also be cut out to allow the chart area to be enlarged. This should give you a plot similar to Figure 6.3, which clearly shows how this function cuts the horizontal axis twice.

Table 6.1

CELL	Enter	Explanation
A1	Ex.6.2	Label to remind you what example this is
B1	QUADRATIC SOLUTION TO y = aq^2+bq+c = 0	Title of spreadsheet (Note that this label is not an actual Excel formula.)
B2	a =	These are labels that tell you that the actual
D2	b =	parameter values will go in the cells next
F2	c =	to them. Right justify these labels.
C2	2	These are the actual parameter values for
E2	-85	this example.
G2	200	
A4	q	Column heading label
B4	y	Column heading label
A5	0	Initial value for *q*
A6	=A5+0.5	Calculates a 0.5 unit increment in *q*
A7 to A90	*Copy formula from cell A6 down column A*	Calculates a series of values of *q* in 0.5 unit increments
B5	=$C$2*A5^2+$E$2*A5+$G$2	This formula calculates the value of the function corresponding to the value of *q* in cell A5 and the parameter values in cells C2, E2 and G2. Note that the $ sign is used so that these cell references do not change when this function is copied down the *y* column.
B6 to B92	*Copy formula from cell B5 down column B*	Calculates values for *y* in each row corresponding to values of *q* in column A.

This spreadsheet can easily be amended to calculate values and plot graphs of other quadratic functions by entering different values for the parameters a, b and c in cells C2, E2 and G2. For example, to calculate values for the function from Example 6.2

$$y = 2q^2 - 85q + 1,500$$

the value in cell G2 should be changed to 1,500. A computer plot of this function should produce the shape shown in Figure 6.2 above, confirming again that this function will not cut the horizontal axis and that there is no solution to the quadratic equation

$$0 = 2q^2 - 85q + 1,500$$

6.3 Factorization

In Chapter 3 factorization was explained, i.e. how some expressions can be broken down into terms which when multiplied together give the original expression. For example,

$$a^2 - 2ab + b^2 = (a - b)(a - b)$$

If a quadratic function which has been rearranged to equal zero can be factorized in this way then one or the other of the two factors must equal zero. (Remember that if $A \times B = 0$ then either A or B, or both, must be zero.)

Table 6.2

	A	B	C	D	E	F	G	H
1	Ex 6.2	QUADRATIC	SOLUTION TO		y = aq^2+bq+c = 0			
2		a =	2	b =	-85	c =	200	
3	q	y	q	y	q	y	q	y
4	0	200	11	-493	22	-702	33	-427
5	0.5	158	11.5	-513	22.5	-700	33.5	-403
6	1	117	12	-532	23	-697	34	-378
7	1.5	77	12.5	-550	23.5	-693	34.5	-352
8	2	38	13	-567	24	-688	35	-325
9	**2.5**	**0**	13.5	-583	24.5	-682	35.5	-297
10	3	-37	14	-598	25	-675	36	-268
11	3.5	-73	14.5	-612	25.5	-667	36.5	-238
12	4	-108	15	-625	26	-658	37	-207
13	4.5	-142	15.5	-637	26.5	-648	37.5	-175
14	5	-175	16	-648	27	-637	38	-142
15	5.5	-207	16.5	-658	27.5	-625	38.5	-108
16	6	-238	17	-667	28	-612	39	-73
17	6.5	-268	17.5	-675	28.5	-598	39.5	-37
18	7	-297	18	-682	29	-583	**40**	**0**
19	7.5	-325	18.5	-688	29.5	-567	40.5	38
20	8	-352	19	-693	30	-550	41	77
21	8.5	-378	19.5	-697	30.5	-532	41.5	117
22	9	-403	20	-700	31	-513	42	158
23	9.5	-427	20.5	-702	31.5	-493	42.5	200
24	10	-450	21	-703	32	-472	43	243
25	10.5	-472	21.5	-703	32.5	-450	43.5	287

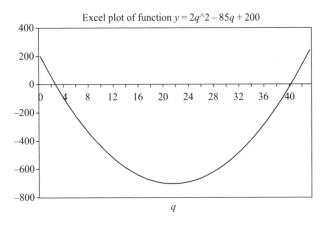

Excel plot of function $y = 2q^2 - 85q + 200$

Figure 6.3

Example 6.3

Solve by factorization the quadratic equation

$$2q^2 - 85q + 200 = 0$$

Solution

This expression can be factorized as

$$(2q - 5)(q - 40) = 2q^2 - 85q + 200$$

Therefore

$$(2q - 5)(q - 40) = 0$$

This means that

$$2q - 5 = 0 \quad \text{or} \quad q - 40 = 0$$

giving solutions

$$q = 2.5 \quad \text{or} \quad q = 40$$

As expected, these are the same solutions as those found by the graphical method.

It may be the case that mathematically a quadratic equation has one or more solutions with a negative value that will not apply in an economic problem. One cannot have a negative output, for example.

Example 6.4

Solve by factorization the quadratic equation

$$2x^2 - 6x - 20 = 0$$

Solution

By factorization $(2x - 10)(x + 2) = 0$

Therefore

$$2x - 10 = 0 \quad \text{or} \quad x + 2 = 0$$
$$x = 5 \quad \text{or} \quad x = -2$$

If x represented output, then $x = 5$ would be the only answer we would use.

If a quadratic equation cannot be factorized then the formula method in Section 6.5 below must be used. The formula method can also be used, however, when an equation can be factorized. Therefore, if you cannot quickly see a way of factorizing then you should use the formula method. Factorization is only useful as a short-cut way of solving certain quadratic equations. It defeats the object of the exercise if you spend half an hour trying to find a way of factorizing an expression when it would be quicker to use the formula.

It should also go without saying that quadratic equations for which no solutions exist cannot be factorized. For example, it is not possible to factorize the equation

$$2q^2 - 85q + 1,500 = 0$$

which we have already shown to have no solution.

Test Yourself, Exercise 6.1

1. Solve for x in the equation $x^2 - 5x + 6 = 0$.
2. Find the output at which total revenue is £600 if a firm's demand schedule is

$$p = 70 - q$$

3. A firm faces the average cost function

$$AC = 40x^{-1} + 10x$$

where x is output. When will average cost be 40?
4. Is there a positive solution for x when

$$0 = 12x^2 + 90x - 48?$$

5. A firm faces the total cost schedule

$$TC = 6 - 2q + 2q^2$$

when $q > 2$. At what output level will TC = £150?

6.4   The quadratic formula

Any quadratic equation expressed in the form

$$ax^2 + bx + c = 0$$

where a, b and c are given parameters and for which a solution exists can be solved for x by using the quadratic formula

$$x = \frac{-b \pm \sqrt{b^2 - 4ac}}{2a}$$

(The sign $\pm$ means $+$ or $-$.) There is no need for you to understand how the formula is derived. You just need to know that it works.

Example 6.5

Use the quadratic formula to solve the quadratic equation

$$2q^2 - 85q + 200 = 0$$

Solution

In the quadratic formula applied to this example $a = 2, b = -85$ and $c = 200$ (and, of course, $x = q$). Note that the minus signs for any negative coefficients must be included. One also needs to take special care to remember to use the rules for arithmetic operations using negative numbers. Substituting these values for a, b and c into the formula we get

$$
\begin{aligned}
q &= \frac{-(-85) \pm \sqrt{(-85)^2 - 4 \times 2 \times 200}}{2 \times 2} \\
&= \frac{85 \pm \sqrt{7,225 - 1,600}}{4} \\
&= \frac{85 \pm \sqrt{5,625}}{4} \\
&= \frac{85 \pm 75}{4} = \frac{160}{4} \text{ or } \frac{10}{4} = 40 \text{ or } 2.5
\end{aligned}
$$

These are, of course, the same as the solutions found by factorization in Example 6.3 above.

What happens if you try to use the quadratic formula when no solution exists? We can find out by applying the formula to the quadratic equation in Example 6.2 above, where a sketch graph showed that there was no solution.

Example 6.6

Use the quadratic formula to try to solve

$$2q^2 - 85q + 1,500 = 0$$

Solution

In this example $a = 2, b = -85$ and $c = 1,500$. Therefore

$$
\begin{aligned}
q &= \frac{-(-85) \pm \sqrt{(-85)^2 - 4 \times 2 \times 1,500}}{2 \times 2} \\
&= \frac{85 \pm \sqrt{7,225 - 12,000}}{4} \\
&= \frac{85 \pm \sqrt{-4,775}}{4}
\end{aligned}
$$

We are now stuck! It is impossible to find the square root of a negative number. In other words, no solution exists.

It will always be the case that the quadratic formula will require the square root of a negative number if no solution exists.

Test Yourself, Exercise 6.2

(Use the quadratic formula to try to solve these problems.)

1. Solve for x if $0 = x^2 + 2.5x - 125$.
2. A firm faces the demand schedule $q = 400 - 2p - p^2$. What price does it need to charge to sell 100 units?
3. If a firm's demand function is $p = 100 - q$, what quantities need to be sold to bring in a total revenue of

    (a) £100   (b) £1,000   (c) £10,000?

    (Give answers to 2 decimal places, where they exist.)
4. Make up your own quadratic equation and then find whether a solution exists.

6.5   Quadratic simultaneous equations

If one or more equations in a simultaneous equation system are quadratic then it may be possible to eliminate all but one unknown and to reduce the problem to a single quadratic equation. If this can be solved then the other unknowns can be found by substitution.

Example 6.7

Find the equilibrium values of p and q in a competitive market where the demand schedule is

$$p = 200q^{-1}$$

and the supply function is

$$p = 30 + 2q$$

Solution

In equilibrium, demand price equals supply price. Therefore

$$200q^{-1} = 30 + 2q$$

Multiplying through by q,

$$200 = 30q + 2q^2$$
$$0 = 2q^2 + 30q - 200$$
$$0 = (2q - 10)(q + 20)$$

Therefore $2q - 10 = 0$ or $q + 20 = 0$
$$q = 5 \quad \text{or} \quad q = -20$$

We can ignore the second solution as negative quantities cannot exist. Thus the equilibrium quantity is 5.

Substituting this value into the supply function gives equilibrium price

$$p = 30 + 2 \times 5 = 40$$

You should now be able to link the different mathematical techniques you have learned so far to tackle more complex problems. If you have covered the theory of perfect competition in your economics course, then you should be able to follow the analysis in the example below.

Example 6.8

An industry is made up of 100 firms, all with the cost schedules

$$AC = 40q^{-1} + 0.4q^2 \qquad TC = 40 + 0.4q^3 \qquad MC = 1.2q^2$$

They sell in a market where the demand schedule is

$$p = 70 - 0.08Q$$

where Q is industry output (and q is an individual firm's output).

 (i) What will be the short-run price, industry output and profit for each firm?
 (ii) What will happen to price, industry output and the number of firms in the long run? (Assume new entrants have the same cost structure.)

Solution

(i) The industry supply schedule is the horizontal sum of the individual firms' marginal cost schedules. Given the marginal cost function

$$MC = 1.2q^2$$

$$\left(\frac{MC}{1.2}\right)^{0.5} = q$$

There are 100 firms, and so the amount supplied by the whole industry is

$$Q = 100q = 100\left(\frac{MC}{1.2}\right)^{0.5} \tag{1}$$

In perfect competition MC corresponds to the price at which any given quantity will be supplied, and so (1) can be rewritten as

$$Q = 100\left(\frac{p}{1.2}\right)^{0.5}$$

Therefore

$$0.01Q = \left(\frac{p}{1.2}\right)^{0.5}$$

$$(0.01Q)^2 = \frac{p}{1.2}$$

$$0.0001Q^2 = \frac{p}{1.2}$$

$$0.00012Q^2 = p \tag{2}$$

The function (2) will be the industry supply schedule. The demand schedule given in the question is

$$p = 70 - 0.08Q \tag{3}$$

In equilibrium, demand price equals supply price. Thus equating (3) and (2) we get

$$70 - 0.08Q = 0.00012Q^2$$

$$0 = 0.00012Q^2 + 0.08Q - 70 \tag{4}$$

Using the quadratic formula to solve (4)

$$Q = \frac{-0.08 \pm \sqrt{0.0064 + 0.0336}}{0.00024}$$

$$= \frac{-0.08 \pm \sqrt{0.04}}{0.00024}$$

$$= \frac{-0.08 + 0.2}{0.00024} \quad \text{or} \quad \frac{-0.08 - 0.2}{0.00024}$$

$$= \frac{0.12}{0.00024} \quad \text{or} \quad \frac{-0.28}{0.00024}$$

$$= 500 \quad \text{(ignoring the negative answer)}$$

Substituting this value of Q into the demand schedule (3) gives

$$p = 70 - 0.08(500) = 70 - 40 = £30$$

Each of the 100 firms produces the same amount q. Therefore,

$$q = \frac{Q}{100} = \frac{500}{100} = 5$$

Each firm's profit will be

$$\text{TR} - \text{TC} = pq - (40 + 0.4q^3)$$

$$= 30(5) - (40 + 50)$$

$$= 150 - 90 = £60$$

(ii) If existing firms are making a profit then in the long run new entrants will be attracted into the industry. This will shift the supply schedule to the right and price will be driven down until each firm is only just breaking even, when price equals the lowest value on the firm's U-shaped average cost schedule.

How do we find when AC is at its minimum point? The MC and AC functions are given in the question. From cost theory you should know that MC always cuts AC at its minimum point. Therefore

$$MC = AC$$

$$1.2q^2 = 40q^{-1} + 0.4q^2$$

$$0.8q^2 = 40q^{-1}$$

$$q^3 = 50$$

$$q = 3.684 \quad \text{(to 3 dp)}$$

When $q = 3.684$, then

$$AC = 40q^{-1} + 0.4q^2$$

$$= 40(3.684)^{-1} + 0.4(3.684)^2$$

$$= 16.2865$$

Therefore $\quad p = £16.29 \quad$ (to the nearest penny)

The old supply schedule does not now apply because of the increased number of firms in the industry. Therefore, substituting this price into the demand schedule (3) to get total output gives

$$16.29 = 70 - 0.08Q$$

$$0.08Q = 53.71$$

$$Q = 671.375$$

We already know that each firm produces 3.684 units in long-run equilibrium. Therefore, the new number of firms in the industry is

$$\frac{Q}{q} = \frac{671.375}{3.684} = 182.24 = 182 \text{ firms}$$

Given that there were originally 100 firms, the number of new entrants is therefore 82. Note that the fraction is rounded down to the nearest whole number. Any extra firms would bring price below the break-even level.

Test Yourself, Exercise 6.3

1. If $y = 255 - x - x^2$ and $y = 180 + (2/3)x^2 - 21x$, find x and y.
2. Find x and y given the functions

$$2y + 4x^2 + 10x - 36 = 0$$

and

$$4y - 10x^2 + 24x = 24$$

3. A monopoly faces the marginal cost function $MC = 0.5q^2$
   and the marginal revenue function $MR = 200 - 4q$
   What output will maximize profits?

4. A price-discriminating monopoly sells in two markets whose demand schedules are

$$p_1 = 200 - 20q_1 \quad \text{and} \quad p_2 = 120 - 5q_2$$

Total output $q = q_1 + q_2$ and the firm's marginal cost schedule is

$$MC = 40 + 0.5q^2$$

How much should it sell in each market, and at what price, in order to maximize profit?

5. A firm's marginal cost schedule is $MC = 2.3 + 0.00012q^2$ and it sells its output in two separate markets with demand schedules

$$p_1 = 25 - 0.125q_1 \quad \text{and} \quad p_2 = 12 - 0.05q_2$$

What prices and quantities will maximize profits if this firm is a price-discriminating monopoly?

6.6 Polynomials

Quadratic equations are a special case of polynomial equations. The general format of a polynomial function is

$$y = a_0 + a_1x + a_2x^2 + a_3x^3 + \cdots + a_nx^n$$

where n is any non-negative integer. Linear equations contain polynomials where $n = 1$. Quadratic equations contain polynomials where $n = 2$.

When n is greater than 2 the solution of a polynomial equation by algebraic means becomes complex and time-consuming. For practical purposes, however, you may use a spreadsheet to find a solution by the iterative method. This means calculating values of a function for different values of the unknown variable until a solution or a good approximation to it is found. As a spreadsheet can quickly perform the necessary calculations, it is an ideal tool for the calculation of polynomial solutions.

The format of the spreadsheet will depend on the problem tackled. Below are some examples of how problems can be approached.

Example 6.9

A firm's total costs (TC) are given by the function

$$TC = 420 + 32.5q - 6.25q^2 + 0.8q^3$$

where q is output level and TC is measured in pounds. If the firm's management is given a budget of £43,000, what output can it produce?

Solution

A spreadsheet needs to be constructed that will calculate TC for different values of *q*. You can then experiment with different ranges of q until the solution is found.

The method for constructing the spreadsheet is basically the same as that used for quadratic equations as set out in Table 6.1 earlier. This time the spreadsheet calculates the cubic TC function that corresponds to the parameters entered. The instructions for doing this are set out in Table 6.3.

Although AC and MC may initially fall as a firm's output increases, its TC function should never fall. It would therefore be useful to have a check that this cubic TC function always increases as *q* increases and MC is never negative. To do this the spreadsheet also includes a third column where values of MC are calculated. (See Section 8.4 for further analysis of cubic functions with this property.)

Table 6.3

CELL	Enter	Explanation
A1	Ex.6.9	Label to remind you what example this is
B2	CUBIC POLYNOMIAL SOLUTION TO	Title of spreadsheet
B3	TC =a + bq + cq^2 + dq^3	(Note that this is not an actual Excel formula.)
F2	Parameter	Labels that tell you that the parameter values
F3	Values	will be shown below
E4	a =	These are labels that tell you that the
E5	b =	parameter values will go in the cells next to
E6	c =	them.
E7	d =	Right justify these cells.
F4	420	These are the actual parameter values for
F5	32.5	*a,b,c* and *d*, respectively, for this example.
F6	-6.25	
F7	0.8	
A3	q	Column heading labels
B3	TC	
C3	MC	
A4	0	Initial value for *q*
A5	=A4+1	Calculates a one unit increment in *q*
A6 to A45	*Copy formula from cell A5 down column A*	Calculates a series of values of *q* in one unit increments
B4	=F$4+F$5*A4+F$6*A4^2+ F$7*A4^3	Formula to calculate value of *TC* corresponding to value of *q* in cell A4 and parameter values in cells F4, F5, F6 and F7. Note the $ sign used to anchor row references for when this formula is copied down row B.
B5 to B45	*Copy formula from cell B4 down column B*	Calculates values for TC in each row corresponding to values of *q* in column A.
C5	=B5-B4	Calculates values MC as the change in TC from a one unit increment in *q*.
C6 to C45	*Copy formula from cell C5 down column C*	Calculates MC of a unit of *q* corresponding to increment in TC shown in column B.
B4 to C45	*Highlight these columns and format to 2 decimal places*	TC and MC are both monetary values measured in £ so use numerical format 0.00

Table 6.4

	A	B	C	D	E	F
1	Ex 6.9	CUBIC POLYNOMIAL SOLUTION TO				
2		TC =a + bq + cq^2 + dq^3				Parameter
3	q	TC	MC			Values
4	0	420.00			a =	420
5	1	447.05	27.05		b =	32.5
6	2	466.40	19.35		c =	-6.25
7	3	482.85	16.45		d =	0.8
8	4	501.20	18.35			
9	5	526.25	25.05			
10	6	562.80	36.55			
11	7	615.65	52.85			
12	8	689.60	73.95			
13	9	789.45	99.85			
14	10	920.00	130.55			
15	11	1086.05	166.05			
16	12	1292.40	206.35			
17	13	1543.85	251.45			
18	14	1845.20	301.35			
19	15	2201.25	356.05			
20	16	2616.80	415.55			
21	17	3096.65	479.85			
22	18	3645.60	548.95			
23	19	4268.45	622.85			
24	20	4970.00	701.55			
25	21	5755.05	785.05			
26	22	6628.40	873.35			
27	23	7594.85	966.45			
28	24	8659.20	1064.35			
29	25	9826.25	1167.05			
30	26	11100.80	1274.55			
31	27	12487.65	1386.85			
32	28	13991.60	1503.95			
33	29	15617.45	1625.85			
34	30	17370.00	1752.55			
35	31	19254.05	1884.05			
36	32	21274.40	2020.35			
37	33	23435.85	2161.45			
38	34	25743.20	2307.35			
39	35	28201.25	2458.05			
40	36	30814.80	2613.55			
41	37	33588.65	2773.85			
42	38	36527.60	2938.95			
43	39	39636.45	3108.85			
44	40	42920.00	3283.55			
45	41	46383.05	3463.05			

Your spreadsheet should now look like Table 6.4, which shows that when q is 40, TC will be 42,920. Thus, if output is constrained to whole units, 40 is the maximum output that the firm's management can produce for a budget of £43,000.

This spreadsheet also confirms that MC declines in value then increases, but is never negative. This is what we would expect. Save your spreadsheet for use with other examples.

This example was constructed for a range of values of q that contained the answer we were seeking. If you had no idea where the solution to this cubic polynomial lay then you could get a 'ball park' estimate by producing a range of values in jumps of 10 in the column headed q by entering the formula $= A4 + 10$ in cell A5 and then copying it down the column for a few dozen rows. This would tell you that when $q = 31$, TC = £19,254.05, and when $q = 41$, TC = £46,383.05. Therefore, TC = £43,000 must lie somewhere between these values of q. Once you have a rough idea of where the solution value for q will lie, you can change the q column so that values increase in only one unit increments, or smaller units if necessary, until the actual solution is pinpointed.

To solve other cubic polynomials, one simply enters the corresponding parameters into the spreadsheet set up for Example 6.9 above and adjusts the range of the independent variable (q) until the solution is found.

Example 6.10

If TC $= 880 + 72q - 14.5q^2 + 1.5q^3$, at what value of q will TC = £9,889?

Table 6.5

	A	B	C	D	E	F
1	Ex 6.10	CUBIC POLYNOMIAL SOLUTION TO				
2		TC =a + bq + cq^2 + dq^3				Parameter
3	q	TC	MC			Values
4	0	880.00			a =	880
5	1	939.00	59.00		b =	72
6	2	978.00	39.00		c =	-14.5
7	3	1006.00	28.00		d =	1.5
8	4	1032.00	26.00			
9	5	1065.00	33.00			
10	6	1114.00	49.00			
11	7	1188.00	74.00			
12	8	1296.00	108.00			
13	9	1447.00	151.00			
14	10	1650.00	203.00			
15	11	1914.00	264.00			
16	12	2248.00	334.00			
17	13	2661.00	413.00			
18	14	3162.00	501.00			
19	15	3760.00	598.00			
20	16	4464.00	704.00			
21	17	5283.00	819.00			
22	18	6226.00	943.00			
23	19	7302.00	1076.00			
24	20	8520.00	1218.00			
25	21	9889.00	1369.00			
26	22	11418.00	1529.00			

Solution

Entering the new values for a, b, c and d into the spreadsheet constructed for Example 6.9 and adjusting the range of q, one should get a spreadsheet similar to Table 6.5. This shows that $q = 21$ when TC = £9,889.

This spreadsheet can also be adjusted to cope with more complex polynomials. Its crucial part is the formula in cell B4. This needs to be amended to calculate the value of the new polynomial function if more terms are added. Note, however, that large *polynomial equations may have several solutions*. In particular, if there are both positive and negative coefficients, a polynomial function may equal zero at more than two values of the independent variable. You can usually deduce the number of solutions from the format of the equation, or get a plot from your spreadsheet to see how many times the function crosses the horizontal axis. On the other hand, no solutions may exist, in which case a graph will not cut the axis; e.g. there is no positive value of x which will satisfy the equation

$$0 = 8 + 32x + 6x^2 + 0.9x^3$$

although in this case there will be a negative solution.

Example 6.11

Assuming $x < 1,000$, is there a positive value of x that is a solution to the function

$$0 = -770,077.6 + 262x - 74x^2 + 12x^3 + 2x^4 - 0.05x^5?$$

Solution

To solve this equation we need to calculate values of the polynomial function

$$y = -770,077.6 + 262x - 74x^2 + 12x^3 + 2x^4 - 0.05x^5$$

and find the value(s) of x where this function equals zero. To do this, call up the spreadsheet created for Example 6.9 and then follow the instructions in Table 6.6 to add two new terms so that it will be able to calculate values for polynomials in the format

$$y = a + bx + cx^2 + dx^3 + ex^4 + fx^5$$

Once the basic spreadsheet has been created, the ball-park method explained above can be used to narrow down the possible solution range to between 30 and 40. This should give you a spreadsheet that looks like Table 6.7. This clearly shows that y is zero when x is 38 and so this is the solution. (If you try increasing the range of x you will see that there are no other solutions in the range $0 < x < 1,000$.)

Table 6.6

(Only shows changes needed to adapt *Table 6.3* for this example.)

CELL	Enter	Explanation
A1	Ex.6.11	Label for example number
B2	y =a + bx + cx^2 + dq^3 + ex^4 + fx^5	Title of new formula.
E8	e =	Additional labels for the two extra
E9	f =	parameter values. (Right justify.)
A3	x	New labels for column headings.
B3	y	
Column C	*Highlight and hit Edit-Clear- All*	This column can be cleared as MC not calculated for this example.
F4	-770077.6	These are the actual parameter values for
F5	262	a,b,c, d, e and f, respectively, for this
F6	-74	example.
F7	12	
F8	2	
F9	-0.05	
A4	30	Initial value for x (Ball-park range found.)
Rows 15 onward	*Highlight-Edit-Delete*	Delete extra rows as only need range of x from 30 to 40 for this example.
B4	=F$4+F$5*A4+F$6*A4^2+F$7*A4^3+F$8*A4^4+F$9*A4^5	Calculates the value of y function corresponding to value of x in cell A4 and the parameter values in cells F4 to F9. Note $ sign used to anchor row references.
B5 to B14	*Copy formula from cell B4 down column B*	Calculates values for y in each row corresponding to values of x in column A.

Table 6.7

	A	B	C	D	E	F
1	Ex 6.11	CUBIC POLYNOMIAL SOLUTION TO				
2		y =a + bx + cx^2 + dq^3 + ex^4 + fx^5				Parameter
3	x	y				Values
4	30	-99817.60			a =	-770077.6
5	31	-59993.15			b =	262
6	32	-24823.20			c =	-74
7	33	4298.75			d =	12
8	34	25835.20			e =	2
9	35	38098.65			f =	-0.05
10	36	39245.60				
11	37	27270.55				
12	38	0.00				
13	39	-44913.55				
14	40	-109997.60				

Test Yourself, Exercise 6.4

(You will need to use a spreadsheet to tackle these questions.)

1. How much of q can be produced for £60,000 if the total cost function is

    $$TC = 86 + 152q - 12q^2 + 0.6q^3?$$

2. What output can be produced for £150,000 if

    $$TC = 130 + 62q - 3.5q^2 + 0.15q^3?$$

3. Solve for x when

    $$0 = -1,340 + 14x + 2x^2 - 1.5x^3 + 0.2x^4 + 0.005x^5 - 0.0002x^6$$

7 Financial mathematics

Series, time and investment

Learning objectives

After completing this chapter students should be able to:

- Calculate the final sum, the initial sum, the time period and the interest rate for an investment.
- Calculate the Annual Equivalent Rate for part year investments and compare this with the nominal annual rate of return.
- Calculate the Net Present Value and Internal Rate of Return on an investment, constructing relevant spreadsheets when required.
- Use the appropriate investment appraisal method to decide if an investment project is worthwhile.
- Find the sum of finite and infinite geometric series.
- Calculate the value of an annuity.
- Calculate monthly repayments and the APR for a loan.
- Apply appropriate mathematical methods to solve problems involving the growth and decline over discrete time periods of other economic variables, including the depletion of natural resources.

7.1 Discrete and continuous growth

In economics we come across many variables that grow, or decline, over time. A sum of money invested in a deposit account will grow as interest accumulates on it. The amount of oil left in an oilfield will decline as production continues over the years. This chapter explains how mathematics can help answer certain problems concerned with these variables that change over time. The main area of application is finance, including methods of appraising different forms of investment. Other applications include the management of natural resources, where the implications of different depletion rates are analysed.

The interest earned on money invested in a deposit account is normally paid at set regular intervals. Calculations of the return are therefore usually made with respect to specific time intervals. For example, Figure 7.1(a) shows the amount of money in a deposit account at any given moment in time assuming an initial deposit of £1,000 and interest credited at the end of each year at a rate of 10%. There is not a continuous relationship between time and the total sum in the deposit account. Instead there is a 'jump' at the end of each year when the interest

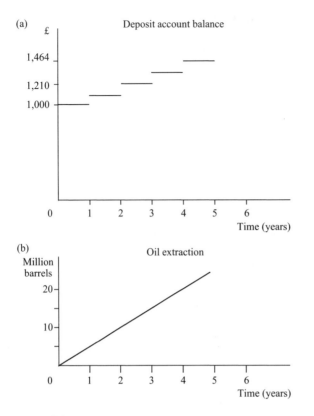

Figure 7.1

on the account is paid. This is an example of a 'discrete' function. Between the occasions when interest is added there is no change in the value of the account.

A *discrete function* can therefore be defined as one where the value of the dependent variable is known for specific values of the independent variable but does not continuously change between these values. Hence one gets a series of values rather than a continuum. For example, teachers' salaries are based on scales with series of increments. A hypothetical scale linking completed years of service to salary might be:

0 yrs = £20,000, 1 yr = £21,800, 2 yrs = £23,600, 3 yrs = £25,400

The relationship between salary and years of service is a discrete function. At any moment in time one knows what a teacher's salary will be but there is not a continuous relationship between time and salary level.

An example of a continuous function is illustrated in Figure 7.1(b). This shows the cumulative total amount of oil extracted from an oilfield when there is a steady 5 million barrels per year extraction rate. There is a continuous smooth function showing the relationship between the amount of oil extracted and the time elapsed.

In this chapter we analyse a number of discrete-variable problems. Algebraic formulae are developed to solve some applications of discrete functions and methods of solution using spreadsheets are explained where appropriate, for investment appraisal analysis in particular. The analysis of continuous growth requires the use of the exponential function and will be explained in Chapter 14.

7.2   Interest

Time is money. If you borrow money you have to pay interest on it. If you invest money in a deposit account you expect to earn interest on it. From an investor's viewpoint the interest rate can be looked on as the 'opportunity cost of capital'. If a sum of money is tied up in a project for a year then the investor loses the interest that could have been earned by investing the money elsewhere, perhaps by putting it in a deposit account.

Simple interest is the interest that accrues on a given sum in a set time period. It is *not* reinvested along with the original capital. The amount of interest earned on a given investment each time period will be the same (if interest rates do not change) as the total amount of capital invested remains unaltered.

Example 7.1

An investor puts £20,000 into a deposit account and has the annual interest paid directly into a separate current account and then spends it. The deposit account pays 8.5% interest. How much interest is earned in the fifth year?

Solution

The interest paid each year will remain constant at 8.5% of the original investment of £20,000. Thus in year 5 the interest will be

$$0.085 \times £20{,}000 = £1{,}700$$

Most investment decisions, however, need to take into account the fact that any interest earned can be reinvested and so compound interest, explained below, is more relevant. The calculation of simple interest is such a basic arithmetic exercise that the only mistake you are likely to make is to transform a percentage figure into a decimal fraction incorrectly.

Example 7.2

How much interest will be earned on £400 invested for a year at 0.5%?

Solution

To convert any percentage figure to a decimal fraction you must divide it by 100. Therefore

$$0.5\% = \frac{0.5}{100} = 0.005$$

and so

$$0.5\% \text{ of } £400 = 0.005 \times £400 = £2$$

If you can remember that $1\% = 0.01$ then you should be able to transform any interest rate specified in percentage terms into a decimal fraction in your head. Try to do this for the following interest rates:

(i) 1.5%   (ii) 30%   (iii) 0.075%   (iv) 1.02%   (v) 0.6%

Now check your answers with a calculator. If you got any wrong you really ought to go back and revise Section 2.5 before proceeding. Converting decimal fractions back to percentage interest rates is, of course, simply a matter of multiplying by 100;

e.g. $0.02 = 2\%, 0.4 = 40\%, 1.25 = 125\%, 0.008 = 0.8\%$.

Compound interest is interest which is added to the original investment every time it accrues. The interest added in one time period will itself earn interest in the following time period. The total value of an investment will therefore grow over time.

Example 7.3

If £600 is invested for 3 years at 8% interest compounded annually at the end of each year, what will the final value of the investment be?

Solution

	£
Initial sum invested	600.00
Interest at end of year $1 = 0.08 \times 600$	48.00
Total sum invested for year 2	648.00
Interest at end of year $2 = 0.08 \times 648$	51.84
Total sum invested for year 3	699.84
Interest at end of year $3 = 0.08 \times 699.84$	55.99
Final value of investment	755.83

Example 7.4

If £5,000 is invested at an annual rate of interest of 12% how much will the investment be worth after 2 years?

Solution

	£
Initial sum invested	5,000
Year 1 interest $= 0.12 \times 5,000$	600
Sum invested for year 2	5,600
Year 2 interest $= 0.12 \times 5,600$	672
Final value of investment	6,272

The above examples only involved the calculation of interest for a few years and did not take too long to solve from first principles. To work out the final sum of an investment after longer time periods one could construct a spreadsheet, but an even quicker method is to use the formula explained below.

Calculating the final value of an investment

Consider an investment at compound interest where:

A is the initial sum invested,
F is the final value of the investment,
i is the interest rate per time period (as a decimal fraction) and
n is the number of time periods.

The value of the investment at the end of each year will be $1 + i$ times the sum invested at the start of the year. For instance, the £648 at the start of year 2 is 1.08 times the initial investment of £600 in Example 7.3 above. The value of the investment at the start of year 3 is 1.08 times the value at the start of year 2, and so on. Thus, for any investment

Value after 1 year $= A(1 + i)$

Value after 2 years $= A(1 + i)(1 + i) = A(1 + i)^2$

Value after 3 years $= A(1 + i)^2(1 + i) = A(1 + i)^3$ etc.

We can see that each value is A multiplied by $(1 + i)$ to the power of the number of years that the sum is invested. Thus, after n years the initial sum A is multiplied by $(1 + i)^n$.

The **formula for the final value F of an investment** of £A for n time periods at interest rate i is therefore

$$F = A(1 + i)^n$$

Let us rework Examples 7.3 and 7.4 using this formula just to check that we get the same answers.

Example 7.3 (reworked)

If £600 is invested for 3 years at 8% then the known values for the formula will be

$$A = £300 \qquad n = 3 \qquad i = 8\% = 0.08$$

Thus the final sum will be

$$F = A(1 + i)^n = 600(1.08)^3 = 600(1.259712) = £755.83$$

Example 7.4 (reworked)

£5,000 invested for 2 years at 12% means that

$$A = £5,000 \qquad n = 2 \qquad i = 12\% = 0.12$$

$$F = A(1 + i)^n = 5,000(1.12)^2 = 5,000(1.2544) = £6,272$$

Having satisfied ourselves that the formula works we can now tackle some more difficult problems.

Example 7.5

If £4,000 is invested for 10 years at an interest rate of 11% per annum what will the final value of the investment be?

Solution

$$A = £4,000 \qquad n = 10 \qquad i = 11\% = 0.11$$

$$F = A(1 + i)^n = 4,000(1.11)^{10}$$

$$= 4,000(2.8394205)$$

$$= £11,357.68$$

(Refer back to Chapter 2, Section 8 if you cannot remember how to use the $[y^x]$ function key on your calculator to work out large powers of numbers.)

Sometimes a compound interest problem may be specified in a rather different format, but the method of solution is still the same.

Example 7.6

You estimate that you will need £8,000 in 3 years' time to buy a new car, assuming a reasonable trade-in price for your old car. You have £7,000 which you can put into a fixed interest building society account earning 4.5%. Will you have enough to buy the car?

Solution

You need to work out the final value of your savings to see whether it will be greater than £8,000. Using the usual notation,

$$A = £7,000 \qquad n = 3 \qquad i = 0.045$$

$$F = A(1 + i)^n = 7,000(1.045)^3 = 7,000(1.141166) = £7,988.16$$

So the answer is 'almost'. You will have to find another £12 to get to £8,000, but perhaps you can get the dealer to knock this off the price.

Changes in interest rates

What if interest rates are expected to change before the end of the investment period? The final sum can be calculated by slightly adjusting the usual formula.

Example 7.7

Interest rates are expected to be 14% for the next 2 years and then fall to 10% for the following 3 years. How much will £2,000 be worth if it is invested for 5 years?

Solution

After 2 years the final value of the investment will be

$$F = A(1 + i)^n = 2{,}000(1.14)^2 = 2{,}000(1.2996) = \text{£}2{,}599.20$$

If this sum is then invested for a further 3 years at the new interest rate of 10% then the final sum is

$$F = A(1 + i)^n = 2{,}599.20(1.1)^3 = 2{,}599.20(1.331) = \text{£}3{,}459.54$$

This could have been worked out in one calculation by finding

$$F = 2{,}000(1.14)^2(1.1)^3 = \text{£}3{,}459.54$$

Therefore the formula for the final sum F that an initial sum A will accrue to after n time periods at interest rate i_n and q time periods at interest rate i_q is

$$F = A(1 + i_n)^n(1 + i_q)^q$$

If more than two interest rates are involved then the formula can be adapted along the same lines.

Example 7.8

What will £20,000 invested for 10 years be worth if the expected rate of interest is 12% for the first 3 years, 9% for the next 2 years and 8% thereafter?

Solution

$$F = 20{,}000(1.12)^3(1.09)^2(1.08)^5 = \text{£}49{,}051.90$$

Test Yourself, Exercise 7.1

1. If £4,000 is invested at 5% interest for 3 years what will the final sum be?
2. How much will £200 invested at 12% be worth at the end of 4 years?
3. A parent invests £6,000 for a 7-year-old child in a fixed interest scheme which guarantees 8% interest. How much will the child have at the age of 21?
4. If £525 is invested in a deposit account that pays 6% interest for 6 years, what will the final sum be?
5. What will £24,000 invested at 11% be worth at the end of 5 years?
6. Interest rates are expected to be 10% for the next 3 years and then to fall to 8% for the following 3 years. How much will an investment of £3,000 be worth at the end of 6 years?

7.3 Part year investment and the annual equivalent rate

If the duration of an investment is less than a year the usual final sum formula does not always apply. It is usually the custom to specify interest rates on an annual basis for part year investments, but two different types of annual interest rates can be used:

(a) the nominal annual interest rate, and
(b) the Annual Equivalent Rate (AER).

The ways that these annual interest rates relate to part year investments differ. They are also used in different circumstances.

Nominal annual interest rates

For large institutional investors on the money markets, and for some forms of individual savings accounts, a nominal annual interest rate is quoted for part year investments. To find the interest that will actually be paid, this nominal annual rate is multiplied by the fraction of the year that it is quoted for.

Example 7.9

What interest is payable on a £100,000 investment for 6 months at a nominal annual interest rate of 6%?

Solution

6 months is 0.5 of one year and so the interest rate that applies is

$$0.5 \times \text{(nominal annual rate)} = 0.5 \times 6\% = 3\%$$

Therefore interest earned is

$$3\% \text{ of } £100,000 = £3,000$$

and the final sum is

$$F = (1.03)100{,}000 = £103{,}000$$

If this nominal annual interest rate of 6% applied to a 3-month investment then the actual interest payable would be a quarter of 6% which is 1.5%. If it applied to an investment for one month then the interest payable would be 6% divided by 12 which gives 0.5%.

The calculation of part year interest payments on this basis can actually give investors a total annual return that is greater than the nominal interest rate if they can keep reinvesting through the year at the same part year interest rate. The total final value of the investment can be calculated with reference to these new time periods using the $F = A(1 + i)^n$ formula as long as the interest rate i and the number of time periods n refer to the same time periods.

For example, if £100,000 can be invested for four successive three month periods at a nominal annual interest rate of 6% then, letting i represent the effective quarterly interest rate and n represent the number of three month periods, we get

$$A = £100{,}000 \qquad n = 4 \qquad i = 0.25 \times 6\% = 1.5\% = 0.015$$

$$F = A(1 + i)^n = 100{,}000(1.015)^4 = £106{,}136.35$$

This final sum gives a 6.13635% return on the initial £100,000 sum invested. (Although in practice interest rates are usually only specified to 2 dp.)

The more frequently that interest based on the nominal annual rate is paid the greater will be the total annual return when all the interest is compounded. For example, if a nominal annual interest rate of 6% is paid monthly at 0.5% a month and £100,000 is invested for 12 months then

$$A = £100{,}000 \qquad n = 12 \qquad i = 0.5\% = 0.005$$

$$F = A(1 + i)^n = 100{,}000(1.005)^{12} = £106{,}167.78$$

This new final sum is greater than that achieved from quarterly interest payment and is equivalent to an annual rate of 6.17%.

The Annual Equivalent Rate (AER) and Annual Percentage Rate (APR)

Although some part year investments on money markets may earn a return which is not equivalent to the nominal annual interest rate, individual investors are usually quoted an annual equivalent rate (AER) which is an accurate reflection of the interest that they earn on investments. For example, interest on the money you may have in a building society will normally be worked out on a daily basis although you will only be told the AER and the interest on your account may only be credited once a year. For loan repayments the annual equivalent rate is usually referred to as the annual percentage rate (APR). If you take out a bank loan you will usually be quoted an APR even though you will be asked to make monthly repayments.

The examples above have already demonstrated that the AER is not simply 12 times the monthly interest rate. To determine the relationship between part year interest rates and their true AER, consider another example.

Example 7.10

If interest is credited monthly at a monthly rate of 0.9% how much will £100 invested for 12 months accumulate to?

Solution

Using the standard investment formula where the time period n is measured in months:

$$A = £100 \qquad n = 12 \qquad i = 0.9\% = 0.009$$

$$F = A(1 + i)^n = 100(1.009)^{12} = 100(1.1135) = £111.35$$

This final sum of £111.35 after investing £100 for one year corresponds to an annual rate of interest of 11.35%. This is greater than 12 times the monthly rate of 0.9%, since

$$12 \times 0.9\% = 10.8\%$$

The calculations in the above example that tell us that the ratio of the final sum to the initial sum invested is $(1.009)^{12}$. Using the same principle, the **corresponding AER for any given monthly rate of interest** i_m can be found using the formula

$$\text{AER} = (1 + i_m)^{12} - 1$$

Because $(1 + i_m)^{12}$ gives the ratio of the final sum F to the initial amount A the -1 has to be added to the formula in order to get the proportional increase in F over A. The APR on loans is the same thing as the annual equivalent rate and so the same formula applies.

Example 7.11

If the monthly rate of interest on a loan is 1.75% what is the corresponding APR?

Solution

$$i_m = 1.75\% = 0.0175$$
$$\text{APR} = (1 + i_m)^{12} - 1$$
$$= (1.0175)^{12} - 1$$
$$= 1.2314393 - 1$$
$$= 0.2314393 = 23.14\%$$

If you have a credit card you can check out this formula by referring to the leaflet on interest rates that the credit card company should give you. For example, in October 2001 the

LloydsTSB Trustcard had an interest rate per month of 1.527% and quoted the APR as 19.9%. We can check this using the formula

$$APR = (1 + i_m)^{12} - 1$$
$$= (1.01527)^{12} - 1$$
$$= 1.19944 - 1$$
$$= 0.19944 = 19.9\%$$

The calculation of monthly loan repayments from a given APR will be explained later, in Section 7.9.

Savers may put money into deposit accounts with banks and building societies, or may make withdrawals, at any time throughout the year and so different amounts will remain in their accounts for different periods of time. The interest on these accounts is therefore usually calculated on a daily basis. However, only the AER is widely publicized as this is much more useful to savers to help them make comparisons between different possible investment opportunities. The relationship between the daily interest rate i_d on a deposit account and the AER can be formulated as

$$AER = (1 + i_d)^{365} - 1$$

For example, if a building society tells you that it will pay you an AER of 6% on a savings account, what it actually will do is credit interest at a rate of 0.015954% a day. We can check this out using the formula

$$AER = (1 + i_d)^{365} - 1$$
$$= (1.00015954)^{365} - 1$$
$$= 1.06 - 1 = 0.06 = 6\%$$

To derive a formula for the daily interest rate i_d that corresponds to a given AER, we start with the AER formula. Thus

$$AER = (1 + i_d)^{365} - 1$$
$$AER + 1 = (1 + i_d)^{365}$$
$$\sqrt[365]{AER + 1} = 1 + i_d$$
$$\left(\sqrt[365]{AER + 1} \right) - 1 = i_d$$

Example 7.12

A building society account pays interest on a daily basis at an AER of 4.5%. If you deposited £2,750 in such an account on 1st October how much would you get back if you closed the account 254 days later?

Solution

First we find the daily interest rate using the formula derived above. Thus

$$i_\text{d} = \left(\sqrt[365]{\text{AER} + 1} \right) - 1 = \left(\sqrt[365]{0.045 + 1} \right) - 1 = 1.0001206 - 1$$
$$= 0.0001206 = 0.01206\%$$

The final sum accumulated when £2,750 is invested at this daily rate for 254 days will therefore be

$$F = 2,750(1 + 0.0001206)^{254} = 2,750(1.0311045) = £2,835.54$$

(So you could earn about £85 interest if you put a student loan of this amount in a building society and didn't touch it for the whole academic year – not a very likely scenario!)

Interest rates on Treasury Bills

A government Treasury Bill, like certain other forms of bond, guarantees the owner a fixed some of money payable at a fixed date in the future. So, for example, a 3-month Treasury Bill for £100,000 is effectively a promise from the government that it will pay £100,000 to the owner on a date 3 months from when it was issued. The prices that the institutional investors who trade in these bills will pay for them will reflect the returns that can be made on other similar investments.

Suppose that investors are currently willing to pay £95,000 for 12-month Treasury Bills when they are issued. This would mean that they consider an annual return of £5,000 on their £95,000 investment to be acceptable.

An annual return of £5,000 on a £95,000 investment is equivalent to an interest rate of

$$i = \frac{5,000}{95,000} = 0.0526316 = 5.26\%$$

However, in the financial press the interest rates quoted relate to a nominal annual rate of return based on the final sum paid out when the Treasury Bill matures. Thus, in the example above the 12-month Treasury Bill rate quoted would be 5%, because this is the discount the price of £95,000 yields on the final maturity sum of £100,000. This is why they are called discount rates.

Although the above example considered a 12-month Treasury Bill so that the equivalent annual rate of return could be easily compared, in practice UK government Treasury Bills are normally issued for shorter periods. Also, the nominal annual rates are quoted using fractions, such as $4\frac{5}{16}\%$, rather than in decimal format.

Example 7.13

If an annual discount rate of $4\frac{7}{8}\%$ is quoted for 3-month Treasury Bills, what would it cost to buy a tranch of these bills with redemption value of £100,000? What would be the annual equivalent rate of return on the sum paid for them?

Solution

A nominal annual rate of $4\frac{7}{8}\%$ corresponds to a 3-month rate of

$$\frac{4\frac{7}{8}}{4} = 1\frac{7}{32} = 1.21875\%$$

As this rate is actually the discount on the maturity sum then the cost of 3-month Treasury Bills with redemption value of £100,000 of would be

$$£100,000(1 - 0.0121875) = £98,781.25$$

and the amount of the discount is £1,218.75.

   Therefore, the rate of return on the sum of £98,781.25 invested for 3 months is

$$\frac{1,218.75}{98,781.25} = 0.012338 = 1.2338\%$$

If this investment could be compounded for four 3-month periods at this quarterly rate of 1.2338% then the annual equivalent rate calculated using the standard formula would be

$$\text{AER} = (1.012338)^4 - 1 = 1.050273 - 1 = 0.050273 = 5.0273\%$$

Test Yourself, Exercise 7.2

1. If £40,000 is invested at a monthly rate of 1% what will it be worth after 9 months? What is the corresponding AER?
2. A sum of £450,000 is invested at a monthly interest rate of 0.6%. What will the final sum be after 18 months? What is the corresponding AER?
3. Which is the better investment for someone wishing to invest a sum of money for two years:

    (a) an account which pays 0.9% monthly, or
    (b) an account which pays 11% annually?

4. If £1,600 is invested at a quarterly rate of interest of 4.5% what will the final sum be after 18 months? What is the corresponding AER?
5. How much interest is earned on £50,000 invested for three months at a nominal annual interest rate of 5%? If money can be reinvested each quarter at the same rate, what is the AER?
6. If a credit card company charges 1.48% a month on any outstanding balance, what APR is it charging?
7. A building society pays an AER of 5.5% on an investment account, calculated on a daily basis. What daily rate of interest will it pay?
8. If 3-month government Treasury Bills are offered at an annual discount rate of $4\frac{7}{16}\%$, what would it cost to buy bills with redemption value of £500,000? What would the AER be for this investment?

7.4   Time periods, initial amounts and interest rates

The formula for the final sum of an investment contains the four variables F, A, i and n. So far we have only calculated F for given values of A, i and n. However, if the values of any three of the variables in this equation are given then one can usually calculate the fourth.

Initial amount

A formula to calculate A, when values for F, i and n are given, can be derived as follows. Since the final sum formula is

$$F = A(1 + i)^n$$

then, dividing through by $(1 + i)^n$, we get the initial sum formula

$$\frac{F}{(1 + i)^n} = A$$

or

$$A = F(1 + i)^{-n}$$

Example 7.14

How much money needs to be invested now in order to accumulate a final sum of £12,000 in 4 years' time at an annual rate of interest of 10%?

Solution

Using the formula derived above, the initial amount is

$$A = F(1 + i)^{-n}$$

$$= 12{,}000(1.1)^{-4}$$

$$= \frac{12{,}000}{1.4641} = £8{,}196.16$$

What we have actually done in the above example is find the sum of money that is equivalent to £12,000 in 4 years' time if interest rates are 10%. An investor would therefore be indifferent between (a) £8,196.16 now and (b) £12,000 in 4 years' time. The £8,196.16 is therefore known as the 'present value' (PV) of the £12,000 in 4 years' time. We shall come back to this concept in the next few sections when methods of appraising different types of investment project are explained.

Time period

Calculating the time period is rather more tricky than the calculation of the initial amount. From the final sum formula

$$F = A(1 + i)^n$$

Then

$$\frac{F}{A} = (1+i)^n$$

If the values of F, A and i are given and one is trying to find n this means that one has to work out to what power $(1+i)$ has to be raised to equal F/A. One way of doing this is via logarithms.

Example 7.15

For how many years must £1,000 be invested at 10% in order to accumulate £1,600?

Solution

$$A = £1,000 \quad F = £1,600 \quad i = 10\% = 0.1$$

Substituting these values into the formula

$$\frac{F}{A} = (1+i)^n$$

we get $\dfrac{1,600}{1,000} = (1+0.1)^n$

$$1.6 = (1.1)^n \tag{1}$$

If equation (1) is specified in logarithms then

$$\log 1.6 = n \log 1.1 \tag{2}$$

since to find the nth power of a number its logarithm must be multiplied by n. Finding logs, this means that (2) becomes

$$0.20412 = n \, 0.0413927$$

$$n = \frac{0.20412}{0.0413927} = 4.93 \text{ years}$$

If investments must be made for whole years then the answer is 5 years. This answer can be checked using the final sum formula

$$F = A(1+i)^n = 1,000(1.1)^5 = 1,610.51$$

If the £1,000 is invested for a full 5 years then it accumulates to just over £1,600, which checks out with the answer above.

A general formula to solve for n can be derived as follows from the final sum formula:

$$F = A(1+i)^n$$

$$\frac{F}{A} = (1+i)^n$$

Taking logs

$$\log\left(\frac{F}{A}\right) = n\log(1+i)$$

Therefore the time period formula is

$$\frac{\log(F/A)}{\log(1+i)} = n \tag{3}$$

An alternative approach is to use the iterative method and plot different values on a spreadsheet. To find the value of n for which

$$1.6 = (1.1)^n$$

this entails setting up a formula to calculate the function $y = (1.1)^n$ and then computing it for different values of n until the answer 1.6 is reached. Although some students who find it difficult to use logarithms will prefer to use a spreadsheet, logarithms are used in the other examples in this section. Logarithms are needed to analyse other concepts related to investment and so you really need to understand how to use them.

Example 7.16

How many years will £2,000 invested at 5% take to accumulate to £3,000?

Solution

$$A = 2,000 \qquad F = 3,000 \qquad i = 5\% = 0.05$$

Using these given values in the time period formula derived above gives

$$\begin{aligned} n &= \frac{\log(F/A)}{\log(1+i)} \\ &= \frac{\log 1.5}{\log 1.05} \\ &= \frac{0.1760913}{0.0211893} = 8.34 \text{ years} \end{aligned}$$

Example 7.17

How long will any sum of money take to double its value if it is invested at 12.5%?

Solution

Let the initial sum be A. Therefore the final sum is

$$F = 2A$$

and $i = 12.5\% = 0.125$

Substituting these value for F and i into the final sum formula

$$F = A(1 + i)^n$$

gives

$$2A = A(1.125)^n$$
$$2 = (1.125)^n$$

Taking logs of both sides

$$\log 2 = n \log 1.125$$
$$n = \frac{\log 2}{\log 1.125} = \frac{0.30103}{0.0511525} = 5.9 \text{ years}$$

Interest rates

A method of calculating the interest rate on an investment is explained in the following example.

Example 7.18

If £4,000 invested for 10 years is projected to accumulate to £6,000, what interest rate is used to derive this forecast?

Solution

$$A = 4{,}000 \qquad F = 6{,}000 \qquad n = 10$$

Substituting these values into the final sum formula

$$F = A(1 + i)^n$$

Gives $6{,}000 = 4{,}000(1 + i)^{10}$

$$1.5 = (1 + i)^{10}$$
$$1 + i = \sqrt[10]{(1.5)}$$
$$= 1.0413797$$
$$i = 0.0414 = 4.14\%$$

A general formula for calculating the interest rate can be derived. Starting with the familiar final sum formula

$$F = A(1 + i)^n$$

$$\frac{F}{A} = (1 + i)^n$$

$$\sqrt[n]{(F/A)} = 1 + i$$

$$\sqrt[n]{(F/A)} - 1 = i \tag{4}$$

This **interest rate formula** can also be written as

$$i = \left(\frac{F}{A}\right)^{1/n} - 1$$

Example 7.19

At what interest rate will £3,000 accumulate to £10,000 after 15 years?

Solution

Using the interest rate formula (4) above

$$i = \sqrt[n]{\left(\frac{F}{A}\right)} - 1 = \sqrt[15]{\left(\frac{10,000}{3,000}\right)} - 1$$

$$= \sqrt[15]{(3.3333)} - 1 = 1.083574 - 1$$

$$= 0.083574 = 8.36\%$$

Example 7.20

An initial investment of £50,000 increases to £56,711.25 after 2 years. What interest rate has been applied?

Solution

$$A = 50,000 \qquad F = 56,711.25 \qquad n = 2$$

Therefore

$$\frac{F}{A} = \frac{56,711.25}{50,000} = 1.134225$$

Substituting these values into the interest rate formula gives

$$i = \sqrt[n]{\left(\frac{F}{A}\right)} - 1 = \sqrt[2]{(1.13455)} - 1 = 1.065 - 1 = 0.065$$

$$i = 6.5\%$$

Test Yourself, Exercise 7.3

1. How much needs to be invested now in order to accumulate £10,000 in 6 years' time if the interest rate is 8%?
2. What sum invested now will be worth £500 in 3 years' time if it earns interest at 12%?
3. Do you need to invest more than £10,000 now if you wish to have £65,000 in 15 years' time and you have a deposit account which guarantees 14%?
4. You need to have £7,500 on 1 January next year. How much do you need to invest at 1.3% per month if your investment is made on 1 June?
5. How much do you need to invest now in order to earn £25,000 in 10 years' time if the interest rate is
   (a) 10%   (b) 8%   (c) 6.5%?
6. How many complete years must £2,400 be invested at 5% in order to accumulate a minimum of £3,000?
7. For how long must £5,000 be kept in a deposit account paying 8% interest before it accumulates to £7,500?
8. If it can earn 9.5% interest, how long would any given sum of money take to treble its value?
9. If one needs to have a final sum of £20,000, how many years must one wait if £12,500 is invested at 9%?
10. How long will £70,000 take to accumulate to £100,000 if it is invested at 11%?
11. If £6,000 is to accumulate to £10,000 after being invested for 5 years, what rate must it earn interest at?
12. What interest rate will turn £50,000 into £60,000 after 2 years?
13. At what interest rate will £3,000 accumulate to £4,000 after 4 years?
14. What monthly rate of interest must be paid on a sum of £2,800 if it is to accumulate to £3,000 after 8 months?
15. What rate of interest would turn £3,000 into £8,000 in 10 years?
16. At what rate of interest will £600 accumulate to £900 in 5 years?
17. Would you prefer (a) £5,000 now or (b) £8,000 in 4 years' time if money can be borrowed or lent at 11%?

7.5 Investment appraisal: net present value

Assume that you have £10,000 to invest and that someone offers you the following proposal: pay £10,000 now and get £11,000 back in 12 months' time. Assume that the returns on this investment are guaranteed and there are no other costs involved. What would you do? Perhaps you would compare this return of 10% with the rate of interest your money could earn in a deposit account, say 4%. In a simple example like this the comparison of rates of return, known as the internal rate of return (IRR) method, is perhaps the most intuitively obvious method of judging the proposal.

This is not the preferred method for investment appraisal, however. The net present value (NPV) method has several advantages over the IRR method of comparing the project rate of return with the market interest rate. These advantages are explained more fully in the following section, but first it is necessary to understand what the NPV method involves.

We have already come across the concept of present value (PV) in Section 7.4. If a certain sum of money will be paid to you at some given time in the future its PV is the amount of money that would accumulate to this sum if it was invested now at the ruling rate of interest.

Example 7.21

What is the present value of £1,500 payable in 3 years' time if the relevant interest rate is 4%?

Solution

Using the initial amount investment formula, where

$$F = £1,500 \qquad i = 0.08 \qquad n = 3$$

$$A = F(1 + i)^{-n}$$

$$= \frac{1,500}{(1.04)^3} = 1,500(1.04)^{-3}$$

$$= \frac{1,500}{1.124864} = £1,333.49$$

An investor would be indifferent between £1,333.49 now and £1,500 in 3 years' time. Thus £1,333.49 is the PV of £1,500 in 3 years' time at 4% interest.

In all the examples in this chapter it is assumed that future returns are assured with 100% certainty. Of course, in reality some people may place greater importance on earlier returns just because the future is thought to be more risky. If some form of measure of the degree of risk can be estimated then more advanced mathematical methods exist which can be used to adjust the investment appraisal methods explained in this chapter. However, here we just assume that estimated future returns and costs, are correct. An investor has to try to make the most rational decision based on whatever information is available.

The **net present value** (NPV) of an investment project is defined as the PV of the future returns minus the cost of setting up the project.

Example 7.22

An investment project involves an initial outlay of £600 now and a return of £1,000 in 5 years' time. Money can be invested at 9%. What is the NPV?

Solution

The PV of £1,000 in 5 years' time at 9% can be found using the initial amount formula as

$$A = F(1 + i)^{-n} = 1,000(1.09)^{-5} = £649.93$$

Therefore NPV = £649.93 − £600 = £49.93.

This project is clearly worthwhile. The £1,000 in 5 years' time is equivalent to £649.93 now and so the outlay required of only £600 makes it a bargain. In other words, one is being asked to pay £600 for something which is worth £649.93.

Another way of looking at the situation is to consider what alternative sum could be earned by the investor's £600. If £649.93 was invested for 5 years at 9% it would accumulate to £1,000. Therefore the lesser sum of £600 must obviously accumulate to a smaller sum. Using the final sum investment formula this can be calculated as

$$F = A(1 + i)^n = 600(1.09)^5 = 600(1.538624) = £923.17$$

The investor thus has the choice of

(a) putting £600 into this investment project and securing £1,000 in 5 years' time, or
(b) investing £600 at 9%, accumulating £923.17 in 5 years.

Option (a) is clearly the winner.

If the outlay is less than the PV of the future return an investment must be a profitable venture. The basic **criterion for deciding whether or not an investment project is worthwhile** is therefore

NPV > 0

As well as deciding whether specific projects are profitable or not, an investor may have to decide how to allocate limited capital resources to competing investment projects. The *rule for choosing between projects* is that they should be ranked according to their NPV. If only one out of a set of possible projects can be undertaken then the one with the largest NPV should be chosen, as long as its NPV is positive.

Example 7.23

An investor can put money into any one of the following three ventures:

Project A costs £2,000 now and pays back £3,000 in 4 years
Project B costs £2,000 now and pays back £4,000 in 6 years
Project C costs £3,000 now and pays back £4,800 in 5 years

The current interest rate is 10%. Which project should be chosen?

Solution

NPV of project A $= 3,000(1.1)^{-4} - 2,000$
$$= 2,049.04 - 2,000 = £49.04$$

NPV of project B $= 4,000(1.1)^{-6} - 2,000$
$$= 2,257.90 - 2,000 = £257.90$$

NPV of project C $= 4,800(1.1)^{-5} - 3,000$
$$= 2,980.42 - 3,000 = -£19.58$$

Project B has the largest NPV and is therefore the best investment. Project C has a negative NPV and so would not be worthwhile even if there was no competition.

The investment examples considered so far have only involved a single return payment at some given time in the future. However, most real investment projects involve a stream of returns occurring over several time periods. The same principle for calculating NPV is used to assess these projects, the initial outlay being subtracted from the sum of the PVs of the different future returns.

Example 7.24

An investment proposal involves an initial payment now of £40,000 and then returns of £10,000, £30,000 and £20,000 respectively in 1, 2 and 3 years' time. If money can be invested at 10% is this a worthwhile investment?

Solution

PV of £10,000 in 1 year's time $= \dfrac{£10,000}{1.1} = £9,090.91$

PV of £30,000 in 2 years' time $= \dfrac{£30,000}{1.1^2} = £24,793.39$

PV of £20,000 in 3 years' time $= \dfrac{£20,000}{1.1^3} = £15,026.30$

Total PV of future returns	£48,910.60
less initial outlay	−£40,000
NPV of project	£8,910.60

This NPV is greater than zero and so the project is worthwhile. At an interest rate of 10% one would need to invest a total of £48,910.60 to get back the projected returns and so £40,000 is clearly a bargain price.

The further into the future the expected return occurs the greater will be the discounting factor. This is made obvious in Example 7.25 below, where the returns are the same each time period. The PV of each successive year's return is smaller than that of the previous year because it is multiplied by $(1 + i)^{-1}$.

Example 7.25

An investment project requires an initial outlay of £7,500 and will pay back £2,000 at the end of the next 5 years. Is it worthwhile if capital can be invested elsewhere at 12%?

Solution

$$\text{PV of £2,000 in 1 year's time} = \frac{£2,000}{1.12} = £1,785.71$$

$$\text{PV of £2,000 in 2 years' time} = \frac{£2,000}{1.12^2} = £1,594.39$$

$$\text{PV of £2,000 in 3 years' time} = \frac{£2,000}{1.12^3} = £1,423.56$$

$$\text{PV of £2,000 in 4 years' time} = \frac{£2,000}{1.12^4} = £1,271.04$$

$$\text{PV of £2,000 in 5 years' time} = \frac{£2,000}{1.12^5} = £1,134.85$$

Total PV of future returns	£7,209.55
less initial outlay	−£7,500.00
NPV of project	− £290.45

The NPV < 0 and so this is not a worthwhile investment.

Investment appraisal using a spreadsheet

From the above examples one can see that the mathematics involved in calculating the NPV of a project can be quite time-consuming. For this type of problem a spreadsheet program can be a great help. Although Excel has a built in NPV formula, this does not take the initial outlay into account and so care has to be taken when using it. We shall therefore construct a spreadsheet to calculate NPV from first principles.

To derive an algebraic formula for calculating NPV assume that R_j is the net return in year j, i is the given rate of interest, n is the number of time periods in which returns occur and C is the initial cost of the project. Then

$$\text{NPV} = \frac{R_1}{1+i} + \frac{R_2}{(1+i)^2} + \cdots + \frac{R_n}{(1+i)^n} - C$$

Using the Σ notation this becomes

$$\text{NPV} = \sum_{j=1}^{n} \frac{R_j}{(1+i)^j} - C \tag{1}$$

If the initial outlay C is considered as a negative return at time 0 (i.e. $R_0 = -C$) the formula can be more neatly stated as

$$\text{NPV} = \sum_{j=0}^{n} \frac{R_j}{(1+i)^j} \tag{2}$$

There will be no discounting of the initial outlay in the first term

$$\frac{R_0}{(1+i)^0}$$

since $(1+i)^0 = 1$. (Remember that $x^0 = 1$ whatever the value of x.)

The following example shows how an Excel spreadsheet program based on this formula can be used to work out the NPV of a project. The answer obtained is then compared with the solution using the Excel built in NPV function.

Example 7.26

An investment project requires an initial outlay of £25,000 with the following expected returns:

£5,000	at the end of year 1
£6,000	at the end of year 2
£10,000	at the end of year 3
£10,000	at the end of year 4
£10,000	at the end of year 5

Is this a viable investment if money can be invested elsewhere at 15%?

Solution

Follow the instructions for creating an Excel spreadsheet set out in Table 7.1, which should give you the spreadsheet in Table 7.2. This calculates the PVs of the returns in each year separately, including the outlay in year 0. It then sums the PVs, giving a total NPV of £1,149.15 which is positive and hence means that the project is a viable investment opportunity.

This can be compared with the answer obtained using the Excel built-in NPV formula. Because this formula always treats the number in the first cell of the range as the return at the end of year 1, the computed answer of £26,149.15 is the total PV of the returns in years 1 to 5 only. To get the overall NPV of the project one has to subtract the initial outlay. (The outlay amount was entered as a negative quantity and so this is actually added in the formula.) This adjusted Excel NPV figure should be the same as the NPV calculated from first principles, which it is. Having an answer computed by two separate methods is a useful check. If you save this spreadsheet and adapt it for other problems then, if you do not get the same answer from both methods, you will know that a mistake has been made somewhere.

The spreadsheet created for the above example can be used to work out the NPV for other projects. The initial cost and returns need to be entered in cells B4 to B9 and the new interest rate goes in cell D2. Obviously if there are more (or less) years when returns occur then rows will need to be added (or deleted or left blank).

As investment appraisal involves the comparison of different projects, as well as the assessment of the financial viability of individual projects, a spreadsheet can be adapted to work

Table 7.1

CELL	Enter	Explanation
A1	Ex.7.26	Label to remind you what example this is
A3	YEAR	Column heading label
B3	RETURN	Column heading label
C3	PV	Column heading label
C1	Interest rate =	Label to tell you interest rate goes in next cell.
D1	15%	Value of interest rate. (NB Excel automatically treats this % format as 0.15 in any calculations.)
A4 to A9	Enter numbers 0 to 5	These are the time periods
B4	-25000	Initial outlay (negative because it is a cost)
B5	5000	Returns at end of years 1 to 5
B6	6000	
B7	10000	
B8	10000	
B9	10000	
C4	=B4/(1+$D$1)^A4	Formula calculates PV corresponding to return in cell B4, time period in cell A4 and interest rate in cell D1. Note the $ to anchor cell D1.
C5 to C9	Copy cell C4 formula down column C	Calculates PV for return in each time period. Format to 2 d.p. as monetary values
B11	NPV =	Label to tell you NPV goes in next cell.
C11	=SUM(C4:C9)	Calculates NPV of project by summing PVs for each year in cells C4-C9, which includes the negative return of the initial outlay.
B13	Excel NPV	Label tells you Excel NPV goes in next cell.
B14	less cost =	Label tells you what goes in next cell.
C13	=NPV(D1,B5:B9)	The Excel NPV formula will calculate NPV based only on the interest rate in D1 and the 5 years of future returns in cells B5 to B9.
C14	=C13+B4	Adjusts the Excel computed NPV in C13 by subtracting initial outlay in B4. (This was entered as a negative number so it is added.)

Table 7.2

	A	B	C	D
1	Ex 7.26		Interest rate=	15%
2				
3	YEAR	RETURN	PV	
4	0	-25000	-25000	
5	1	5000	4347.83	
6	2	6000	4536.86	
7	3	10000	6575.16	
8	4	10000	5717.53	
9	5	10000	4971.77	
10				
11		NPV =	1149.15	
12				
13		Excel NPV	£26,149.15	
14		less cost =	£1,149.15	

out the NPV for more than one project. The following example shows how the spreadsheet created for Example 7.26 can be extended so that two projects can be compared.

Example 7.27

An investor has to choose between two projects A and B whose outlay and returns are set out in Table 7.3. Which is the better investment if the going rate of interest is 10%?

Table 7.3

(All values in £)	Project A	Project B
Initial outlay	30,000	30,000
Return in 1 year's time	6,000	8,000
Return in 2 years' time	10,000	8,000
Return in 3 years' time	10,000	8,000
Return in 4 years' time	10,000	8,000
Return in 5 years' time	8,000	8,000

Table 7.4

CELL	Enter	Explanation
A1	Ex.7.27	New example label
B3	PROJECT A	Changed column heading label
C3	PV A	Changed column heading label
D1	10%	New value of interest rate.
D3	PROJECT B	New column heading label for project B returns.
E3	PV B	New column heading label for project B PVs.
B4	-30000	Project A initial outlay
B5	6000	Project A returns at end of years 1 to 5
B6	10000	
B7	10000	
B8	10000	
B9	8000	
D4	-30000	Project B initial outlay
D5	8000	Project B returns at end of years 1 to 5
D6	8000	
D7	8000	
D8	8000	
D9	8000	
E4	=D4/(1+$D$1)^A4	Formula calculates PV for project B corresponding to return in cell D4.
E5 to E9	*Copy cell E4 formula down column* E	Calculates PV for project B for return in each time period.
E11	=SUM(E4:E9)	Calculates NPV of B by summing PVs for each time period
E13	=NPV(D1,D5:D9)	Excel NPV formula applied to project B
E14	=E13+D4	Adjusts the Excel NPV for project B
C12	=B3	Writes "PROJECT A" under relevant NPV
E12	=D3	Writes "PROJECT B" under relevant NPV

Table 7.5

	A	B	C	D	E
1	Ex 7.27		Interest rate =	10%	
2					
3	YEAR	PROJECT A	PV A	PROJECT B	PV B
4	0	-30000	-30000.00	-30000	-30000.00
5	1	6000	5454.55	8000	7272.73
6	2	10000	8264.46	8000	6611.57
7	3	10000	7513.15	8000	6010.52
8	4	10000	6830.13	8000	5464.11
9	5	8000	4967.37	8000	4967.37
10					
11		NPV =	3029.66		326.29
12			PROJECT A		PROJECT B
13		Excel NPV	£33,029.66		£30,326.29
14		less cost =	£3,029.66		£326.29

Solution

Call up the worksheet which you created for Example 7.26 and make the changes shown in Table 7.4. This should give you a spreadsheet that looks similar to Table 7.5. The computed NPV for project B is £326.29 compared with £3,029.66 for project A. Therefore, although both projects are financially viable, the better investment is project A because it has the greater NPV.

If you do not have access to a spreadsheet program then you can still work out the NPV of different projects from first principles. However, there are now available financial calculators with an NPV function which may be a cheaper alternative than a computer. To assist students without a spreadsheet program or a financial calculator, a set of discounting factors is reproduced in Table 7.6. Although the actual monetary returns will differ from project to project the discounting factor will be the same for a given time period and a given rate of interest. For example, the PV of a sum of money £x payable in 8 years' time when the interest rate is 7% will be

$$\frac{£x}{(1.07)^8} \quad \text{or} \quad £x(1.07)^{-8}$$

The value of $(1.07)^{-8}$ can be read off from Table 7.6 by looking at the column headed 7% and the row corresponding to year 8, giving a figure of 0.582009. If £x was £525 then the PV would be

$$£525(0.582009) = £305.55$$

You can also compute these values on any mathematical calculator with a $[y^x]$ function key. For example, to calculate £$525(1.07)^{-8}$ enter $525\ [\div]\ 1.07\ [y^x]\ 8\ [=]$ or $525\ [\times]\ 1.07\ [y^x]\ 8\ [+/-]\ [=]$.

Table 7.6 Discounting factors for Net Present Value

Rate of interest i	4%	5%	6%	7%	8%	9%	10%	11%	12%	13%	14%	15%
Year 0	1	1	1	1	1	1	1	1	1	1	1	1
1	0.961538	0.952381	0.943396	0.934579	0.925926	0.917431	0.961538	0.952381	0.943396	0.934579	0.925926	0.917431
2	0.924556	0.907029	0.889996	0.873439	0.857339	0.84168	0.924556	0.907029	0.889996	0.873439	0.857339	0.84168
3	0.888996	0.863838	0.839619	0.816298	0.793832	0.772183	0.888996	0.863838	0.839619	0.816298	0.793832	0.772183
4	0.854804	0.822702	0.792094	0.762895	0.73503	0.708425	0.854804	0.822702	0.792094	0.762895	0.73503	0.708425
5	0.821927	0.783526	0.747258	0.712986	0.680583	0.649931	0.821927	0.783526	0.747258	0.712986	0.680583	0.649931
6	0.790315	0.746215	0.704961	0.666342	0.63017	0.596267	0.790315	0.746215	0.704961	0.666342	0.63017	0.596267
7	0.759918	0.710681	0.665057	0.62275	0.58349	0.547034	0.759918	0.710681	0.665057	0.62275	0.58349	0.547034
8	0.73069	0.676839	0.627412	0.582009	0.540269	0.501866	0.73069	0.676839	0.627412	0.582009	0.540269	0.501866
9	0.702587	0.644609	0.591898	0.543934	0.500249	0.460428	0.702587	0.644609	0.591898	0.543934	0.500249	0.460428
10	0.675564	0.613913	0.55839	0.508349	0.463193	0.422411	0.675564	0.613913	0.558395	0.508349	0.463193	0.422411
11	0.649581	0.584679	0.526788	0.475093	0.428883	0.387533	0.649581	0.584679	0.526788	0.475093	0.428883	0.387533
12	0.624597	0.556837	0.496969	0.444012	0.397114	0.355535	0.624597	0.556837	0.496969	0.444012	0.397114	0.355535

Test Yourself, Exercise 7.4

1. The following investment projects all involve an outlay now and a single return at some point in the future. Calculate the NPV and say whether or not each is a worthwhile investment:

    (a) £1,100 outlay, £1,500 return after 3 years, interest rate 8%
    (b) £750 outlay, £1,000 return after 5 years, interest rate 9%
    (c) £10,000 outlay, £12,000 return after 3 years, interest rate 8%
    (d) £50,000 outlay, £75,000 return after 3 years, interest rate 14%
    (e) £50,000 outlay, £100,000 return after 5 years, interest rate 14%
    (f) £5,000 outlay, £7,000 return after 3 years, interest rate 6%
    (g) £5,000 outlay, £7,750 return after 5 years, interest rate 6%
    (h) £5,000 outlay, £8,500 return after 6 years, interest rate 6%

2. An investor has to choose between the following three projects:

    Project A requires an outlay of £35,000 and returns £60,000 after 4 years
    Project B requires an outlay of £40,000 and returns £75,000 after 5 years
    Project C requires an outlay of £25,000 and returns £50,000 after 6 years

    Which project would you advise this investor to put money into if the cost of capital is 10%?

3. A firm has a choice between three investment projects, all of which involve an initial outlay of £36,000. The returns at the end of the next 4 years are given in Table 7.7. If the interest rate is 15%, say (a) whether each project is viable or not, and (b) which is the best investment.

    Table 7.7

Year	*Project A*	*Project B*	*Project C*
1	15,000	5,000	20,000
2	15,000	10,000	15,000
3	15,000	20,000	10,000
4	15,000	25,000	5,000

    Note
    All values are given in £.

4. If money can be invested elsewhere at 6%, is the following project worthwhile?

Initial outlay	£100,000
Return at end of year 1	£10,000
Return at end of year 2	£12,000
Return at end of year 3	£15,000
Return at end of year 4	£18,000
Return at end of year 5	£20,000
Return at end of year 6	£20,000

Return at end of year 7   £20,000
Return at end of year 8   £15,000
Return at end of year 9   £10,000
Return at end of year 10   £5,000

5. Would you put £40,000 into a project which pays back nothing in the first year but then brings annual net returns of £12,000 from the end of year 2 until the end of year 6, assuming an interest rate of 8%?

6. A project requires an initial outlay of £20,000 and will pay back the following returns (in £):

   1,000 at the end of years 1 and 2
   2,000 at the end of years 3 and 4
   5,000 at the end of years 5, 6, 7, 8, 9 and 10

   Is this project a worthwhile investment if the going rate of interest is (a) 9%, (b) 10%?

7. Which of the three projects shown in Table 7.8 is the best investment if the interest rate is 20%?

Table 7.8

	Project A	Project B	Project C
Outlay now	85,000	40,000	40,000
Return after year 1	20,000	15,000	10,000
Return after year 2	24,000	20,000	12,000
Return after year 3	30,000	25,000	12,000
Return after year 4	30,000	0	12,000
Return after year 5	25,000	0	15,000
Return after year 6	20,000	0	15,000

Note
All values are given in £.

7.6   The internal rate of return

The IRR method of investment appraisal involves finding the rate of return (r) on a project and comparing it with the market rate of interest (i). If $r > i$ then the project is viable. Alternative projects can be ranked according to the magnitude of the different rates of return.

Example 7.28

Find the IRR for the three projects in Table 7.9, decide whether they are viable if the market rate of interest is 7%, and then rank them in order of profitability according to the IRR method.

Table 7.9

	Project A	Project B	Project C
Initial outlay	£5,000	£4,000	£8,000
Return after 1 year	£5,750	£4,300	£8,500

Solution

In this simple example it is obvious from basic arithmetic that

$$\text{IRR for A} = r_A = \frac{750}{5,000} = 0.15 = 15\%$$

$$\text{IRR for B} = r_B = \frac{300}{4,000} = 0.075 = 7.5\%$$

$$\text{IRR for C} = r_C = \frac{500}{8,000} = 0.0625 = 6.25\%$$

Only projects A and B produce an IRR of more than the market rate of interest of 7% and so C is not viable.

A is preferred to B because $r_A > r_B$.

From the above example one can see that the IRR is the rate of interest which, if applied to the initial outlay, gives the return in year 1. Put another way, r is the rate of interest at which the PV of the future return equals the initial outlay, thus making the NPV of the whole project zero. This principle can be used to help calculate the IRR for more complex problems.

Example 7.29

Use the IRR method to evaluate the following project given a market rate of interest of 11%.

Initial outlay	£75,000
Return at end of year 1	£15,000
Return at end of year 2	£20,000
Return at end of year 3	£20,000
Return at end of year 4	£25,000
Return at end of year 5	£25,000
Return at end of year 6	£12,000

Solution

One needs to find the value of r for which

$$0 = -75,000 + 15,000(1 + r)^{-1} + 20,000(1 + r)^{-2} + 20,000(1 + r)^{-3}$$
$$+ 25,000(1 + r)^{-4} + 25,000(1 + r)^{-5} + 12,000(1 + r)^{-6}$$

The algebraic method of solution is far too complex and time-consuming to consider using here. The most practical method is to use a spreadsheet. Excel has a built-in IRR formula

Table 7.10

CELL	Enter	Explanation
A1	Ex.7.29	Label to remind you what example this is
A3	YEAR	Column heading label
B3	RETURN	Column heading for project returns
D2	Interest	Column heading label for the range of interest
D3	rate	rates for which NPV will be computed
E3	NPV	Column heading label
A4 to A10	*Enter numbers 0 to 6*	These are the time periods for this example
B4	-75000	Initial outlay (negative because it is a cost)
B5	15000	Project returns at end of years 1 to 6
B6	20000	
B7	20000	
B8	25000	
B9	25000	
B10	12000	
D4	4%	Interest rate to start range used
D5	=D4+0.01	Calculates a 1% rise in interest rate.
D6 to D20	*Copy cell D5 formula down column* D	Calculates a series of interest rates with increments of 1%.
E4	=NPV(D4,B$5:B$10) +B$4	Calculates project NPV corresponding to interest rate in D4 using Excel NPV formula less outlay in B4. Note the $ to anchor rows.
E5 to E20	*Copy cell E4 formula down column* E	Calculates NPV corresponding to interest rates in column D.
A12	IRR =	Label to tell you IRR calculated in next cell.
B12	=IRR(B4:B10)	Excel IRR formula calculates IRR of project returns in cells B4 to B10, which includes the negative return of the initial outlay.

which can immediately calculate r. You could also find r by using Excel to calculate the project NPV for a range of interest rates and then identifying the interest rate at which NPV is zero. Instructions for constructing a spreadsheet to solve this problem by both methods are shown in Table 7.10. Note that because the Excel NPV function does not take into account the initial outlay in cell B4 this is subtracted to get the true NPV of the project in column E.

The resulting spreadsheet should look like Table 7.11. This shows that the rate of interest that corresponds to an NPV of zero will lie somewhere between 14% and 15%, which checks out with the precise value for the IRR of 14.14% computed in cell B12. The market rate of interest given in the question is 11% and so, as the calculated IRR of 14.14% exceeds this, the project is worthwhile according to the IRR criterion.

Deficiencies of the IRR method

Although the IRR method may appear to be the most obvious and easily understood criterion for deciding on investment projects, and is still frequently used, it has several deficiencies which make it less useful than the NPV method.

First, it ignores the total value of the profit, as illustrated in the following example.

Table 7.11

	A	B	C	D	E
1	Ex .7.29				
2				Interest	
3	YEAR	RETURN		Rate	NPV
4	0	-75000		4%	27096.19
5	1	15000		5%	23813.36
6	2	20000		6%	20686.58
7	3	20000		7%	17706.57
8	4	25000		8%	14864.67
9	5	25000		9%	12152.86
10	6	12000		10%	9563.64
11				11%	7090.04
12	IRR =	14.14%		12%	4725.54
13				13%	2464.06
14				14%	299.94
15				15%	-1772.14
16				16%	-3757.14
17				17%	-5659.71
18				18%	-7484.20
19				19%	-9234.68
20				20%	-10914.99

Example 7.30

A firm has to choose between projects A and B. Project A involves an initial outlay of £18,000 and a return in 1 year's time of £20,000. Project B involves an initial outlay of £2,000 and a return in 1 year's time of £2,500. The interest rate is 6%. Which would be the better investment?

Solution

The IRR method ranks B as the best investment opportunity, since

$$r_A = \frac{20,000}{18,000} - 1 = 1.11 - 1 = 0.11 = 11\%$$

$$r_B = \frac{2,500}{2,000} - 1 = 1.25 - 1 = 0.25 = 25\%$$

The NPV method, however, would rank A as the better investment since

$$\text{NPV}_A = -18,000 + \frac{20,000}{1.06} = -18,000 + 18,867.92 = £867.92$$

$$\text{NPV}_B = -2,000 + \frac{2,500}{1.06} = -2,000 + 2,358.49 = £358.49$$

If the firm has a straightforward choice between A and B, then A is clearly the better investment. (The possibility of the firm using its initial £18,000 for investing in nine separate projects all with the same returns as B is ruled out.)

Some students may still not be convinced that the IRR method is faulty in the above example as one so often sees the rate of return used as a measure of the success of an investment in the press and other sources. Let us therefore work from first principles and consider the total assets of the firm after one year.

Assume that the firm has up to £18,000 at its disposal. If it puts this all into project A, then at the end of the year total assets will be £20,000.

If its puts £2,000 into project B, then it can also invest the remaining £16,000 elsewhere at the going rate of interest of 6%. Its total assets will therefore be as follows:

Return on project B	£2,500
plus £16,000 invested at 6% = 16,000 × 1.06	£16,960
Total assets	£19,460

Thus the firm is in a better financial position overall at the end of the year if it chooses project A, which is what the NPV method recommends but what the IRR method advises against.

Another way of reinforcing this point is to consider a third project C. Assume that this involves an investment now of £1 giving a return in one year's time of £1.90. This has a very high IRR of 90% but the small sum involved does not make it an attractive investment, which is why the NPV method should be used.

The second advantage that the NPV investment appraisal method has over the IRR method is that it can easily cope with forecasts of variable interest rates. The IRR method just involves comparing the computed IRR from an investment project with one given interest rate and so it could not be applied to Example 7.31 below.

Example 7.31

An investment project involves an initial outlay of £25,000 and net annual returns as follows:

£6,000	at the end of year 1
£8,000	at the end of year 2
£8,000	at the end of year 3
£10,000	at the end of year 4
£6,000	at the end of year 5

Interest rates are currently 15% but are forecast to fall to 12% next year and 10% the following year. They will then rise by 1 percentage point each year. Is the project worthwhile?

Solution

The variation in interest rates means that one cannot simply use the Excel NPV formula. To compute the answer manually we have to adjust the basic discounting formula to allow for the different discount rates each year. Thus

$$NPV = -25,000 + \frac{6,000}{1.15} + \frac{8,000}{1.15 \times 1.12} + \frac{8,000}{1.15 \times 1.12 \times 1.1}$$
$$+ \frac{10,000}{1.15 \times 1.12 \times 1.1 \times 1.11} + \frac{6,000}{1.15 \times 1.12 \times 1.1 \times 1.11 \times 1.12}$$

$$= -25,000 + \frac{6,000}{1.15} + \frac{8,000}{1.288} + \frac{8,000}{1.4168} + \frac{10,000}{1.572648} + \frac{6,000}{1.7613657}$$

$$= -25,000 + 5,217.39 + 6,211.18 + 5,646.53 + 6,358.70 + 3,406.45$$

$$= -25,000 + 26,840.25$$

$$= £1,840.25$$

This is positive and so the investment is worthwhile.

Although this is not a straightforward NPV calculation, a spreadsheet can be constructed to do the calculations. One suggested format for solving Example 7.31 is shown in Table 7.12, which shows the formulae to enter in relevant cells. This should produce the figures shown in Table 7.13, which confirm that NPV is £1,840.25.

A third drawback of the IRR method is that there may not be one unique solution for *r* when there are several negative terms in the polynomial to be solved. This point was made in Chapter 6 when the solution of polynomial equations was discussed. Apart from the initial outlay, negative returns may occur if further investment is required, or if a company has to pay to dismantle a project and return it to an environmentally acceptable state and the end of its useful life. However, investment project multiple solutions for the IRR are unusual and you are unlikely to come across them.

Table 7.12

	A	B	C	D	E	F
1	Ex 7.31	NPV	WITH	VARIABLE	INTEREST	RATES
2						
3	YEAR	i	DISCOUNT	FACTOR	RETURN	PV
4	0	0	=1/(1 + B4)	1	−25000	=D4*E4
5	1	0.15	=1/(1 + B5)	=D4*C5	6000	=D5*E5
6	2	0.12	=1/(1 + B6)	=D5*C6	8000	=D6*E6
7	3	0.1	=1/(1 + B7)	=D6*C7	8000	=D7*E7
8	4	0.11	=1/(1 + B8)	=D7*C8	10000	=D8*E8
9	5	0.12	=1/(1 + B9)	=D8*C9	6000	=D9*E9
10						
11				TOTAL	NPV	=SUM(F4.F9)

Table 7.13

	A	B	C	D	E	F
1	Ex 7.31	NPV	WITH	VARIABLE	INTEREST	RATES
2						
3	YEAR	i	DISCOUNT	FACTOR	RETURN	PV
4	0	0	1	1	-25000	-25000.00
5	1	0.15	0.8695652	0.869565	6000	5217.39
6	2	0.12	0.8928571	0.776398	8000	6211.18
7	3	0.1	0.9090909	0.705816	8000	5646.53
8	4	0.11	0.9009009	0.63587	10000	6358.70
9	5	0.12	0.8928571	0.567741	6000	3406.45
10						
11				TOTAL	NPV =	1840.25

Test Yourself, Exercise 7.5

1. Calculate the IRR for the projects in Table 7.14 and then say whether or not the IRR ranking is consistent with the NPV ranking for these projects if the market rate of interest is 15%.

    Table 7.14

	Project A	Project B	Project C	Project D
Outlay now (All values in £)	20,000	6,000	25,000	10,000
Return after 1 year (All values in £)	24,000	8,500	30,000	12,000

2. Two projects A and B each involve an initial outlay of £40,000 and guarantee the returns (in £) given in Table 7.15. The market rate of interest is 18%. Which is the better investment according to (a) the IRR criterion, (b) the NPV criterion?

    Table 7.15

	Project A	Project B
End of year 1	15,000	10,000
End of year 2	20,000	12,000
End of year 3	25,000	12,000
End of year 4	0	12,000
End of year 5	0	15,000
End of year 6	0	15,000

3. Using a spreadsheet, find the IRR and show that the NPV of the following project is zero when the discount rate used is approximately equal to this IRR.

    Outlay now: £25,000
    Annual returns: (1) £4,000   (2) £6,000   (3) £7,500
                      (4) £7,500   (5) £10,000   (6) £10,000

7.7   Geometric series and annuities

You may have noted that in some of the examples in Sections 7.5 and 7.6 above the return on the investment was the same in each time period. Although not many actual industrial investment projects give such a constant stream of returns there are other forms of financial investments which are designed to. These are called 'annuities'. For example, someone might pay a fixed sum for a guaranteed pension payment of £14,000 a year for the next 5 years.

The present value of a steady stream of a fixed return of £a per year for the next n years when interest rates are i% will be

$$\text{PV} = a(1+i)^{-1} + a(1+i)^{-2} + \cdots + a(1+i)^{-n}$$

This sequence of terms is a special case of what is known as a 'geometric series'. There exists a mathematical formula for the sum of such sequences of numbers so that one does not have to calculate each of the terms separately before summing. This would obviously be useful if you did not have access to a computer with an NPV program, but is this formula of any use otherwise? Well, there are some forms of annuities that are called 'perpetual annuities' which promise a fixed annual monetary return forever. For example, a bond that pays a fixed 6% return on a nominal price of £100 is a perpetual annuity of £6. The NPV of such an annuity at a rate of interest i would be

$$\frac{6}{1+i} + \frac{6}{(1+i)^2} + \cdots + \frac{6}{(1+i)^n} + \cdots$$

as n continues to infinity. Each successive term gets smaller and smaller but the sum of this sequence continues to grow as n gets bigger. You cannot sum such an infinite series of numbers without using the formula for the sum of an infinite geometric series. The next section deals with infinite geometric series and perpetual annuities but first we shall look at some more general features of geometric series and the appraisal of investments in annuities with a finite life span.

Geometric series

A geometric series is a sequence of terms where each successive term is the previous term multiplied by a *common ratio*. The series starts with a given initial term. Any number of terms may be in a series.

Example 7.32

If the given initial term is 24 and the common ratio is 5, what is the corresponding geometric series? (Find up to six terms.)

Solution

The series will be

$$24 \quad 24 \times 5 \quad 24 \times 5^2 \quad 24 \times 5^3 \quad 24 \times 5^4 \quad 24 \times 5^5$$

or

$$24 \quad 120 \quad 600 \quad 3{,}000 \quad 15{,}000 \quad 75{,}000$$

Example 7.33

A firm's sales revenue is initially £40,000 and then grows by 20% each successive year. What is the pattern of sales revenue over 5 years?

Solution

Each year's sales are 120% of the previous year's. The time profile of sales revenue is therefore a geometric series with an initial term of £40,000 and a common ratio of 1.2. Thus (in £) the series is

$$40{,}000 \quad 40{,}000 \times 1.2 \quad 40{,}000 \times 1.2^2 \quad 40{,}000 \times 1.2^3 \quad 40{,}000 \times 1.2^4$$

If we use the algebraic notation a for the initial term, k for the common ratio and n for the number of terms then the general form of a geometric series will be

$$a, ak, ak^2, \ldots, ak^{n-1}$$

Note that the last (nth) term is ak^{n-1} and not ak^n, because the initial term a is not multiplied by k.

Sum of a geometric series

The sum of a geometric series can be found by simply adding all the terms together. This is easy enough to do using a pocket calculator for the examples above. More complex series are more difficult to sum in this way, however, and so we need to derive a formula for summing them.

The general format for the sum of a geometric series with n terms will be

$$GP_n = a + ak + ak^2 + \cdots + ak^{n-1} \tag{1}$$

Multiplying each term by k gives

$$kGP_n = ak + ak^2 + \cdots + ak^{n-1} + ak^n$$

Subtracting (1) $GP_n = a + ak + ak^2 + \cdots + ak^{n-1}$

gives $(k-1)GP_n = -a + ak^n$

Therefore

$$GP_n = \frac{-a + ak^n}{k-1} = \frac{-a(1-k^n)}{k-1} = \frac{(-1)a(1-k^n)}{(-1)(1-k)} = \frac{a(1-k^n)}{1-k}$$

Thus the formula for the sum of a geometric series is

$$GP_n = \frac{a(1-k^n)}{1-k}$$

The examples below illustrate how this formula can be used to sum some simple numerical sequences of numbers.

Example 7.34

Use the geometric series sum formula to sum the geometric series

15   45   135   405   1,215   3,645

Solution

In this geometric series with six terms, each number except the first is 3 times the previous one. Thus

$$a = 15 \qquad k = 3 \qquad n = 6$$

Substituting these values into the geometric series sum formula we get

$$GP_n = \frac{a(1-k^n)}{1-k} = \frac{15(1-3^6)}{-2}$$

$$= \frac{15(1-729)}{-2} = \frac{15(-728)}{-2}$$

$$= 15 \times 364 = 5{,}460$$

You can check that this formula gives the same answer as that found using a calculator. In fact, in simple examples like this using the calculator may be the quicker method, but in other more complex cases the formula will provide the quickest method of solution.

Example 7.35

A firm expects its sales to grow by 12% per month. If its January sales figure is £9,200 per month what will its expected total annual sales be?

Solution

The firm's total annual sales will be the sum of the geometric series

$$9{,}200 + 9{,}200(1.12) + 9{,}200(1.12)^2 + \cdots + 9{,}200(1.12)^{11}$$

In this example $a = 9{,}200$, $k = 1.12$ and $n = 12$. Therefore, the sum is

$$GP_n = \frac{9{,}200(1-1.12^{12})}{1-1.12} = \frac{9{,}200(1-3.895976)}{1-1.12} = \frac{-26{,}642.979}{-0.12} = £222{,}024.83$$

Example 7.36

A 5-year saving scheme requires investors to pay in £5,000 now followed by £5,000 at 12-month intervals. Interest is credited at 14% at the end of each year of the investment. What will the final sum be at the end of the fifth year?

Solution

The £5,000 invested at the start of year 5 will be worth $5{,}000(1.14)$. The £5,000 invested at the start of year 4 will be worth $5{,}000(1.14)^2$ etc. Therefore the final sum will be

$$5{,}000(1.14) + 5{,}000(1.14)^2 + \cdots + 5{,}000(1.14)^5$$

This is a geometric series with $a = 5{,}000(1.14)$, $k = 1.14$ and $n = 5$. The sum will therefore be

$$GP_n = \frac{a(1-k^n)}{1-k} = \frac{5{,}000(1.14)(1-1.14^5)}{1-1.14}$$

$$= \frac{5{,}700(1-1.9254146)}{-0.14} = \frac{-5{,}274.8626}{-0.14} = £37{,}677.59$$

The formula for the sum of a geometric series can be used in Present Value (PV) calculations for a constant stream of returns. However, one has to be very careful not to get the algebraic terminology mixed up when, as in the previous example, the initial payback figure includes the constant ratio.

Example 7.37

An annuity will pay £8,000 at the end of each year for 5 successive years, the first payment being 12 months from the initial purchase date. What is the maximum price any rational investor would pay for such an annuity if the opportunity cost of capital is 10%?

Solution

The maximum purchase price will be the PV of the stream of returns, using 10% as the discount rate. Therefore (in £):

$$PV = \frac{8,000}{1.1} + \frac{8,000}{1.1^2} + \frac{8,000}{1.1^3} + \frac{8,000}{1.1^4} + \frac{8,000}{1.1^5}$$

This is a geometric series with five terms. The first term a is $\frac{8,000}{1.1}$ (*not* 8,000). The constant ratio k is $\frac{1}{1.1}$. Therefore

$$PV = \frac{a(1 - k^n)}{1 - k} = \frac{\dfrac{8,000}{1.1}\left[1 - \left(\dfrac{1}{1.1}\right)^5\right]}{1 - \dfrac{1}{1.1}}$$

$$= \frac{8,000(1 - 0.6209211)}{1.1(1 - 1/1.1)} = \frac{8,000(0.3790789)}{1.1 - 1}$$

$$= \frac{3,032.6312}{0.1} = £30,326.31$$

In the above example some of the terms cancelled out. The same terms will cancel in any annuity PV calculations and so a simplified general formula for the PV of an annuity can be derived.

Assuming an annual payment of R for n years and an interest rate of i, then

$$PV = \frac{R}{1 + i} + \frac{R}{(1 + i)^2} + \cdots + \frac{R}{(1 + i)^n}$$

In this geometric series the initial term is

$$a = \frac{R}{1 + i}$$

And the constant ratio is

$$k = \frac{1}{1 + i}$$

Therefore

$$PV = \frac{a(1-k^n)}{1-k} = \frac{\frac{R}{1+i}\left[1-\left(\frac{1}{1+i}\right)^n\right]}{1-\frac{1}{1+i}}$$

$$= \frac{R\left[1-\frac{1}{(1+i)^n}\right]}{1+i-1} = \frac{R[1-(1+i)^{-n}]}{i}$$

Thus for any annuity

$$PV = \frac{R[1-(1+i)^{-n}]}{i}$$

We can use this formula to check the answer to Example 7.37 above. Given $R = 8,000$, $i = 0.1$ and $n = 5$, then

$$PV = \frac{8,000[1-(1.1)^{-5}]}{0.1} = £30,326.31$$

This is the same answer as that derived from first principles.

Example 7.38

An annuity will pay £2,000 a year for the next 5 years, with the first payment in 12 months' time. Capital can be invested elsewhere at an interest rate of 14%. Is £6,000 a reasonable price to pay for this annuity?

Solution

For this annuity (in £)

$$PV = 2,000(1.14)^{-1} + 2,000(1.14)^{-2} + 2,000(1.14)^{-3} + 2,000(1.14)^{-4} + 2,000(1.14)^{-5}$$

In this example the annual payment $R = 2,000$, $i = 0.14$ and $n = 5$. Therefore, using the annuity formula

$$PV = \frac{R[1-(1+i)^{-n}]}{i} = \frac{2,000[1-(1.14)^{-5}]}{0.14}$$

$$= \frac{2,000(1-0.5193686)}{0.14} = £6,866.16$$

The PV of this annuity is greater than its purchase price of £6,000 and so it is clearly a worthwhile investment.

Example 7.39

What would you pay for an annuity that promises to pay £450 a year for the next 10 years given an interest rate of 8%?

Solution

In this example $R = £450$, $i = 8\% = 0.08$ and $n = 10$. The present value of the stream of returns will be

$$PV = \frac{R[1 - (1+i)^{-n}]}{i} = \frac{450[1 - (1.08)^{-10}]}{0.08}$$

$$= \frac{450(1 - 0.4631935)}{0.08} = £3,019.54$$

Thus any price less than £3,019.54 would make this annuity a worthwhile purchase.

Test Yourself, Exercise 7.6

1. In the geometric series below (i) identify the constant ratio, (ii) say what the sixth term will be and (iii) calculate the sum of each series up to ten terms using the formula for summation of a geometric series.

   (a) $8, 20, 50, \ldots$
   (b) $0.5, 1.5, 4.5, \ldots$
   (c) $2, 2.8, 3.92, \ldots$
   (d) $60, 48, 38.4, \ldots$
   (e) $2.4, 1.8, 1.35, \ldots$

2. A firm starts producing a new product. It sells 420 units in January and then sales increase by 10% each month. What will total demand be in the last 6 months of the year?

3. What would be the maximum price you would pay for the following annuities if money can be invested elsewhere at 8%?

   Annuity A pays £200 a year for the next 8 years
   Annuity B pays £900 a year for the next 4 years
   Annuity C pays £6,000 a year for the next 12 years

4. Would you pay £3,500 for an annuity which guarantees to pay you £750 annually for the next 7 years if you can invest money elsewhere at 9%?

5. What would be a reasonable price to pay for a pension plan which guarantees to pay £200 a month for the next 2 years if you can earn 1.2% a month on your bank deposit account?

7.8   Perpetual annuities

We now return to the problem of how to calculate the worth of an annuity that promises to pay a fixed annual return indefinitely. The PV of the stream of returns from a perpetual annuity is an infinite geometric progression. Whether or not one can find the sum of an infinite geometric progression depends on whether the progression is convergent or divergent. Before looking at the formal mathematical conditions for convergence or divergence these concepts are illustrated with some simple examples.

When you were at school you may have come across the teaser about the frog jumping across a pond, which goes something like this. 'A frog is sitting on a leaf in the middle of a circular pond. The pond is 10 metres in radius and the frog jumps 5 metres with its first jump. Its second jump is 2.5 m, its third jump 1.25 m and so on. How many jumps will it take for the frog to reach the edge of the pond? Assume that each time it jumps it lands on a leaf.'

The correct answer is, of course, 'never'. Each time the frog manages to jump half of the remaining distance to the edge of the pond. The total distance the frog travels in n jumps is given by the sum of the geometric series

$$5 + (0.5)5 + (0.5)^2 5 + \cdots + (0.5)^{n-1} 5$$

As n gets larger the sum of this series continues to increase but never actually reaches 10 metres. Only if an infinite number of jumps can be made will the total distance travelled be 10 metres. Thus in this example we have a geometric series which converges on 10 metres.

Geometric series may also be divergent. For example the sequence

40   60   90   135 ... etc.

can be written as the geometric series

40   40(1.5) $40(1.5)^2$ $40(1.5)^3$ ... $40(1.5)^n$

It is intuitively obvious that each successive term is larger than the previous one. Therefore, as the number of terms approaches infinity the sum of the series will also become infinitely large. The last term $40(1.5)^n$ will itself become infinitely large. There is thus no set quantity towards which the sum of the series converges.

You will probably have already guessed by now that the way to distinguish a convergent and a divergent geometric series is to look at the value of the common ratio k.

If $|k| > 1$ then successive terms become larger and larger and the series *diverges*.

If $|k| < 1$ then successive terms become smaller and smaller and the series *converges*.

The absolute value is used because it is possible to have a negative common ratio.

To find the **sum of a convergent geometric series** (such as the case of a perpetual annuity) let us look again at the general formula for the sum of a geometric series:

$$GP_n = \frac{a(1 - k^n)}{1 - k}$$

This can be rewritten as

$$GP_n = \frac{a}{1 - k} - \left(\frac{a}{1 - k}\right) k^n \tag{1}$$

If $-1 < k < 1$ then $k^n \to 0$ as $n \to \infty$ (i.e. the value of k^n approaches zero as n approaches infinity) and so the second term in (1) will disappear and the sum to infinity will be

$$GP_n = \frac{a}{1 - k} \tag{2}$$

We can now use formula (2) for the frog example. The total distance jumped is

$$\sum_{n=0}^{\infty} 5(0.5)^n$$

In this geometric series $k = 0.5$ and $a = 5$. The sum for an infinite number of terms will thus be

$$\frac{a}{1-k} = \frac{5}{1-0.5} = \frac{5}{0.5} = 10 \text{ metres}$$

The PV of a perpetual annuity can also be found using this formula although care must be taken to include the discounting factor in the initial term, as explained in the following example.

Example 7.40

What is the PV of an annuity which will pay £6 a year *ad infinitum*, with the first payment due in 12 months' time? Assume that capital can be invested elsewhere at 15%.

Solution

$$PV = \frac{6}{1.15} + \frac{6}{1.15^2} + \cdots + \frac{6}{1.15^n}$$

where $n \to \infty$.

In this geometric series $a = \frac{6}{1.5}$ and $k = \frac{1}{1.5}$. This is clearly convergent as $|k| < 1$. The sum to infinity is therefore

$$NPV = \frac{a}{1-k} = \frac{\dfrac{6}{1.15}}{1 - \dfrac{1}{1.15}} = \frac{6}{1.15\left(1 - \dfrac{1}{1.15}\right)}$$

$$= \frac{6}{1.15 - 1} = \frac{6}{0.15} = £40$$

A simplified formula for the PV of a perpetual annuity can be derived as certain terms will always cancel out, as Example 7.40 above illustrates.

Assume that an annuity pays a fixed return R each year, starting in 12 months' time, and the opportunity cost of capital is $i\%$. For this annuity

$$PV = R(1+i)^{-1} + R(1+i)^{-2} + \cdots + R(1+i)^{-n} \quad \text{where } n \to \infty$$

In this geometric series the initial value $a = R(1+i)^{-1}$ and constant ratio $k = (1+i)^{-1}$. Therefore, using the formula for the sum of an infinite converging geometric series

$$PV = \frac{a}{1-k} = \frac{R(1+i)^{-1}}{1-(1+i)^{-1}} = \frac{R}{(1+i)[1-(1+i)^{-1}]} = \frac{R}{1+i-1} = \frac{R}{i}$$

Thus the **formula for the PV of a perpetual annuity** is

$$PV = \frac{R}{i}$$

Reworking Example 7.40 above using this formula we get

$$PV = \frac{6}{0.15} = £40$$

which is identical to the answer derived from first principles, although the formula obviously makes the calculations much easier.

Example 7.41

An investment opportunity involves an initial outlay of £50,000 and gives a £5,000 annual return, starting in 12 months' time and continuing indefinitely. Capital can be invested elsewhere at 8%. Is this worth considering?

Solution

The PV of the annual income stream can be calculated using the formula for the PV of a perpetual annuity as

$$PV = \frac{R}{i} = \frac{5,000}{0.08} = £62,500$$

This is greater than the initial outlay of £50,000 and so this investment is clearly an attractive proposition.

Example 7.42

What would be the maximum price you would pay for a perpetual annuity that will pay £900 per annum, starting in 12 months' time, given an interest rate of 15%?

Solution

For this stream of returns

$$PV = \frac{R}{i} = \frac{900}{0.15} = £6,000$$

This is the maximum price a rational investor would pay for this annuity.

Test Yourself, Exercise 7.7

1. Identify which of the following geometric series are convergent and then calculate the sum to which these series converge as the number of terms approaches infinity:

   (a) 4, 6, 9, ...
   (b) 120, 96, 76.8, ...

    (c) $0.8, -1.2, 1.8, \ldots$
    (d) $36, 12, 4, \ldots$
    (e) $500, 500(0.48), 500(0.48)^2, \ldots$
    (f) $850, 850(1.2)^{-1}, 850(1.2)^{-2}, \ldots$

2. What is the maximum price you would pay for a perpetual annuity that will commence annual payment of £400 in 12 months' time if the market rate of interest is 13%?

3. Is it worth paying £40,000 for a perpetual annuity of £1,500 per annum, commencing payments in 12 months' time, if money can be invested elsewhere at 3%?

4. What would you calculate the price of an annuity paying £12,000 per annum (starting in 12 months' time) to be if the market rate of interest is
    (a) 5%,   (b) 10%,   (c) 15%,   (d) 20%?

5. A government bond guarantees an annual payment of £140 in perpetuity; what will it be priced at, given a market rate of interest of 4%?

7.9 Loan repayments

If someone takes out a loan now, to be paid off in regular equal instalments over a given time period, how can these payments be calculated? The following example shows how the formula for calculating the PV of an annuity can be adapted for this type of problem. As most loans are paid off monthly we shall mainly use monthly rates of interest in this section.

Example 7.43

If a £2,000 loan is taken out now to be paid back over the next 12 months at a monthly interest rate of 2% what will the monthly payments be?

Solution

From the lender's viewpoint the repayments can be viewed as a monthly annuity which pays £R per month for the following 12 months, where R is the monthly repayment. If the lender is willing to exchange the loan of £2,000 for this stream of payments then this must be the PV of this 'annuity'. Therefore, for a loan of amount L we can adapt the formula for the PV of an annuity as

$$\text{PV} = \frac{R[1 - (1+i)^{-n}]}{i} = L$$

However, instead of calculating PV we need to find the value of R for a given size of loan L. Since

$$\frac{R[1 - (1+i)^{-n}]}{i} = L$$

$$R[1 - (1+i)^{-n}] = iL$$

Table 7.16 Powers for loan repayments: calculation of $(1 + i)^{-n}$

i (%)	$n = 12$	$n = 24$	$n = 36$	$n = 60$	$n = 240$	$n = 300$
0.50	0.941905	0.887185	0.835644	0.741372	0.302096	0.223965
0.55	0.936300	0.876658	0.820815	0.719574	0.268103	0.192920
0.60	0.930731	0.866260	0.806255	0.698427	0.237949	0.166190
0.65	0.925197	0.855991	0.791961	0.677911	0.211199	0.143174
0.70	0.919700	0.845848	0.777927	0.658008	0.187467	0.123355
0.75	0.914238	0.835831	0.764148	0.638699	0.166412	0.106287
0.80	0.908811	0.825937	0.750621	0.619966	0.147731	0.091588
0.85	0.903418	0.816165	0.737339	0.601791	0.131154	0.078927
0.90	0.898061	0.806514	0.724299	0.584157	0.116444	0.068022
0.95	0.892738	0.796981	0.711495	0.567049	0.103390	0.058627
1.00	0.887449	0.787566	0.698924	0.550449	0.091805	0.050534
1.05	0.882194	0.778266	0.686582	0.534343	0.081523	0.043561
1.10	0.876972	0.769081	0.674463	0.518717	0.072397	0.037553
1.15	0.871784	0.760008	0.662564	0.503554	0.064296	0.032376
1.20	0.866630	0.751048	0.650880	0.488842	0.057105	0.027915
1.25	0.861508	0.742197	0.639409	0.474567	0.050721	0.024070
1.30	0.856419	0.733454	0.628145	0.460715	0.045053	0.020757
1.35	0.851363	0.724819	0.617084	0.447275	0.040022	0.017900
1.40	0.846339	0.716290	0.606224	0.434232	0.035554	0.015438
1.45	0.841347	0.707865	0.595560	0.421577	0.031586	0.013316
1.50	0.836387	0.699543	0.585089	0.409295	0.028064	0.011486
1.55	0.831459	0.691324	0.574807	0.397378	0.024935	0.009908
1.60	0.826562	0.683204	0.564711	0.385813	0.022156	0.008584
1.65	0.821696	0.675185	0.554797	0.374590	0.019689	0.007375
1.70	0.816861	0.667263	0.545061	0.363699	0.017497	0.006363
1.75	0.812057	0.659438	0.535501	0.353130	0.015550	0.005491
1.80	0.807284	0.651708	0.526114	0.342873	0.013820	0.004738
1.85	0.802541	0.644073	0.516895	0.332918	0.012284	0.004089
1.90	0.797828	0.636531	0.507842	0.323257	0.010919	0.003529
1.95	0.793146	0.629080	0.498953	0.313881	0.009706	0.003046
2.00	0.788493	0.621721	0.490223	0.304782	0.008628	0.002629

$$R = \frac{iL}{1 - (1 + i)^{-n}}$$

which is the general formula for calculating loan repayments.

The known values for this example are $L = 2{,}000$, $i = 2\% = 0.02$ and $n = 12$. Substituting these into the loan repayment formula gives

$$R = \frac{0.02 \times 2{,}000}{1 - (1.02)^{-12}}$$

$$= \frac{40}{1 - 0.7884934}$$

$$= \pounds189.12 \text{ per month}$$

To work out loan repayments using the above formula you need to use the power function key on a calculator. Table 7.16 shows some calculated values that may be useful for this type of problem for those of you who do not have such a calculator to hand.

Example 7.44

What will be the monthly repayments on a loan of £6,000 taken out for 5 years at a monthly interest rate of 0.7%?

Solution

$$L = £6,000 \qquad i = 0.7\% = 0.007 \qquad n = 5 \times 12 = 60$$

Using the loan repayment formula

$$R = \frac{iL}{1 - (1 + i)^{-n}} = \frac{0.007 \times 6,000}{1 - (1.007)^{-60}}$$

$$= \frac{0.007 \times 6,000}{1 - 0.658008} = \frac{42}{0.342}$$

$$= £122.81 \text{ monthly repayment}$$

(Note: To use Table 7.16 to find the value of 1.007^{-60} in the above example, just read along the row for 0.7% until you get to the column for 60.)

Example 7.45

What are the monthly payments on a repayment mortgage of £60,000 taken out for 25 years if the monthly rate of interest is 0.75%?

Solution

$$L = 60,000 \qquad i = 0.75\% = 0.0075 \qquad n = 25 \times 12 = 300 \text{ months}$$

Using the loan repayment formula the monthly payments will be

$$R = \frac{iL}{1 - (1 + i)^{-n}} = \frac{0.0075 \times 60,000}{1 - (1.0075)^{-300}}$$

$$= \frac{450}{1 - 0.106287} = £503.32$$

If only the APR for a loan is quoted, then it will be necessary to calculate the equivalent monthly interest rate before working out monthly repayments.

Example 7.46

If a loan of £4,200 is taken out over a period of 3 years at an APR of 6.8% what will the monthly repayments be?

Solution

First we need to convert the APR of 6.8% to its equivalent monthly rate i_m. We know that

$$1 + \text{APR} = (1 + i_m)^{12}$$

and so

$$\sqrt[12]{(1 + \text{APR})} = 1 + i_m$$
$$\sqrt[12]{(1 + \text{APR})} - 1 = i_m$$

Substituting in the value of APR = 6.8% = 0.068

$$
\begin{aligned}
i_m &= \sqrt[12]{(1 + 0.068)} - 1 \\
&= \sqrt[12]{(1.068)} - 1 \\
&= 1.0054974 - 1 \\
&= 0.0055 = 0.55\%
\end{aligned}
$$

The values to be entered into the loan repayment formula are therefore

$$L = 4{,}200 \qquad i = 0.0055 \qquad n = 3 \times 12 = 36$$

Giving
$$
R = \frac{iL}{1 - (1+i)^{-n}} = \frac{0.0055 \times 4{,}000}{1 - (1.0055)^{-36}}
$$
$$
= \frac{23.1}{1 - 0.820815} = £128.92 \text{ monthly payment}
$$

From an individual consumer's viewpoint you may be more interested in finding out the interest rate you have to pay on a loan. All lenders now have to quote their APR by law, but you may still wish to check this.

Example 7.47

A car dealer offers you a £12,000 car for a £4,000 deposit now followed by 24 monthly payments of £400. What is the APR on this effective loan of £8,000?

Solution

As in the examples above, treat the stream of repayments as an annuity for the lender. Referring again to the formula for loan repayments

$$R = \frac{iL}{1 - (1+i)^{-n}}$$

Table 7.17

CELL	Enter	Explanation
A1	Ex.7.47	Label to remind you what example this is
B1	LOAN =	Label to tell you loan value goes in next cell.
C1	8000	Loan value for this example.
D1	n MONTHS=	Label to tell you number of months for repayment goes in next cell.
E1	24	Number of months for this example.
A3	APR	Column heading labels
B3	MONTHLY i	
C3	REPAYMENT	
A4	15.00%	Start of (guessed) interest rate for APR range
A5	=A4+0.0025	Gives increment of 0.25%
A6 to A24	*Copy cell A5 formula down column A*	Gives a range of values for APR in 0.25% increments. (Format to 2 d.p.)
B4	=(1+A4)^(1/12)-1	Formula calculates monthly interest rate corresponding to APR in cell A4.
B5 to B24	*Copy cell B4 formula down column B*	Calculates monthly interest rates corresponding to APR in column A.
C4	=B4*C$1/(1-(1+B4)^-E$1)	Formula calculates repayment corresponding to value of total loan in cell C1, number of months in cell E1 and the monthly interest rate in cell B4, which is determined by APR in column A.
C5 to C24	*Copy cell C4 formula down column C*	Calculates repayment corresponding to different APR values.

we can see that even if we know that $L = 8,000, R = 400$ and $n = 24$ this still leaves us with the awkward equation

$$400 = \frac{i \times 8,000}{1 - (1 + i)^{-24}}$$

to solve for i (the monthly interest rate) which can then be used to calculate the APR.

The quickest way to solve this is to use a spreadsheet. Instructions for constructing an appropriate Excel format are shown in Table 7.17, which should give the actual spreadsheet shown in Table 7.18. This calculates the repayment values that correspond to a range of monthly interest rates which, in turn, will correspond to specific values for the APR. Once a repayment equal (or very close) to £400 has been identified then the corresponding monthly interest rate and APR can be read off. Near the bottom of Table 7.18 we can see that a £400.01 repayment corresponds to a 1.51% monthly interest rate and a 19.75% APR, which is the solution to this problem.

This spreadsheet format can be used to solve similar types of problems. You only need to change the total loan figure in cell C1 and the time period in cell E2 to compute a new set of repayment figures for a range of monthly interest rates and, in some cases, you may have to extend the interest rate range or just change the initial trial value in cell A4.

Example 7.48

A loan company will require 36 monthly payments of £438.25 in return for a loan of £12,500. What APR is it charging?

Table 7.18

	A	B	C	D	E
1	Ex 7.47	LOAN =	8000	n MONTHS=	24
2					
3	APR	MONTHLY i	REPAYMENT		
4	15.00%	1.17%	384.32		
5	15.25%	1.19%	385.15		
6	15.50%	1.21%	385.98		
7	15.75%	1.23%	386.81		
8	16.00%	1.24%	387.64		
9	16.25%	1.26%	388.47		
10	16.50%	1.28%	389.30		
11	16.75%	1.30%	390.13		
12	17.00%	1.32%	390.95		
13	17.25%	1.33%	391.78		
14	17.50%	1.35%	392.61		
15	17.75%	1.37%	393.43		
16	18.00%	1.39%	394.26		
17	18.25%	1.41%	395.08		
18	18.50%	1.42%	395.90		
19	18.75%	1.44%	396.73		
20	19.00%	1.46%	397.55		
21	19.25%	1.48%	398.37		
22	19.50%	1.50%	399.19		
23	19.75%	1.51%	400.01	<< Solution	
24	20.00%	1.53%	400.83		

Solution

Using the spreadsheet constructed for Example 7.47 above, enter the new values for the loan and time period in cells C1 and E1. In row 12 you should then be able then read off the values:

APR	MONTHLY i	REPAYMENT
17.00%	1.32%	438.25

The APR this company charges is therefore 17%.

Test Yourself, Exercise 7.8

1. What will be the monthly repayments on a loan of £6,500 taken out over 5 years at a monthly interest rate of 1.2%?
2. You wish to buy a car priced at £6,000 by putting down a cash deposit of £2,000 and borrowing £4,000, the loan being paid back in monthly instalments over 2 years. How much will you have to budget to pay out of your salary if the monthly interest rate is 1.4%?
3. A loan company will lend you £5,000, repayable over the next 3 years in monthly payments. What will these payments be if the APR on the loan is 24.6%?
4. What will be the monthly payments on a repayment mortgage of £75,000 taken out over 20 years if the interest rate is fixed at 0.95% per month?

5. A loan of £800 is taken out. What APR is being charged if the monthly payments are

   (a) £27.00　over 36 months?
   (b) £31.13　over 36 months?
   (c) £25.00　over 48 months?
   (d) £21.78　over 48 months?

6. A car dealer has on offer a special '0% finance' deal on the advertised price of £8,671 for a particular model. This requires an initial deposit of £1,734 followed by 24 monthly payments of £289.00. If you could get the price reduced to £8,095 if you paid cash and can earn 9% per annum on money invested in a building society, which method would you use to purchase this car?

7.10　Other applications of growth and decline

In the previous sections of this chapter, various mathematical methods have been explained in order to solve a variety of problems concerned with finance and investment. Rather than introducing even more mathematical methods, this section now considers how the techniques already explained can be adapted to some non-financial problems. A series of different types of problem are presented and the most appropriate method of solution is explained. Note that in all these examples growth and decline are still treated as discrete processes.

Example 7.49

There are limited world reserves of mineral M. The current rate of extraction is 45 million tonnes a year, with all mined material being used up by manufacturing industry. This extraction rate is expected to increase at 3% per annum. Total estimated reserves are 1,200 million tonnes. When will they be expected to run out if this 3% growth rate continues?

Solution

The annual pattern of consumption will be (in millions of tonnes) the geometric series

$$45, \ 45(1.03), \ 45(1.03)^2, \ldots \ 45(1.03)^{n-1}$$

where n is the number of years that mining continues. The initial term $a = 45$ and the common ratio $k = 1.03$. The sum of this geometric series is therefore

$$\frac{a(1 - k^n)}{1 - k} = \frac{45(1 - 1.03^n)}{1 - 1.03}$$

which must sum to 1,200 if all reserves are used up. Therefore

$$1,200 = \frac{45(1 - 1.03^n)}{-0.03}$$

$$-0.8 = 1 - 1.03^n$$

$$1.03^n = 1.8$$

Putting this in logarithmic form we get

$$n \log 1.03 = \log 1.8$$

$$n = \frac{\log 1.8}{\log 1.03} = \frac{0.2552725}{0.0128372} = 19.885$$

Therefore mineral M is expected to run out within 20 years at the current rate of extraction.

Example 7.50

A developing country currently produces 3,600 million units of food per annum and this rate of production is expected to increase by 4% a year. Its population is currently 2.5 million and expected to grow by 6% per annum. The minimum recommended average intake of food is 1,200 units of food per person per year. Assuming no changes in production or population growth rates, no imports and exports and no foreign aid, when will food production fall below the subsistence level?

Solution

The demand for food is 1,200 × population.

Initial demand is therefore 1,200 × 2.5 million = 3,000 million units.

The rate of growth of the population is 6% and so total demand for food after n years will be $3,000(1.06)^n$ million units.

Initial production is 3,600 million units of food. The rate of growth of production is 4% and so total production after n years will be $3,600(1.04)^n$ million units.

The subsistence level is reached in year n when

$$\text{food demand} = \text{food production}$$

$$3,000(1.06)^n = 3,600(1.04)^n$$

$$\left(\frac{1.06}{1.04}\right)^n = \frac{3,600}{3,000}$$

$$(1.0192307)^n = 1.2$$

putting this in log form,

$$n \log 1.01923 = \log 1.2$$

$$n = \frac{\log 1.2}{\log 1.01923} = \frac{0.079182}{0.0082722} = 9.572$$

Therefore food production will fall below the subsistence level in 10 years' time.

Example 7.51

Estimated reserves of an oil field are 84 million barrels. What annual growth in the rate of extraction will exhaust the oil in 12 years given that this year's production is 6 million barrels?

Solution

Total oil extraction will be the sum of the geometric series with 12 terms:

$$6 + 6(1 + r) + 6(1 + r)^2 + \cdots + 6(1 + r)^{11}$$

where r is the growth in the extraction rate. The oilfield will be exhausted when this totals to 84. Therefore, given the initial term $a = 6$ and the constant ratio $k = 1 + r$, and employing the formula for the sum of a geometric series

$$84 = GP_n = \frac{a(1 - k^n)}{1 - k} = \frac{6[1 - (1 + r)^{12}]}{1 - (1 + r)}$$

$$84 = \frac{6[1 - (1 + r)^{12}]}{-r}$$

The easiest way to solve for r in this equation is to set up a spreadsheet to calculate different values of GP_n for different values of r and then see which one is closest to 84.

An Excel spreadsheet which does this is shown in Table 7.19. To construct this spreadsheet yourself, you should now be able to enter the labels, given parameter values and the range of interest rates without any difficulty. The crucial calculation is the formula that calculates the total amount of extraction corresponding to the given initial extraction rate, the time period and the interest rate. This is achieved by entering the formula $=C\$2^* - (1 + A6)\wedge C\$3/- A6$ in cell B6 and then copying it down the column.

You can now read off the interest rate that corresponds to the total extraction amount which is closest to 84, which is 2.75% giving total extraction of 83.953. If you wanted to get a more precise answer, you could make the interest rate increments smaller close to this approximate solution. This should show that a growth rate of 2.76% in the annual level of extraction will exhaust the oil reserves in 12 years.

Example 7.52

In a water authority's area the current river flows allow a maximum extraction rate of 100 million gallons per day and current usage is 25 million gallons per day. When will a crisis point be reached if consumption grows by 4% per annum? What rate of growth would allow current supply sources to be sufficient for the next 100 years?

Solution

Consumption rate in n years' time (in millions of gallons) will be $25(1.04)^n$.

The crisis point will be reached when consumption equals the maximum extraction rate and so

$$100 = 25(1.04)^n$$
$$4 = 1.04^n$$

Table 7.19

	A	B	C	D
1	Ex 7.51	OIL	RESERVES	
2	INITIAL	EXTRACTION =	6	m barrels
3		TIME PERIOD =	12	years
4	GROWTH	TOTAL		
5	r	EXTRACTION		
6	2.00%	80.473		
7	2.05%	80.699		
8	2.10%	80.927		
9	2.15%	81.155		
10	2.20%	81.384		
11	2.25%	81.613		
12	2.30%	81.844		
13	2.35%	82.075		
14	2.40%	82.307		
15	2.45%	82.540		
16	2.50%	82.773		
17	2.55%	83.008		
18	2.60%	83.243		
19	2.65%	83.479		
20	2.70%	83.715		
21	2.75%	83.953	<< solution	
22	2.80%	84.191		
23	2.85%	84.430		
24	2.90%	84.670		
25	2.95%	84.911		
26	3.00%	85.152		

Putting this in log form this gives

$$\log 4 = n \log 1.04$$

$$n = \frac{\log 4}{\log 1.04} = \frac{0.60206}{0.017033} = 35.346$$

Therefore a growth rate of 4% can be sustained for 35 years.

If current water supplies are to last another 100 years at growth rate r, then the current maximum supply rate of 100 million gallons per day will equal demand when

$$100 = 25(1 + r)^{100}$$

$$4 = (1 + r)^{100}$$

$$\sqrt[100]{4} = 1 + r$$

$$1.0139595 = 1 + r$$

$$0.0139595 = r$$

Therefore current water supplies will be sufficient for the next 100 years with a growth rate of 1.4%.

Example 7.53

A retailer has to order stock of a particular summer seasonal product in one batch at the start of the season. The first week's sales are expected to be 200 units. Past years' sales suggest that demand will then grow by 5% a week for the next 14 weeks and then fall by 10% a week for the remaining 10 weeks of the season. How much stock needs to be ordered to meet the anticipated sales for the whole 25-week season?

Solution

This problem involves the summing of two separate geometric series.
   Sales over the first 15 weeks are expected to be

$$200 + 200(1.05) + 200(1.05)^2 + \cdots + 200(1.05)^{14}$$

In this geometric series $a = 200$, $k = 1.05$, $n = 15$ and so its sum will be

$$\frac{a(1-k^n)}{1-k} = \frac{200(1 - 1.05^{15})}{1 - 1.05}$$
$$= \frac{200(1 - 2.0789282)}{-0.05}$$
$$= \frac{-215.78564}{-0.05}$$
$$= 4{,}315.7128$$

Thus, to the nearest whole unit over the first 15 weeks, sales will be 4,316.
   In the fifteenth week (which is part of the first geometric series) the sales will be

$$200(1.05)^{14} = 395.98632$$
$$= 396 \text{ (to nearest whole unit)}$$

Over the remaining 10 weeks the total sales will therefore be

$$396(0.9) + 396(0.9)^2 + \cdots + 396(0.9)^{10}$$

In this geometric series $a = 396(0.9)$, $k = 0.9$ and $n = 10$. Its sum will therefore be

$$\frac{a(1-k^n)}{1-k} = \frac{396(0.9)(1 - 0.9^{10})}{1 - 0.9}$$
$$= \frac{356.4(0.6513216)}{0.1}$$
$$= \frac{232.131}{0.1} = 2{,}321.31$$

Thus, to the nearest whole unit, expected total sales over the whole season sales will be

$$4{,}316 + 2{,}321 = 6{,}637 \text{ units}$$

Example 7.54

Average annual income in a developing country is $420, and average annual expenditure on food is $280. If average income rises at 3% per annum and income elasticity of demand for food is 0.8, when will average expenditure on food reach $340?

Solution

For every 1% rise in average income Y, there will be a 0.8% rise in food consumption F, given an income elasticity of demand of 0.8. Therefore a 3% growth in Y will mean a 2.4% growth in F. The mathematical problem then becomes 'how long will it take $280 to grow to $340 at a growth rate of 2.4%?' and we need to solve for n in the equation

$$340 = 280(1.024)^n$$

$$\frac{340}{280} = 1.024^n$$

$$1.2142857 = 1.024^n$$

Putting this in log form

$$\log 1.2142857 = n \log 1.024$$

$$n = \frac{\log 1.2142857}{\log 1.024} = \frac{0.0843209}{0.0103} = 8.1865$$

Therefore, average food expenditure will reach $340 after approximately 8.19 years.

Test Yourself, Exercise 7.9

1. Total reserves of mineral Z are 140 million tonnes. Current annual consumption is 18 million tonnes. If consumption is expected to grow by 4.5% a year, how long will these reserves last?

2. A country's gross national product (GNP) is forecast to grow at 3% per annum and its population is expected to expand at 1.5% per annum. GNP is currently $12,000 million and the population is 15 million, giving a GNP per capita of $800. When will GNP per capita reach $1,000?

3. What rate of growth of consumption will allow the current reserves of a natural resource to last for the next 50 years if this year's consumption is forecast to be 8 million tonnes and total reserves are 1,220 million tonnes?

4. World annual usage of mineral M is actually declining by 5% a year. The current rate of extraction is 65 million tonnes per year. Total reserves in existence amount to 1,500 million tonnes. Will they last forever if the 5% per annum decline persists?

5. The estimated reserves of resource R are 1,650 million tonnes. Current annual consumption is 80 million tonnes. What percentage reduction in the annual consumption rate will ensure that the resource never runs out, assuming that consumption falls each year by the same percentage?

8 Introduction to calculus

Learning objectives

After completing this chapter students should be able to:

- Differentiate functions with one unknown variable.
- Find the slope of a function using differentiation.
- Derive marginal revenue and marginal cost functions using differentiation and relate them to the slopes of the corresponding total revenue and cost functions.
- Calculate point elasticity for non-linear demand functions.
- Use calculus to find the sales tax that will maximize tax yield.
- Derive the Keynesian multiplier using differentiation.

8.1 The differential calculus

This chapter introduces some of the basic techniques of calculus and their application to economic problems. We shall be concerned here with what is known as the 'differential calculus'.

Differentiation is a method used to find the slope of a function at any point. Although this is a useful tool in itself, it also forms the basis for some very powerful techniques for solving optimization problems, which are explained in this and the following chapters.

The basic technique of differentiation is quite straightforward and easy to apply. Consider the simple function that has only one term

$$y = 6x^2$$

To derive an expression for the slope of this function for any value of x *the basic rules of differentiation* require you to:

(a) multiply the whole term by the value of the power of x, and
(b) deduct 1 from the power of x.

In this example there is a term in x^2 and so the power of x is reduced from 2 to 1. Using the above rule the expression for the slope of this function therefore becomes

$$2 \times 6x^{2-1} = 12x$$

This is known as the derivative of y with respect to x, and is usually written as dy/dx, which is read as 'dy by dx'.

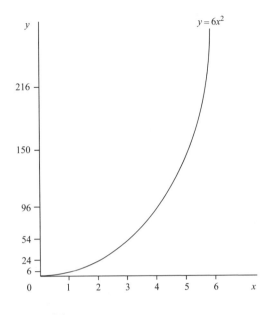

Figure 8.1

We can check that this is approximately correct by looking at the graph of the function $y = 6x^2$ in Figure 8.1. Any term in x^2 will rise at an ever increasing rate as x is increased. In other words, the slope of this function must increase as x increases. The slope is the derivative of the function with respect to x, which we have just worked out to be $12x$. As x increases the term $12x$ will also obviously increase and so we can confirm that the formula derived for the slope of this function does behave in the expected fashion.

To determine the actual value of the slope of the function $y = 6x^2$ for any given value of x, one simply enters the given value of x into the formula

   Slope $= 12x$

When $x = 4$, then slope $= 48$; when $x = 5$, then slope $= 60$; etc.

Example 8.1

What is the slope of the function $y = 4x^2$ when x is 8?

Solution
By differentiating y we get

$$\text{Slope} = \frac{dy}{dx} = 2 \times 4x^{2-1} = 8x$$

When $x = 8$, then slope $= 8(8) = 64$.

Example 8.2

Find a formula that gives the slope of the function $y = 6x^3$ for any value of x.

Solution

$$\text{Slope} = \frac{dy}{dx} = 3 \times 6x^{3-1} = 18x^2 \quad \text{for any value of } x.$$

Example 8.3

What is the slope of the function $y = 45x^4$ when $x = 10$?

Solution

$$\text{Slope} = \frac{dy}{dx} = 180x^3$$

When $x = 10$, then slope $= 180(1,000) = 180,000$.

Test Yourself, Exercise 8.1

1. Derive an expression for the slope of the function $y = 12x^3$.
2. What is the slope of the function $y = 6x^4$ when $x = 2$?
3. What is the slope of the function $y = 0.2x^4$ when $x = 3$?
4. Derive an expression for the slope of the function $y = 52x^3$.
5. Make up your own single-term function and then differentiate it.

8.2   Rules for differentiation

The rule for differentiation can be formally stated as:
   If $y = ax^n$ where a and n are given parameters then

$$\frac{dy}{dx} = nax^{n-1}$$

When there are several terms in x added together or subtracted in a function then this rule for differentiation is applied to each term individually. (The special rules for differentiating functions where terms are multiplied or divided are explained in Chapter 12.)

Example 8.4

Differentiate the function $y = 3x^2 + 10x^3 - 0.2x^4$.

Solution

$$\frac{dy}{dx} = 2 \times 3x^{2-1} + 3 \times 10x^{3-1} - 4 \times 0.2x^{4-1} = 6x + 30x^2 - 0.8x^3$$

Example 8.5

Find the slope of the function $y = 6x^2 - 0.5x^3$ when $x = 10$.

Solution

$$\text{Slope} = \frac{dy}{dx} = 12x - 1.5x^2$$

When $x = 10$, slope $= 120 - 1.5(100) = 120 - 150 = -30$.

Example 8.6

Derive an expression for the slope of the function $y = 4x^2 + 2x^3 - x^4 + 0.1x^5$ for any value of x.

Solution

$$\text{Slope} = \frac{dy}{dx} = 8x + 6x^2 - 4x^3 + 0.5x^4$$

In using the formula for differentiation, one has to remember that $x^1 = x$ and $x^0 = 1$.

Example 8.7

Differentiate the function $y = 8x$.

Solution

$$y = 8x = 8x^1$$

$$\frac{dy}{dx} = 1 \times 8x^{1-1} = 8x^0 = 8$$

Example 8.8

Derive an expression for the slope of the function $y = 30x - 0.5x^2$ for any value of x.

Solution

$$\text{Slope} = \frac{dy}{dx} = 30x^0 - 2(0.5)x = 30 - x$$

Example 8.9

Differentiate the function $y = 14x$.

Solution

$$\frac{dy}{dx} = 14x^{1-1} = 14x^0 = 14$$

The example above illustrates the point that the derivative of any term in x (to the power of 1) is simply the value of the parameter that x is multiplied by.

Any constant terms always disappear when a function is differentiated. To understand why, consider a function with one constant such as the function $y = 5$. This could be written as $y = 5x^0$. Differentiating this function gives

$$\frac{dy}{dx} = 0(5x^{-1}) = 0$$

Example 8.10

Differentiate the function $y = 20 + 4x - 0.5x^2 + 0.01x^3$.

Solution

$$\frac{dy}{dx} = 4 - x + 0.03x^2$$

Example 8.11

Derive an expression for the slope of the function $y = 6 + 3x - 0.1x^2$.

Solution

$$\text{Slope} = \frac{dy}{dx} = 3 - 0.2x$$

Even when the power of x in a function is negative or not a whole number, the same rules for differentiation still apply.

Example 8.12

What is the slope of the function $y = 4x^{0.5}$ when $x = 4$?

Solution

$$\text{Slope} = \frac{dy}{dx} = 0.5 \times 4x^{0.5-1} = 2x^{-0.5}$$

When $x = 4$, slope $= 2 \times 4^{-0.5} = 2 \times \left(\frac{1}{2}\right) = 1$.

Example 8.13

Differentiate the function $y = x^{-1} + x^{0.5}$.

Solution

$$\frac{dy}{dx} = -1 \times x^{-1-1} + 0.5x^{0.5-1} = -x^{-2} + 0.5x^{-0.5}$$

Test Yourself, Exercise 8.2

1. Differentiate the function $y = x^3 + 60x$.
2. What is the slope of the function $y = 12 + 0.5x^4$ when $x = 5$?
3. Derive a formula for the slope of the function $y = 4 + 4x^{-1} - 4x$.
4. What is the slope of the function $y = 4x^{0.5}$ when $x = 4$?
5. Differentiate the function $y = 25 - 0.1x^{-2} + 2x^{0.3}$.
6. Make up your own function with at least three different terms in x and then differentiate it.

8.3   Marginal revenue and total revenue

What differentiation actually does is look at the effect of an infinitely small change in the independent variable x on the dependent variable y in a function $y = f(x)$. This may seem a strange concept, and the rest of this section tries to explain how it works, but first consider the following example which shows how a function can be differentiated from first principles.

Example 8.14

Differentiate the function $y = 6x + 2x^2$ from first principles.

Solution

Assume that x is increased by the small amount Δx (Δ is the Greek letter 'delta' which usually signifies a change in a variable). This will produce a small change Δy in y.

Given the original function

$$y = 6x + 2x^2 \tag{1}$$

the new value of y (i.e. $y + \Delta y$) can be found by substituting the new value of x (i.e. $x + \Delta x$) into the function. Thus

$$y + \Delta y = 6(x + \Delta x) + 2(x + \Delta x)^2$$

$$y + \Delta y = 6x + 6\Delta x + 2x^2 + 4x\Delta x + 2(\Delta x)^2$$

Subtracting (1) $\qquad \underline{y \qquad\quad = 6x \qquad\quad + 2x^2 \qquad\qquad\qquad\qquad\qquad}$

gives $\qquad\qquad\quad \Delta y = \qquad 6\Delta x \qquad\quad + 4x\Delta x + 2(\Delta x)^2$

Dividing through by Δx,

$$\frac{\Delta y}{\Delta x} = 6 + 4x + 2\Delta x$$

If Δx becomes infinitely small, then the last term disappears and

$$\frac{\Delta y}{\Delta x} = 6 + 4x \tag{2}$$

By definition, dy/dx is the effect of an infinitely small change in x on y. Thus, from (2),

$$\frac{dy}{dx} = 6 + 4x$$

This is the same result for dy/dx that would be obtained using the basic rules for differentiation explained in Section 8.2. It is obviously quicker to use these rules than to differentiate from first principles. However, Example 8.14 should now help you to understand how the differential calculus can be applied to economics.

Up to this point we have been using the usual algebraic notation for a single variable function, assuming that y is dependent on x. Changing the notation so that we can look at some economic applications does not alter the rule for differentiation as long as functions are specified in a form where one variable is dependent on another.

In introductory economics texts, marginal revenue (MR) is sometimes defined as the increase in total revenue (TR) received from sales caused by an increase in output by 1 unit. This is not a precise definition though. It only gives an approximate value for marginal revenue and it will vary if the units that output is measured in are changed. A more precise definition of marginal revenue is that it is the rate of change of total revenue relative to increases in output.

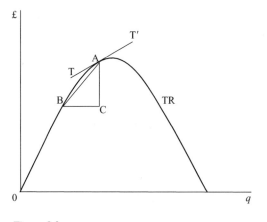

Figure 8.2

In Figure 8.2 the rate of change of total revenue between points B and A is

$$\frac{\Delta \text{TR}}{\Delta Q} = \frac{\text{AC}}{\text{BC}} = \text{the slope of the line AB}$$

which is an approximate value for marginal revenue over this output range.

Now suppose that the distance between B and A gets smaller. As point B moves along TR towards A the slope of the line AB gets closer to the value of the slope of TT′, which is the tangent to TR at A. (A tangent to a curve at any point is a straight line having the slope at that point.) Thus for a very small change in output, MR will be almost equal to the slope of TR at A. If the change becomes infinitesimally small, then the slope of AB will exactly equal the slope of TT′. Therefore, MR will be equal to the slope of the TR function at any given output.

We know that the slope of a function can be found by differentiation and so it must be the case that

$$\text{MR} = \frac{d\text{TR}}{dq}$$

Example 8.15

Given that TR $= 80q - 2q^2$, derive a function for MR.

Solution

$$\text{MR} = \frac{d\text{TR}}{dq} = 80 - 4q$$

This result helps to explain some of the properties of the relationship between TR and MR. The linear demand schedule D in Figure 8.3 represents the function

$$p = 80 - 2q \tag{1}$$

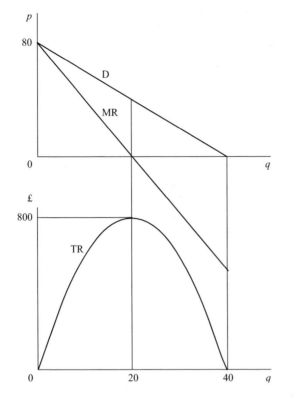

Figure 8.3

We know that by definition TR $= pq$. Therefore, substituting (1) for p,

$$TR = (80 - 2q)q = 80q - 2q^2$$

which is the same as the TR function in Example 8.15 above. This TR function is plotted in the lower section of Figure 8.3 and the function for MR, already derived, is plotted in the top section.

You can see that when TR is rising, MR is positive, as one would expect, and when TR is falling, MR is negative. As the rate of increase of TR gets smaller so does the value of MR. When TR is at its maximum, MR is zero.

With the function for MR derived above it is very straightforward to find the exact value of the output at which TR is a maximum. The TR function is horizontal at its maximum point and its slope is zero and so MR is also zero. Thus when TR is at its maximum

$$MR = 80 - 4q = 0$$
$$80 = 4q$$
$$20 = q$$

One can also see that the MR function has the same intercept on the vertical axis as this straight line demand schedule, but twice its slope. We can show that this result holds for any linear downward-sloping demand schedule.

For any linear demand schedule in the format

$$p = a - bq$$

$$\text{TR} = pq = (a - bq)q = aq - bq^2$$

$$\text{MR} = \frac{\text{dTR}}{\text{d}q} = a - 2bq$$

Thus both the demand schedule and the MR function have a as the intercept on the vertical axis, and the slope of MR is $2b$ which is obviously twice the demand schedule's slope.

It should also be noted that this result does not hold for non-linear demand schedules. If a demand schedule is non-linear then it is best to derive the slope of the MR function from first principles.

Example 8.16

Derive the MR function for the non-linear demand schedule $p = 80 - q^{0.5}$.

Solution

$$\text{TR} = pq = \left(80 - q^{0.5}\right)q = 80q - q^{1.5}$$

$$\text{MR} = \frac{\text{dTR}}{\text{d}q} = 80 - 1.5q^{0.5}$$

In this non-linear case the intercept on the price axis is still 80 but the slope of MR is 1.5 times the slope of the demand function.

For those of you who are still not convinced that the idea of looking at an 'infinitesimally small' change can help find the rate of change of a function at a point, Example 8.17 below shows how a spreadsheet can be used to calculate rates of change for very small increments. This example is for illustrative purposes only though. The main reason for using calculus in the first place is to enable the immediate calculation of rates of change at any point of a function.

Example 8.17

For the total revenue function

$$\text{TR} = 500q - 2q^2$$

find the value of MR when $q = 80$ (i) using calculus, and (ii) using a spreadsheet that calculates increments in q above the given value of 80 that get progressively smaller. Compare the two answers.

Solution

(i) $\text{MR} = \dfrac{d\text{TR}}{dq} = 500 - 4q$

Thus when $q = 80$

$\text{MR} = 500 - 4(80) = 500 - 320 = 180$

(ii) The spreadsheet shown in Table 8.2 can be constructed by following the instructions in Table 8.1. This spreadsheet shows that as increments in q (relative to the initial given value of 80) become smaller and smaller the value of MR (i.e. $\Delta\text{TR}/\Delta q$) approaches 180. This is consistent with the answer obtained by calculus in (i).

Table 8.1

CELL	Enter	Explanation
A1 to B4 and A6 to E6	*Enter labels as shown in Table* 8.2	Labels to indicate where initial values go plus columnheading labels
D2	TR = 500q - 2q^2	Label to remind you what function is used.
D3	80	Given initial value for q.
D4	=500*D3-2*D3^2	Calculates TR corresponding to given q value.
B7	10	Initial size of increment in q.
B8	=B7/10	Calculates an increment in q that is only 10% of the value of the one in cell above.
B9 to B13	*Copy cell* B8 *formula down column* B	Calculates a series of increments in q that get smaller and smaller each time.
A7	=B7+D$3	Calculates new value of q by adding the increment in cell A7 to the given value of 80.
A8 to A13	*Copy cell* A7 *formula down column* A	Calculates a series of values of q that increase by smaller and smaller increments each time.
C7	=500*A7-2*A7^2	Calculates TR corresponding to value of q in cell A7.
C8 to C13	*Copy cell* C7 *formula down column* C	Calculates a series of values of TR corresponding to values of q in row A.
D7	=C7-D$4	Calculates the change in TR relative to the initial given value in cell D4.
D8 to D13	*Copy cell* D7 *formula down column* D	Calculates series of changes in TR corresponding to increments in q in row B.
E7	=D7/B7	Calculates $\Delta\text{TR} / \Delta q$
E8 to E13	*Copy cell* E7 *formula down column* E	Calculates values of $\Delta\text{TR} / \Delta q$ corresponding to decreasing increments in q and TR
A7 to E13	*Widen columns and increase number of decimal places as necessary.*	The point of this example is to show how the value of $\Delta\text{TR} / \Delta q$ converges on dTR/dq so all the decimal places need to be shown.

Table 8.2

	A	B	C	D	E
1	Ex 8.17	DIFFERENTIATION OF TR FUNCTION			
2		GIVEN FUNCTION	TR = 500q - 2q^2		
3		INITIAL q VALUE =	80		
4		INITIAL TR VALUE =	27200		
5					Marginal Revenue
6	q	Delta q	TR	Delta TR	(DeltaTR)/(Delta q)
7	90	10	28800	1600	160
8	81	1	27378	178	178
9	80.1	0.1	27217.98	17.98	179.8
10	80.01	0.01	27201.7998	1.7998	179.98
11	80.001	0.001	27200.18	0.179998	179.998
12	80.0001	0.0001	27200.018	0.01799998	179.9998
13	80.00001	0.00001	27200.0018	0.001800000	179.9999802

Test Yourself, Exercise 8.3

1. Given the demand schedule $p = 120 - 3q$ derive a function for MR and find the output at which TR is a maximum.
2. For the demand schedule $p = 40 - 0.5q$ find the value of MR when $q = 15$.
3. Find the output at which MR is zero when $p = 720 - 4q^{0.5}$ describes the demand schedule.
4. A firm knows that the demand function for its output is $p = 400 - 0.5q$. What price should it charge to maximize sales revenue?
5. Make up your own demand function and then derive the corresponding MR function and find the output level which corresponds to zero marginal revenue.

8.4   Marginal cost and total cost

Just as MR can be shown to be the rate of change of the TR function, so marginal cost (MC) is the rate of change of the total cost (TC) function. In fact, in nearly all situations where one is dealing with the concept of a marginal increase, the marginal function is equal to the rate of change of the original function, i.e. to derive the marginal function one just differentiates the original function.

Example 8.18

Given $TC = 6 + 4q^2$ derive the MC function.

Solution

$$MC = \frac{dTC}{dq} = 8q$$

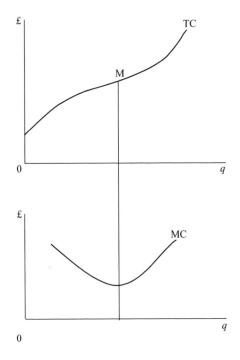

Figure 8.4

The example above is somewhat unrealistic in that it assumes an MC function that is a straight line. This is because the TC function is given as a simple quadratic function, whereas one normally expects a TC function to have a shape similar to that shown in Figure 8.4. This represents a cubic function with certain properties to ensure that:

(a) the rate of change of TC first falls and then rises, and
(b) TC never actually falls as output increases, i.e. MC is never negative. (Although it is quite common to find economies of scale causing *average* costs to fall, no firm is going to find the *total* cost of production falling when output increases.)

The flattest point of this TC schedule is at M, which corresponds to the minimum value of MC.

A cubic total cost function has the above properties if

$$TC = aq^3 + bq^2 + cq + d$$

where a, b, c and d are parameters such that

$$a, c, d > 0, b < 0 \quad \text{and} \quad b^2 < 3ac.$$

This applies to the TC functions in the examples below.

Example 8.19

If $TC = 2.5q^3 - 13q^2 + 50q + 12$ derive the MC function.

Solution

$$MC = \frac{dTC}{dq} = 7.5q^2 - 26q + 50$$

Example 8.20

When will average variable cost be at its minimum value for the TC function.

$$TC = 40 + 82q - 6q^2 + 0.2q^3?$$

Solution

The theory of costs tells us that MC will cut the minimum point of both the average cost (AC) and the average variable cost (AVC) functions. We therefore need to derive the MC and AVC functions and find where they intersect.

It is obvious from this TC function that total fixed costs TFC = 40 and total variable costs TVC = $82q - 6q^2 + 0.2q^3$. Therefore,

$$AVC = \frac{TVC}{q} = 82 - 6q + 0.2q^2$$

and

$$MC = \frac{dTC}{dq} = 82 - 12q + 0.6q^2$$

Setting MC = AVC

$$82 - 12q + 0.6q^2 = 82 - 6q + 0.2q^2$$

$$0.4q^2 = 6q$$

$$q = \frac{6}{0.4} = 15$$

at the minimum point of AVC.

(When you have covered the analysis of maximization and minimization in the next chapter, come back to this example and see if you can think of another way of solving it.)

Test Yourself, Exercise 8.4

1. If $TC = 65 + q^{1.5}$ what is MC when $q = 25$?
2. Derive a formula for MC if $TC = 4q^3 - 20q^2 + 60q + 40$.
3. If $TC = 0.5q^3 - 3q^2 + 25q + 20$ derive functions for: (a) MC, (b) AC, (c) the slope of AC.
4. What is special about MC if $TC = 25 + 0.8q$?
5. Make up your own TC function and then derive the corresponding MC function.

8.5   Profit maximization

We are now ready to see how calculus can help a firm to maximize profits, as the following examples illustrate. At this stage we shall just use the MC = MR rule for profit maximization. The second condition (MC cuts MR from below) will be dealt with in the next chapter.

Example 8.21

A monopoly faces the demand schedule $p = 460 - 2q$

and the cost schedule $TC = 20 + 0.5q^2$

How much should it sell to maximize profit and what will this maximum profit be? (All costs and prices are in £.)

Solution

To find the output where MC = MR we first need to derive the MC and MR functions.

Given $TC = 20 + 0.5q^2$

then $MC = \dfrac{dTC}{dq} = q$ \hfill (1)

As $TR = pq = (460 - 2q)q = 460q - 2q^2$

then $MR = \dfrac{dTR}{dq} = 460 - 4q$ \hfill (2)

To maximize profit MR = MC. Therefore, equating (1) and (2),

$$460 - 4q = q$$
$$460 = 5q$$
$$92 = q$$

The actual maximum profit when the output is 92 will be

$$
\begin{aligned}
TR - TC &= (460q - 2q^2) - (20 + 0.5q^2) \\
&= 460q - 2q^2 - 20 - 0.5q^2 \\
&= 460q - 2.5q^2 - 20 \\
&= 460(92) - 2.5(8,464) - 20 \\
&= 42,320 - 21,160 - 20 = £21,140
\end{aligned}
$$

Example 8.22

A firm faces the demand schedule $p = 184 - 4q$

and the TC function $TC = q^3 - 21q^2 + 160q + 40$

What output will maximize profit?

Solution

Given $TR = pq = (184 - 4q)q = 184q - 4q^2$

then $MR = \dfrac{dTR}{dq} = 184 - 8q$

$MC = \dfrac{dTC}{dq} = 3q^2 - 42q + 160$

To maximize profits MC = MR. Therefore,

$3q^2 - 42q + 160 = 184 - 8q$

$3q^2 - 34q - 24 = 0$

$(q - 12)(3q + 2) = 0$

$q - 12 = 0$ or $3q + 2 = 0$

$q = 12$ or $q = -\dfrac{2}{3}$

One cannot produce a negative quantity and so the firm must produce 12 units of output in order to maximize profits.

Test Yourself, Exercise 8.5

1.   A monopoly faces the following TR and TC schedules:

$TR = 300q - 2q^2$

$TC = 12q^3 - 44q^2 + 60q + 30$

What output should it sell to maximize profit?

2.   A firm faces the demand function $p = 190 - 0.6q$
and the total cost function $TC = 40 + 30q + 0.4q^2$

(a) What output will maximize profit?
(b) What output will maximize total revenue?
(c) What will the output be if the firm makes a profit of £4,760?

3. A firm's total revenue and total cost functions are

$$TR = 52q - q^2$$

$$TC = \frac{q^3}{3} - 2.5q^2 + 34q + 4$$

At what output will profit be maximized?

8.6 Respecifying functions

Many of the examples considered so far have included a demand schedule in the format

$$p = a + bq$$

although, as was explained in Chapter 4, economic theory normally defines a demand function in the format $q = f(p)$, with q being the dependent variable rather than p. However, because the usual convention is to have p on the vertical axis in supply and demand graphical analysis, and also because cost functions have q as the independent variable, it usually helps to work with the inverse demand function $p = f(q)$. The examples below show how to derive the relationship between MR and q by finding the inverse demand function.

Example 8.23

Derive the MR function for the demand function $q = 400 - 0.1p$.

Solution

Given $\qquad q = 400 - 0.1p$

$$10q = 4{,}000 - p$$

$$p = 4{,}000 - 10q$$

Using this inverse demand function we can now derive

$$TR = pq = (4{,}000 - 10q)q = 4{,}000q - 10q^2$$

$$MR = \frac{dTR}{dq} = 4{,}000 - 20q$$

Example 8.24

A firm faces the demand schedule $\qquad q = 200 - 4p$

and the cost schedule $\quad TC = 0.1q^3 - 0.5q^2 + 2q + 8$

What price will maximize profit?

Solution

The demand function $q = 200 - 4p$

can be rewritten as $p = 50 - 0.25q$

This is a linear demand schedule and so MR has the same intercept and twice the slope. Thus

$\text{MR} = 50 - 0.5q$

From the TC function

$\text{MC} = \dfrac{\text{dTC}}{\text{d}q} = 0.3q^2 - q + 2$

To maximize profits MC = MR. Therefore, equating the MR and MC functions already derived

$0.3q^2 - q + 2 = 50 - 0.5q$

$0.3q^2 - 0.5q - 48 = 0$

Using the formula for the solution of quadratic equations

$q = \dfrac{-(-0.5) \pm \sqrt{(-0.5)^2 - 4 \times 0.3 \times (-48)}}{2 \times 0.3}$

$= \dfrac{0.5 \pm \sqrt{0.25 + 57.6}}{0.6}$

$= \dfrac{0.5 \pm \sqrt{57.85}}{0.6}$

Disregarding the negative solution as output cannot be negative

$q = \dfrac{0.5 + 7.6}{0.6} = \dfrac{8.1}{0.6} = 13.5$

Substituting this output into the demand function

$p = 50 - 0.25q = 50 - 3.375 = 46.625$

Test Yourself, Exercise 8.6

1. Given the demand function $q = 150 - 3p$, derive a function for MR.
2. A firm faces the demand schedule $q = 40 - p^{0.5}$ (where $p^{0.5} \geq 0, q \leq 40$) and the cost schedule $\text{TC} = q^3 - 2.5q^2 + 50q + 16$. What price should it charge to maximize profit?
3. Find the MR function corresponding to the demand schedule $q = (60 - 2.5p)^{0.5}$.

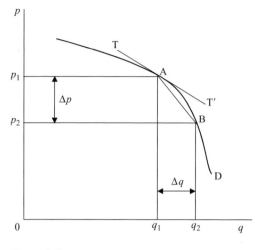

Figure 8.5

8.7   Point elasticity of demand

Price elasticity of demand is defined as

$$e = (-1)\frac{\text{percentage change in quantity}}{\text{percentage change in price}}$$

However, looking at the changes in price and quantity between points A and B on the demand schedule D in Figure 8.5, the question you may ask is 'percentage of what'? Clearly the change in quantity Δq is a much larger percentage of q_1 than of the larger quantity q_2. Although arc elasticity gives an approximate 'average' measure, a more precise measure can be obtained by finding the elasticity of demand at a single point on the demand schedule. In Chapter 4 some simple examples of point elasticity based on linear demand schedules were considered. With the aid of calculus we can now also derive point elasticity for non-linear demand schedules.

If the movement along D from A to B in Figure 8.5 is very small then we can assume $p_1 = p_2 = p$ and $q_1 = q_2 = q$ and so

$$e = (-1)\frac{\Delta q/q}{\Delta p/p} = (-1)\frac{p}{q}\frac{1}{\Delta p/\Delta q} \tag{1}$$

As B gets nearer to A the value of $\Delta p/\Delta q$, which is the slope of the straight line AB, gets closer to the slope of the tangent TT′ at A. (Note that, as price falls in this example, Δp is negative, giving a negative value for the relevant slopes.) Thus for an infinitesimally small movement from A

$$\frac{\Delta p}{\Delta q} = \frac{\mathrm{d}p}{\mathrm{d}q} = \text{slope of D at A}$$

Thus, substituting this result into (1) above, the formula for point elasticity of demand becomes

$$e = (-1)\frac{p}{q}\frac{1}{\mathrm{d}p/\mathrm{d}q}$$

Example 8.25

What is point elasticity when price is 12 for the demand function $p = 60 - 3q$?

Solution

$$\frac{\mathrm{d}p}{\mathrm{d}q} = -3$$

Given $p = 60 - 3q$, then

$$3q = 60 - p$$

$$q = \frac{60 - p}{3}$$

When $p = 12$, then

$$q = \frac{60 - 12}{3} = \frac{48}{3} = 16$$

Therefore,

$$e = (-1)\frac{p}{q}\frac{1}{\mathrm{d}p/\mathrm{d}q} = (-1)\frac{12}{16} \times \frac{1}{-3} = 0.25$$

Example 8.26

What is elasticity of demand when quantity is 8 if a firm's demand function is $q = 60 - 2p^{0.5}$ (where $p^{0.5} \geq 0, q \leq 60$)?

Solution

Deriving the inverse of the demand function

$$q = 60 - 2p^{0.5}$$

$$2p^{0.5} = 60 - q$$

$$p^{0.5} = 30 - 0.5q$$

$$p = (30 - 0.5q)^2 = 900 - 30q + 0.25q^2$$

Therefore,

$$\frac{dp}{dq} = -30 + 0.5q$$

When $q = 8$,

$$\frac{dp}{dq} = -30 + 0.5(8) = -30 + 4 = -26$$

Also, when $q = 8$,

$$p = 900 - 30(8) + 0.25(8)^2$$
$$= 900 - 240 + 16 = 676$$

Thus

$$e = (-1)\frac{p}{q}\frac{1}{dp/dq} = (-1)\frac{676}{8} \times \frac{1}{-26} = 3.25$$

Test Yourself, Exercise 8.7

1. What is the point elasticity of demand when price is 20 for the demand schedule $p = 45 - 1.5q$?
2. Explain why the point elasticity of demand decreases in value as one moves down a straight line demand schedule.
3. Given the demand function $q = (1,200 - 2p)^{0.5}$, what is elasticity of demand when quantity is 30?
4. Explain why the demand function $q = 265p^{-1}$ will have the same point elasticity of demand at all prices and say what its value is.

8.8 Tax yield

Elementary supply and demand analysis tells us that the effect of a per-unit tax t on a good sold in a competitive market will effectively shift up the supply schedule vertically by the amount of the tax. This will cause the price paid by consumers to rise and the quantity bought to fall. The change in total revenue spent by consumers will depend on the price elasticity of demand.

The Chancellor of the Exchequer, however, is more interested in the total amount of tax raised for the government, or the tax yield (TY), than total consumer expenditure. If a per-unit tax is increased, the quantity bought will always fall. The question, however, is whether or not this fall in quantity will outweigh the effect on TY of the increase in the amount of tax raised on each unit. To answer this we need to know the rate of change of the tax yield with respect to increases in the per-unit tax.

Example 8.27

A market has the demand schedule $p = 92 - 2q$ and the supply schedule $p = 12 + 3q$. What per-unit tax will raise the maximum tax revenue for the government? (All prices are in £.)

Solution

Let the per-unit tax be t. This changes the supply schedule to

$$p = 12 + t + 3q$$

i.e. the intercept on the price axis shifts vertically upwards by the amount t.
We now need to derive a function for q in terms of the tax t. In equilibrium, supply price equals demand price. Therefore,

$$12 + 3q + t = 92 - 2q$$
$$5q = 80 - t$$
$$q = 16 - 0.2t$$

The tax yield is (amount sold) × (per-unit tax). Therefore,

$$\text{TY} = qt = (16 - 0.2t)t = 16t - 0.2t^2$$

and so the rate of change of TY with respect to t is

$$\frac{d\text{TY}}{dt} = 16 - 0.4t$$

If $d\text{TY}/dt > 0$, an increase in t will increase TY. However, from the formula for $d\text{TY}/dt$ derived above, one can see that as the amount of the tax t is increased the value of $d\text{TY}/dt$ falls. Therefore in order to maximize TY, t should be increased until $d\text{TY}/dt = 0$. Any further increases in t would cause $d\text{TY}/dt$ to become negative and cause TY to start to fall. Thus

$$\frac{d\text{TY}}{dt} = 16 - 0.4t = 0$$
$$16 = 0.4t$$
$$40 = t$$

Therefore a per-unit tax of £40 will maximize the tax yield.

Rather than working from first principles, as in the above example, a general formula can be derived for the rate of change of the tax yield with respect to a per-unit tax if both demand and supply schedules are linear. Assume that these schedules are:

$$\text{demand } p = a + bq \qquad \text{supply } p = c + dq$$

where a, b, c and d are parameters (note that we expect $b < 0$).
With a per-unit tax of t, the supply schedule becomes

$$p = c + dq + t$$

Setting supply price equal to demand price we can derive the reduced form equation for TY in terms of the independent variable t and then differentiate it to find the comparative static effect of a change in t.

$$c + dq + t = a + bq$$

$$q(d - b) = a - c - t$$

$$q = \frac{a - c}{d - b} - \frac{t}{d - b}$$

$$\text{TY} = qt = \left(\frac{a - c}{d - b}\right) t - \frac{t^2}{d - b}$$

$$\frac{\text{dTY}}{\text{d}t} = \frac{a - c}{d - b} - \frac{2t}{d - b} \tag{1}$$

We can check this formula using the figures from Example 8.27 above. Given the demand schedule $p = 92 - 2q$ and the supply schedule $p = 12 + 3q$, then

$$a = 92 \quad b = -2 \quad c = 12 \quad d = 3$$

Substituting these values into (1) above

$$\frac{\text{dTY}}{\text{d}t} = \frac{92 - 12}{3 - (-2)} - \frac{2t}{5} = \frac{80}{5} - \frac{2t}{5} = 16 - 0.4t$$

This is the same as the function derived from first principles in Example 8.27.

Test Yourself, Exercise 8.8

1. Given the demand schedule $p = 180 - 8q$ and the supply schedule $p = 25 + 2q$, what level of per-unit tax would maximize the government's tax yield?
2. Change one of the parameters in Question 1 above and work out the new answer.
3. Assume a market has the demand function $q = 40 - 0.5p$ and the supply function $q = 2p - 4$. The government currently imposes a per-unit tax of £3. If this tax is slightly increased will the tax yield rise or fall?

8.9   The Keynesian multiplier

In a simple Keynesian macroeconomic model with no government sector and no foreign trade, it is assumed that

$$Y = C + I \tag{1}$$

$$C = a + bY \tag{2}$$

where Y is national income, C is consumption and I is investment, exogenously fixed, and a and b are parameters.

The marginal propensity to consume (MPC) is the rate of change of consumption as national income increases, which is equal to $\text{d}C/\text{d}Y = b$. The multiplier is the rate of change of national income in response to an increase in exogenously determined investment, i.e. dY/dI. The result that the multiplier is equal to

$$\frac{1}{1 - \text{MPC}}$$

can be easily derived by differentiation.

Substituting (2) into (1) we get

$$Y = a + bY + I$$

$$Y(1 - b) = a + I$$

$$Y = \frac{a + I}{1 - b} = \frac{a}{1 - b} + \frac{I}{1 - b}$$

Therefore

$$\frac{dY}{dI} = \frac{1}{1 - b}$$

which is the formula for the multiplier.

This multiplier can be used to calculate the increase in investment necessary to achieve any specified increase in national income.

Example 8.28

In a basic Keynesian macroeconomic model it is assumed that $Y = C + I$ where $I = 250$ and $C = 0.75Y$. What is the equilibrium level of Y? What increase in I would be needed to cause Y to increase to 1,200?

Solution

$$Y = C + I = 0.75Y + 250$$

$$0.25Y = 250$$

Equilibrium level $Y = 1,000$.

For any increase (ΔI) in I the resulting increase(ΔY) in Y will be determined by the formula

$$\Delta Y = K \Delta I \tag{1}$$

where K is the multiplier. We know that

$$K = \frac{1}{1 - \text{MPC}}$$

In this example, MPC $= dC/dY = 0.75$. Therefore,

$$K = \frac{1}{1 - 0.75} = \frac{1}{0.25} = 4 \tag{2}$$

The required change in Y is

$$\Delta Y = 1,200 - 1,000 = 200 \tag{3}$$

Therefore, substituting (2) and (3) into (1),

$$200 = 4\Delta I$$
$$\Delta I = 50$$

This is the required increase in I.

Multipliers for other exogenous variables in more complex macroeconomic models can be derived using the same method. However, for differentiation with respect to one exogenous variable the other variables must remain constant and so we shall return to this topic in Chapter 10 when partial differentiation is explained.

Test Yourself, Exercise 8.9

1. In a basic Keynesian macroeconomic model it is assumed that $Y = C + I$ where $I = 820$ and $C = 60 + 0.8Y$.

    (a) What is the marginal propensity to consume?
    (b) What is the equilibrium level of Y?
    (c) What is the value of the multiplier?
    (d) What increase in I is required to increase Y to 5,000?
    (e) If this increase takes place will savings $(Y - C)$ still equal I?

9 Unconstrained optimization

9.1 First-order conditions for a maximum

Consider the total revenue function

$$TR = 60q - 0.2q^2$$

This will take an inverted U-shape similar to that shown in Figure 9.1. If we ask the question 'when is TR at its maximum?' the answer is obviously at M, which is the highest point on the curve. At this maximum position the TR schedule is flat. To the left of M, TR is rising and has a positive slope, and to the right of M, the TR schedule is falling and has a negative slope. At M itself the slope is zero.

We can therefore say that for a function *of this shape* the maximum point will be where its slope is zero. This zero slope requirement is a necessary *first-order condition* for a maximum.

Zero slope will not guarantee that a function is at a maximum, as explained in the next section where the necessary additional second-order conditions are explained. However, in this particular example we know for certain that zero slope corresponds to the maximum value of the function.

In Chapter 8, we learned that the slope of a function can be obtained by differentiation. So, for the function

$$TR = 60q - 0.2q^2$$

$$slope = \frac{dTR}{dq} = 60 - 0.4q$$

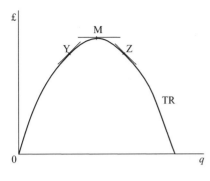

Figure 9.1

The slope is zero when

$$60 - 0.4q = 0$$
$$60 = 0.4q$$
$$150 = q$$

Therefore TR is maximized when quantity is 150.

Test Yourself, Exercise 9.1

1. What output will maximize total revenue if $TR = 250q - 2q^2$?
2. If a firm faces the demand schedule $p = 90 - 0.3q$ how much does it have to sell to maximize sales revenue?
3. A firm faces the total revenue schedule $TR = 600q - 0.5q^2$

    (a) What is the marginal revenue when q is 100?
    (b) When is the total revenue at its maximum?
    (c) What price should the firm charge to achieve this maximum TR?

4. For the non-linear demand schedule $p = 750 - 0.1q^2$ what output will maximize the sales revenue?

9.2 Second-order condition for a maximum

In the example in Section 9.1, it was obvious that the TR function was a maximum when its slope was zero because we knew the function had an inverted U-shape. However, consider the function in Figure 9.2(a). This has a slope of zero at N, but this is its minimum point not its maximum. In the case of the function in Figure 9.2(b) the slope is zero at I, but this is neither a maximum nor a minimum point.

The examples in Figure 9.2 clearly illustrate that although a zero slope is **necessary** for a function to be at its maximum it is not a **sufficient** condition. A zero slope just means that the function is at what is known as a 'stationary point', i.e. its slope is neither increasing nor decreasing. Some stationary points will be turning points, i.e. the slope changes from positive

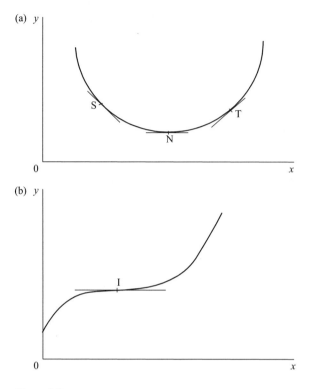

Figure 9.2

to negative (or vice versa) at these points, and will be maximum (or minimum) points of the function.

In order to find out whether a function is at a maximum or a minimum or a point of inflexion (as in Figure 9.2(b)) when its slope is zero we have to consider what are known as the *second-order* conditions. (The first-order condition for any of the three forms of stationary point is that the slope of the function is zero.)

The second-order conditions tell us what is happening to the rate of change of the slope of the function. If the rate of change of the slope is negative it means that the slope decreases as the variable on the horizontal axis is increased. If the slope is decreasing and one is at a point where the actual slope is zero this means that the slope of the function is positive slightly to the left and negative slightly to the right of this point. This is the case in Figure 9.1. The slope is positive at Y, zero at M and negative at Z. Thus, if the rate of change of the slope of a function is negative at the point where the actual slope is zero then that point is a maximum.

This is the second-order condition for a maximum. Until now, we have just assumed that a function is maximized when its slope is zero if a sketch graph suggests that it takes an inverted U-shape. From now on we shall make this more rigorous check of the second-order conditions to confirm whether a function is maximized at any stationary point.

It is a straightforward exercise to find the rate of change of the slope of a function. We know that the slope of a function $y = f(x)$ can be found by differentiation. Therefore if we differentiate the function for the slope of the original function, i.e. dy/dx, we get the rate of change of the slope. This is known as the **second-order derivative** and is written d^2y/dx^2.

Example 9.1

Show that the function $y = 60x - 0.2x^2$ satisfies the second-order condition for a maximum when $x = 150$.

Solution

The slope of this function will be zero at a stationary point. Therefore

$$\frac{dy}{dx} = 60 - 0.4x = 0 \qquad (1)$$

$$x = 150$$

Therefore the first-order condition for a maximum is met when x is 150.

To get the rate of change of the slope we differentiate (1) with respect to x again, giving

$$\frac{d^2y}{dx^2} = -0.4$$

This second-order derivative will always be negative, whatever the value of x. Therefore, the second-order condition for a maximum is met and so y must be a maximum when x is 150.

In the example above, the second-order derivative did not depend on the value of x at the function's stationary point, but for other functions the value of the second-order derivative may depend on the value of the independent variable.

Example 9.2

Show that TR is a maximum when q is 18 for the non-linear demand schedule.

$$p = 194.4 - 0.2q^2$$

Solution

$$TR = pq = (194.4 - 0.2q^2)q = 194.4q - 0.2q^3$$

For a stationary point on this cubic function the slope must be zero and so

$$\frac{dTR}{dq} = 194.4 - 0.6q^2 = 0$$

$$194.4 = 0.6q^2$$

$$324 = q^2$$

$$18 = q$$

When q is 18 then the second-order derivative is

$$\frac{d^2TR}{dq^2} = -1.2q = -1.2(18) = -21.6 < 0$$

Therefore, second-order condition for a maximum is satisfied and TR is a maximum when q is 18. (Note that in this example the second-order derivative $-1.2q < 0$ for any positive value of q.)

Test Yourself, Exercise 9.2

Find stationary points for the following functions and say whether or not they are at their maximum at these points.

1. $TR = 720q - 0.3q^2$
2. $TR = 225q - 0.12q^3$
3. $TR = 96q - q^{1.5}$
4. $AC = 51.2q^{-1} + 0.4q^2$

9.3   Second-order condition for a minimum

By the same reasoning as that set out in Section 9.2 above, if the rate of change of the slope of a function is positive at the point when the slope is zero then the function is at a minimum. This is illustrated in Figure 9.2(a). The slope of the function is negative at S, zero at N and positive at T. As the slope changes from negative to positive, the rate of change of this slope must be positive at the stationary point N.

Example 9.3

Find the minimum point of the average cost function $AC = 25q^{-1} + 0.1q^2$

Solution

The slope of the AC function will be zero when

$$\frac{dAC}{dq} = -25q^{-2} + 0.2q = 0 \tag{1}$$

$$0.2q = 25q^{-2}$$

$$q^3 = 125$$

$$q = 5$$

The rate of change of the slope at this point is found by differentiating (1), giving the second-order derivative

$$\frac{d^2AC}{dq^2} = 50q^{-3} + 0.2$$

$$= \frac{50}{125} + 0.2 \text{ when } q = 5$$

$$= 0.4 + 0.2 = 0.6 > 0$$

Therefore the second-order condition for a minimum value of AC is satisfied when q is 5.

The actual value of AC at its minimum point is found by substituting this value for q into the original AC function. Thus

$$AC = 25q^{-1} + 0.1q^2 = \tfrac{25}{5} + 0.1 \times 25 = 5 + 2.5 = 7.5$$

Test Yourself, Exercise 9.3

Find whether any stationary points exist for the following functions for positive values of q, and say whether or not the stationary points are at the minimum values of the function.

1. $AC = 345.6q^{-1} + 0.8q^2$
2. $AC = 600q^{-1} + 0.5q^{1.5}$
3. $MC = 30 + 0.4q^2$
4. $TC = 15 + 27q - 9q^2 + q^3$
5. $MC = 8.25q$

9.4 Summary of second-order conditions

If $y = f(x)$ and there is a stationary point where $\dfrac{dy}{dx} = 0$, then

(i) this point is a *maximum* if $\dfrac{d^2y}{dx^2} < 0$

(ii) this point is a *minimum* if $\dfrac{d^2y}{dx^2} > 0$

Strictly speaking, (i) and (ii) are conditions for *local* maximums and minimums. It is possible, for example, that a function may take a shape such as that shown in Figure 9.3. This has no true global maximum or minimum, as values of y continue towards plus and minus infinity as shown by the arrows. Points M and N, which satisfy the above second-order conditions for maximum and minimum, respectively, are therefore just local maximum and minimum points. However, for most of the examples that you are likely to encounter in economics any local maximum (or minimum) points will also be global maximum (or minimum) points and so you need not worry about this distinction. If you are uncertain then you can always plot a function using Excel to see the pattern of turning points.

If $d^2y/dx^2 = 0$ there may be an inflexion point that is neither a maximum nor a minimum, such as I in Figure 9.2(b). To check if this is so one really needs to investigate further, looking at the third, fourth and possibly higher order derivatives for more complex polynomial functions. However, we will not go into these conditions here. In all the economic applications given in this text, it will be obvious whether or not a function is at a maximum or minimum at any stationary points.

Some functions do not have maximum or minimum points. Linear functions are obvious examples as they cannot satisfy the first-order conditions for a turning point, i.e. that $dy/dx = 0$, except when they are horizontal lines. Also, the slope of a straight line is always

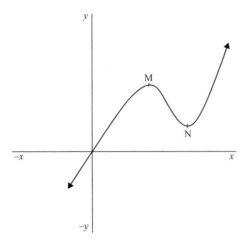

Figure 9.3

a constant and so the second-order derivative, which represents the rate of change of the slope, will always be zero.

Example 9.4

In Chapter 5 we considered an example of a break-even chart where a firm was assumed to have the total cost function TC $= 18q$ and the total revenue function TR $= 240 + 14q$. Show that the profit-maximizing output cannot be determined for this firm.

Solution

The profit function will be

$$\pi = \text{TR} - \text{TC}$$
$$= 240 + 14q - 18q$$
$$= 240 - 4q$$

Its rate of change with respect to q will be

$$\frac{d\pi}{dq} = -4 \tag{1}$$

There is obviously no output level at which the first-order condition that $d\pi/dq = 0$ can be met and so no stationary point exists. Therefore the profit-maximizing output cannot be determined.

End-point solutions

There are some possible exceptions to these first- and second-order conditions for maximum and minimum values of functions. If the domain of a function is restricted, then a maximum or minimum point may be determined by this restriction, giving what is known as an 'end-point' or 'corner' solution. In such cases, the usual rules for optimization set out in this chapter will not apply. For example, suppose a firm faces the total cost function (in £)

$$TC = 45 + 18q - 5q^2 + q^3$$

For a stationary point its slope will be

$$\frac{dTC}{dq} = 18 - 10q + 3q^2 = 0 \tag{1}$$

However, if we try using the quadratic equation formula to find a value of q for which (1) holds we see that

$$q = \frac{-b \pm \sqrt{b^2 - 4ac}}{2a} = \frac{-(-10) \pm \sqrt{10^2 - 4 \times 18 \times 3}}{2 \times 3} = \frac{10 \pm \sqrt{-116}}{6}$$

We cannot find the square root of a negative number and so no solution exists. There is no turning point as no value of q corresponds to a zero slope for this function.

However, if the domain of q is restricted to non-negative values then TC will be at its minimum value of £45 when $q = 0$. Mathematically the conditions for minimization are not met at this point but, from a practical viewpoint, the minimum cost that this firm can ever face is the £45 it must pay even if nothing is produced. This is an example of an end-point solution.

Therefore, when tackling problems concerned with the minimization or maximization of economic variables, you need to ask whether or not there are restrictions on the domain of the variable in question which may give an end-point solution.

Test Yourself, Exercise 9.4

1. A firm faces the demand schedule $p = 200 - 2q$ and the total cost function

$$TC = \tfrac{2}{3}q^3 - 14q^2 + 222q + 50$$

Derive expressions for the following functions and find out whether they have maximum or minimum points. If they do, say what value of q this occurs at and calculate the actual value of the function at this output.

(a) Marginal cost
(b) Average variable cost
(c) Average fixed cost
(d) Total revenue
(e) Marginal revenue
(f) Profit

2. Construct your own example of a function that has a turning point. Check the second-order conditions to confirm whether this turning point is a maximum or a minimum.

3. A firm attempting to expand output in the short-run faces the total product of labour schedule $TP_L = 24L^2 - L^3$. At what levels of L will (a) TP_L, (b) MP_L, and (c) AP_L be at their maximum levels?

4. Using your knowledge of economics to apply appropriate restrictions on their domain, say whether or not the following functions have maximum or minimum points.

    (a) $TC = 12 + 62q - 10q^2 + 1.2q^3$
    (b) $TC = 6 + 2.5q$
    (c) $p = 285 - 0.4q$

9.5   Profit maximization

We have already encountered some problems involving the maximization of a profit function. As profit maximization is one of the most common optimization problems that you will encounter in economics, in this section we shall carefully work through the second-order condition for profit maximization and see how it relates to the different intersection points of a firm's MC and MR schedules.

Consider the firm whose marginal cost and marginal revenue schedules are shown by MC and MR in Figure 9.4. At what output will profit be maximized?

The first rule for profit maximization is that profits are at a maximum when MC = MR. However, there are two points, X and Y, where MC = MR. Only X satisfies the second rule for profit maximization, which is that MC cuts MR from below at the point of intersection. This corresponds to the second-order condition for a maximum required by the differential calculus, as illustrated in the following example.

Example 9.5

Find the profit-maximizing output for a firm with the total cost function

$$TC = 4 + 97q - 8.5q^2 + 1/3q^3$$

and the total revenue function

$$TR = 58q - 0.5q^2.$$

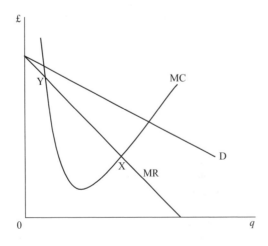

Figure 9.4

Solution

First let us derive the MC and MR functions and see where they intersect.

$$\text{MC} = \frac{\text{dTC}}{\text{d}q} = 97 - 17q + q^2 \tag{1}$$

$$\text{MR} = \frac{\text{dTR}}{\text{d}q} = 58 - q \tag{2}$$

Therefore, when MC = MR

$$97 - 17q + q^2 = 58 - q$$

$$39 - 16q + q^2 = 0 \tag{3}$$

$$(3 - q)(13 - q) = 0$$

Thus $q = 3$ or $q = 13$

These are the two outputs at which the MC and MR schedules intersect, but which one satisfies the second rule for profit maximization? To answer this question, the problem can be reformulated by deriving a function for profit and then trying to find its maximum. Thus, profit will be

$$\pi = \text{TR} - \text{TC} = 58q - 0.5q^2 - (4 + 97q - 8.5q^2 + 1/3q^3)$$

$$= 58q - 0.5q^2 - 4 - 97q + 8.5q^2 - 1/3q^3$$

$$= -39q + 8q^2 - 4 - 1/3q^3$$

Differentiating and setting equal to zero

$$\frac{\text{d}\pi}{\text{d}q} = -39 + 16q - q^2 = 0 \tag{4}$$

$$0 = 39 - 16q + q^2 \tag{5}$$

Equation (5) is the same as (3) above and therefore has the same two solutions, i.e. $q = 3$ or $q = 13$. However, using this method we can also explore the second-order conditions. From (4) we can derive the second-order derivative

$$\frac{\text{d}^2\pi}{\text{d}q^2} = 16 - 2q$$

When $q = 3$ then $\text{d}^2\pi/\text{d}q^2 = 16 - 6 = 10$ and so π is a minimum.

When $q = 13$ then $\text{d}^2\pi/\text{d}q^2 = 16 - 26 = -10$ and so π is a maximum.

Thus only one of the intersection points of MR and MC satisfies the second-order conditions for a maximum and corresponds to the profit-maximizing output. This will be where MC cuts MR from below. We can prove that this must be so as follows:

By differentiating (1) we get

$$\text{slope of MC} = \frac{\text{dMC}}{\text{d}q} = -17 + 2q$$

By differentiating (2) we get

$$\text{slope of MR} = \frac{\text{dMR}}{\text{d}q} = -1$$

When $q = 3$, then the slope of MC is

$$-17 + 2(3) = -17 + 6 = -11 < -1 \text{ (i.e. steeper negative slope than MR)}$$

When $q = 13$, then the slope of MC is

$$-17 + 2(13) = 9 \text{ (i.e. positive slope, greater slope than MR)}$$

Thus, when $q = 3$, the MC schedule has a steeper negative slope than MR and so must cut it from above. When $q = 13$, MC has a positive slope and so must cut MR from below.

Test Yourself, Exercise 9.5

1. A monopoly faces the total revenue schedule $\quad \text{TR} = 300q - 2q^2$
    and the total cost schedule $\quad \text{TC} = 12q^3 - 44q^2 + 60q + 30$
    Are there two output levels at which MC = MR? Which is the profit-maximizing output?

2. If a firm faces the demand schedule $\quad p = 120 - 3q$
    and the total cost schedule $\quad \text{TC} = 120 + 36q + 1.2q^2$

    what output levels, if any, will (a) maximize profit, and (b) minimize profit?

3. Explain why a firm which is a monopoly seller in a market with the demand schedule $p = 66.8 - 0.4q$ and which faces the total cost schedule

    $$\text{TC} = 220 + 120q - 12q^2 + 0.5q$$

    can never make a positive profit.

4. What is the maximum profit a firm can make if it faces the demand schedule $p = 660 - 3q$ and the total cost schedule $\text{TC} = 25 + 240q - 72q^2 + 6q^3$?

5. If a firm faces the demand schedule $p = 53.5 - 0.7q$, what *price* will maximize profits if its total cost schedule is $\text{TC} = 400 + 35q - 6q^2 + 0.1q^3$?

9.6 Inventory control

In Chapter 8, we considered a few applications of differentiation, such as tax yield maximization, without taking second-order conditions into account. We can now look at an application where it is not obvious that a function is maximized or minimized when its slope is zero and where second-order conditions must be fully investigated. This application analyses how the optimum order size can be calculated for a firm wishing to minimize ordering and storage costs.

A manufacturing company has to take into account costs other than the actual purchase price of the components that it uses. These include:

(a) Reorder costs: each order for a consignment of components will involve administration work, delivery, unloading etc.

(b) Storage costs: the more components a firm has in storage the more storage space will be needed. There is also the opportunity cost of the firm's capital which will be tied up in the components it has paid for.

If a firm only makes a few large orders its storage costs will be high but, on the other hand, if it makes lots of small orders its reorder costs will be high. How then can it decide on the optimum order size?

Assume that the total annual demand for components (Q) is evenly spread over the year. Assume that each order is of equal size q and that inventory levels are run down to zero before the next consignment arrives. Also assume that F is the fixed cost for making each order and S is the storage cost per-unit per year. If each consignment of size q is run down at a constant rate, then the average amount of stock held will be $q/2$. (This is illustrated in Figure 9.5 where t represents the time interval between orders.) Thus total storage costs for the year will be $(q/2)S$. The number of orders made in a year will be Q/q. Thus the total order costs for the year will be $(Q/q)F$.

The firm will wish to choose the order size that minimizes the total of order costs plus storage costs, defined as TC. The mathematical problem is therefore to find the value of q that minimizes

$$TC = \left(\frac{Q}{q}\right)F + \left(\frac{q}{2}\right)S$$

As Q, F and S are given constants, and remembering that $1/q$ is q^{-1}, differentiating with respect to q gives

$$\frac{dTC}{dq} = \frac{-QF}{q^2} + \frac{S}{2} \qquad (1)$$

For a stationary point

$$0 = -\frac{QF}{q^2} + \frac{S}{2}$$

$$\frac{QF}{q^2} = \frac{S}{2}$$

$$\frac{2QF}{S} = q^2$$

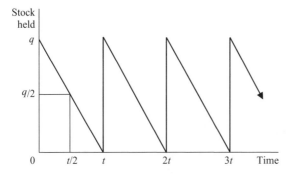

Figure 9.5

Therefore the optimal order size is

$$q = \sqrt{\frac{2QF}{S}} \qquad\qquad (2)$$

Thus q depends on the square root of the total annual demand Q when F and S are exogenously given.

The second-order conditions now need to be inspected to check that this turning point is a minimum. If (1) above is rewritten as

$$\frac{dTC}{dq} = -QFq^{-2} + \frac{S}{2}$$

then we can see that

$$\frac{d^2TC}{dq^2} = 2QFq^{-3} > 0$$

as Q, F and q must all be positive quantities.

Thus any positive value of q that satisfies the first-order condition (2) above must also satisfy the second-order condition for a minimum value of TC.

Example 9.6

A firm uses 200,000 units of a component in a year, with demand evenly spread over the year. In addition to the purchase price, each order placed for a batch of components costs £80. Each unit held in stock over a year costs £8. What is the optimum order size?

Solution

The optimum order size is q and so the average stock held is $q/2$. The number of orders is

$$\frac{Q}{q} = \frac{200,000}{q}$$

As each order made costs £80 and each unit stored for a year costs £8 then

$$TC = \text{order} + \text{stock-holding costs}$$
$$= \frac{200,000(80)}{q} + \frac{8q}{2}$$
$$= 16,000,000q^{-1} + 4q$$

For a stationary point

$$\frac{dTC}{dq} = -16,000,000q^{-2} + 4 = 0$$

$$4 = \frac{16,000,000}{q^2}$$

$$q^2 = \frac{16,000,000}{4} = 4,000,000$$

$$q = \sqrt{(4,000,000)} = 2,000$$

The second-order condition for a minimum is met at this stationary point as

$$\frac{d^2 TC}{dq^2} = 32{,}000{,}000 \, q^{-3} > 0 \quad \text{for any } q > 0$$

Therefore the optimum order size is 2,000 units.

We could, of course, have solved this problem by just substituting the given values into the formula for optimal order size (2) derived earlier. Thus

$$q = \sqrt{\frac{2QF}{S}} = \sqrt{\frac{2 \times 200{,}000 \times 80}{8}} = 2{,}000$$

Test Yourself, Exercise 9.6

In all the questions below assume that demand is spread evenly over the year and stock is run down to zero before a new order is placed.

1. A firm uses 6,000 tonnes of commodity X every year. The fixed transaction costs involved with each order are £80. Each tonne of X held in stock costs £6 per annum. How many separate orders for X should the firm make during the year?
2. If each order for a batch of components costs £700 to make, storage costs per annum per component are £20 and annual usage is 4,480 components, what is the optimal order size?
3. A firm uses 1,280 units of a component each year. The cost of making an order is £540 and each component held in stock for a year costs the firm £6. What average order size would you advise the firm to make? Assume that the demand for this component is steady from year to year and that the same number of orders do not have to be made within each 12-month period.
4. A firm uses 1,400 units per year of component G. Each order costs £350 to make and average storage costs per unit of G are £20. There is also an extra 'capacity' cost given that the firm has to provide warehousing capable of storing a full order size of q even though this warehousing space will be underutilized most of the time. This 'capacity' cost will be £15 per unit of G. Adapt the optimal order size formula to include this extra cost and then find the optimal order size for this firm.

9.7 Comparative static effects of taxes

In Chapter 5, we examined the comparative static effects of taxes on a firm's profit-maximizing output and price when all the relevant functions were linear. Calculus now enables us to extend this analysis to non-linear functions. Having learned how to determine a firm's profit-maximizing output and price by setting up a firm's profit function and then maximizing it, we can now deduce what may happen to these equilibrium values if an exogenous variable changes.

Suppose that a firm operates with the total cost function

$$TC = 50 + 0.4q^2$$

and is a monopoly facing the demand schedule

$$p = 360 - 2.1q$$

There is no independently determined exogenous variable in this economic model as it currently stands and so, if equilibrium was attained, output and price would remain at their profit-maximizing levels. We shall now examine what would happen to these equilibrium values if the following different forms of tax were imposed on the firm:

(a) a per-unit sales tax
(b) a lump sum tax
(c) a percentage profits tax

The approach used in each case is to:

- formulate the firm's objective function for the net (after tax) profit that it will be striving to maximize,
- find the output when the objective function is maximized, checking both first- and second-order conditions,
- specify the profit-maximizing output and price as reduced form functions dependent on the exogenously determined tax and
- differentiate to find the impact of a change in the tax on these optimum values.

It is important for you to learn how to set up objective functions from the economic information available and to understand the different impacts that these different types of taxes will have. A common mistake that students sometimes make in this sort of problem is to try to show the effect of a tax by shifting up the supply schedule by the amount of the tax. That method only applies for a sales tax in a perfectly competitive market. This time we have a firm that operates in a monopolistic market (and so there is no supply schedule as such) and some of these taxes are on profit rather than sales.

(a) Per-unit sales tax

If the firm has to pay the government an amount t on each unit of q that it sells then the total tax it has to pay will be tq. Its total costs, including the tax, will therefore be

$$TC = 50 + 0.4q^2 + tq$$

Given the demand schedule $p = 360q - 2.1q$ the firm's total revenue function will be

$$TR = pq = 360q - 2.1q^2$$

The net profit objective function that the firm will wish to maximize will therefore be

$$\begin{aligned}
\pi &= TR - TC \\
&= 360q - 2.1q^2 - (50 + 0.4q^2 + tq) \\
&= 360q - 2.1q^2 - 50 - 0.4q^2 - tq \\
&= 360q - 2.5q^2 - 50 - tq
\end{aligned}$$

Differentiating with respect to q and setting equal to zero to find the first-order condition for a maximum

$$\frac{d\pi}{dq} = 360 - 5q - t = 0 \tag{1}$$

Before proceeding with the comparative static analysis we need to check the second-order conditions to confirm that this stationary point is indeed a maximum. Differentiating (1) again gives

$$\frac{d^2\pi}{dq^2} = -5 < 0$$

and so the second-order condition for a maximum is met.

Returning to the first-order condition (1) in order to find the optimal level of q in terms of t

$$360 - 5q - t = 0 \tag{1}$$
$$360 - t = 5q$$
$$q = 72 - 0.2t \tag{2}$$

This is the reduced form equation for profit-maximizing output in terms of the independent variable t.

Differentiating (2) with respect to t to find the comparative static effect of a change in t on the optimum value of q gives

$$\frac{dq}{dt} = -0.2$$

This means that a one unit increase in the per-unit sales tax will reduce output by 0.2 units. This comparative static effect is not dependent on any other variable and so at any output level the impact of the tax on q will be the same, as long as it is still profitable for the firm to produce.

The comparative static effect of this tax on price can be found by substituting the function for the optimal level of q

$$q = 72 - 0.2t \tag{2}$$

into the firm's demand schedule

$$p = 360 - 2.1q$$

Thus

$$p = 360 - 2.1(72 - 0.2t)$$
$$= 360 - 151.2 + 0.42t$$

Giving the reduced form

$$p = 208.8 + 0.42t \tag{3}$$

Differentiating

$$\frac{\mathrm{d}p}{\mathrm{d}t} = 0.42$$

This tells us that the comparative static effect of a £1 increase in the per-unit tax t will be a £0.42 increase in the firm's profit-maximizing price.

(b) A lump sum tax

A lump sum tax is a fixed amount that firms are required to pay to the government. The amount of the tax (T) is not related to sales or profit levels.

   Before the tax is introduced the firm in our example faces the total cost and total revenue functions

$$\mathrm{TC} = 50 + 0.4q^2$$

$$\mathrm{TR} = 360q - 2.1q^2$$

The imposition of a lump sum tax T will effectively increase fixed costs by the amount of the tax. The firm's total cost function will therefore become

$$\mathrm{TC} = 50 + 0.4q^2 + T$$

and the net profit objective function that the firm attempts to maximize will become

$$
\begin{aligned}
\pi &= \mathrm{TR} - \mathrm{TC} \\
&= 360q - 2.1q^2 - (50 + 0.4q^2 + T) \\
&= 360q - 2.1q^2 - 50 - 0.4q^2 - T \\
&= 360q - 2.5q^2 - 50 - T
\end{aligned}
$$

Differentiating with respect to q and setting equal to zero to find the first-order conditions for a maximum

$$\frac{\mathrm{d}\pi}{\mathrm{d}q} = 360 - 5q = 0 \tag{4}$$

Differentiating (4) again gives

$$\frac{\mathrm{d}^2\pi}{\mathrm{d}q^2} = -5 < 0$$

and so the second-order condition for a maximum is met.

Returning to (4) to find the optimal level of q

$$-5q + 360 = 0$$
$$360 = 5q$$
$$q = 72 \tag{5}$$

As (5) does not contain any term in T, the firm's profit-maximizing output will always be 72, regardless of the amount of the lump sum tax. Therefore a change in the lump sum tax T will have no effect on output. If the tax has no impact on output then it will also have no effect on price.

This is what economic analysis would predict. If a firm has to pay a fixed sum out of its profits then it would want to be in a position where total gross (before tax) profits are at a maximum in order to maximize net after tax profit. Note, though, that if the lump sum tax was greater than the firm's pre-tax profit then the firm would not be able to pay the tax and might have to close down. It is still possible, though, that the tax might be paid out of accumulated past profits, like the windfall tax that was imposed on some of the UK privatized utility companies in the late 1990s because the government thought that they had earned excessive profits.

(c) A percentage profits tax

If a firm has to pay a proportion of its profits as tax then it will attempt to maximize net profit which will be

$$\pi = (\text{TR} - \text{TC})(1 - c)$$

where c is the rate of profits tax. (Profits tax is called corporation tax in the UK, so we will use the notation c.)

Thus for the firm in this example

$$\pi = (\text{TR} - \text{TC})(1 - c)$$
$$= (360q - 2.1q^2 - 50 - 0.4q^2)(1 - c)$$
$$= (360q - 2.5q^2 - 50)(1 - c)$$

The term $(1 - c)$ can be treated as a constant that multiplies each of the values in the first set of brackets and so differentiating and setting equal to zero to get first-order condition for profit maximization

$$\frac{d\pi}{dq} = (360 - 5q)(1 - c) = 0 \tag{6}$$

Checking the second-order condition for a maximum

$$\frac{d^2\pi}{dq^2} = -5(1 - c) < 0 \quad \text{as long as} \quad 0 < c < 1$$

We would expect a percentage profits tax rate to lie between 0% and 100%. Therefore c will take a value between 0 and 1 and so the second-order condition for a maximum will be met.

Returning to (6) to find the optimal level of q

$$(360 - 5q)(1 - c) = 0$$

Unless there is a 100% profits tax

$$(1 - c) \neq 0$$

and so it must be the case that

$$360 - 5q = 0$$

$$q = 72 \tag{7}$$

As (7) does not contain any term in c, we can say that the firm's profit-maximizing output will always be 72, regardless of the amount of the profits tax. Therefore a change in the rate of profits tax c will have no effect on output. It will therefore also have no effect on price.

This result is what economic analysis would predict and is similar to the case (b) for a lump sum tax. If a firm has to pay a percentage of its profits as tax then it would want to be in a position where total profits before the tax are at a maximum in order to maximize net after tax profit.

Test Yourself, Exercise 9.7

1. Derive reduced form equations for equilibrium price and output in terms of

   (a) a per-unit sales tax t
   (b) a lump sum tax T
   (c) a percentage profits tax c
       in each of the cases below:

       (i) $p = 450 - 2q$ and $TC = 20 + 0.5q^2$
       (ii) $p = 200 - 0.3q$ and $TC = 10 + 0.1q^2$
       (iii) $p = 260 - 4q$ and $TC = 8 + 1.2q^2$

   In each case assume that the demand schedule and total cost function apply to a single firm in an industry.

2. Derive a reduced form equation that will show the comparative static effect of a percentage sales tax on a company that faces the demand schedule $p = 680 - 3q$ and the total cost function $TC = 20 + 0.4q^2$.

10 Partial differentiation

Learning objectives

After completing this chapter students should be able to:

- Derive the first-order partial derivatives of multi-variable functions.
- Apply the concept of partial differentiation to production functions, utility functions and the Keynesian macroeconomic model.
- Derive second-order partial derivatives and interpret their meaning.
- Check the second-order conditions for maximization and minimization of a function with two independent variables using second-order partial derivatives.
- Derive the total differential and total derivative of a multi-variable function.
- Use Euler's theorem to check if the total product is exhausted for a Cobb–Douglas production function.

10.1 Partial differentiation and the marginal product

For the production function $Q = f(K, L)$ with the two independent variables L and K the value of the function will change if one independent variable is increased whilst the other is held constant. If K is held constant and L is increased then we will trace out the total product of labour (TP_L) schedule (TP_L is the same thing as output Q). This will typically take a shape similar to that shown in Figure 10.1.

In your introductory microeconomics course the marginal product of L (MP_L) was probably defined as the increase in TP_L caused by a one-unit increment in L, assuming K to be fixed at some given level. A more precise definition, however, is that MP_L is the rate of change of TP_L with respect to L. For any given value of L this is the slope of the TP_L function. (Refer back to Section 8.3 if you do not understand why.) Thus the MP_L schedule in Figure 10.1 is at its maximum when the TP_L schedule is at its steepest, at M, and is zero when TP_L is at its maximum, at N.

Partial differentiation is a technique for deriving the rate of change of a function with respect to increases in one independent variable when all other independent variables in the function are held constant. Therefore, if the production function $Q = f(K, L)$ is differentiated with respect to L, with K held constant, we get the rate of change of total product with respect to L, in other words MP_L.

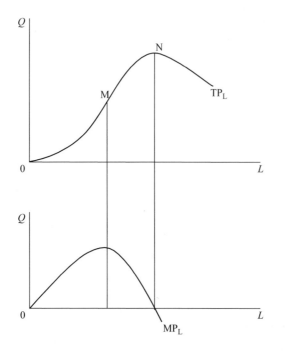

Figure 10.1

The *basic rule for partial differentiation* is that all independent variables, other than the one that the function is being differentiated with respect to, are treated as constants. Apart from this, partial differentiation follows the standard differentiation rules explained in Chapter 8. A curved ∂ is used in a partial derivative to distinguish it from the derivative of a single variable function where a normal letter 'd' is used. For example, the partial derivative of the production function above with respect to L is written $\partial Q/\partial L$.

Example 10.1

If $y = 14x + 3z^2$, find the partial derivatives of this function with respect to x and z.

Solution
The partial derivative of function y with respect to x is

$$\frac{\partial y}{\partial x} = 14$$

(The $3z^2$ disappears as it is treated as a constant. One then just differentiates the term $14x$ with respect to x.)
Similarly, the partial derivative of y with respect to z is

$$\frac{\partial y}{\partial z} = 6z$$

(The $14x$ is treated as a constant and disappears. One then just differentiates the term $3z^2$ with respect to z.)

Example 10.2

Find the partial derivatives of the function $y = 6x^2z$.

Solution

In this function the variable held constant does not disappear as it is multiplied by the other variable. Therefore

$$\frac{\partial y}{\partial x} = 12xz$$

treating z as a constant, and

$$\frac{\partial y}{\partial z} = 6x^2$$

treating x (and therefore x^2) as a constant.

Example 10.3

For the production function $Q = 20K^{0.5}L^{0.5}$

(i) derive a function for MP_L, and
(ii) show that MP_L decreases as one moves along an isoquant by using more L.

Solution

(i) MP_L is found by partially differentiating the production function $Q = 20K^{0.5}L^{0.5}$ with respect to L. Thus

$$MP_L = \frac{\partial Q}{\partial L} = 10K^{0.5}L^{-0.5} = \frac{10K^{0.5}}{L^{0.5}}$$

Note that this MP_L function will continuously slope downward, unlike the MP_L function illustrated in Figure 10.1.

(ii) If the function for MP_L above is multiplied top and bottom by $2L^{0.5}$, then we get

$$MP_L = \left(\frac{2L^{0.5}}{2L^{0.5}}\right)\left(\frac{10K^{0.5}}{L^{0.5}}\right) = \frac{20K^{0.5}L^{0.5}}{2L} = \frac{Q}{2L} \tag{1}$$

An isoquant joins combinations of K and L that yield the same output level. Thus if Q is held constant and L is increased then the function (1) shows us that MP_L will decrease.

(Note that moving along an isoquant entails using more L and less K to keep output constant. Although the amount of K used does therefore change, what this result tells us is that with the new amount of capital MP_L will be lower than it was before.)

We can now see that for any Cobb–Douglas production function in the format $Q = AK^\alpha L^\beta$ the law of diminishing marginal productivity holds for each input as long as $0 < \alpha, \beta < 1$. If K is fixed and L is variable, the marginal product of L is found in the usual way by partial differentiation. Thus, when

$$Q = AK^\alpha L^\beta$$

$$\text{MP}_\text{L} = \frac{\partial Q}{\partial L} = \beta AK^\alpha L^{\beta-1} = \frac{\beta AK^\alpha}{L^{1-\beta}}$$

If K is held constant then, given that α, β and A are also constants, the numerator in this expression βAK^α is constant. In the denominator, as L is increased, $L^{1-\beta}$ gets larger (given $0 < \beta < 1$) and so the whole function for MP_L decreases in value, i.e. the marginal product falls as L is increased.

Similarly, if K is increased while L is held constant,

$$\text{MP}_\text{K} = \frac{\partial Q}{\partial K} = \alpha AK^{\alpha-1} L^\beta = \frac{\alpha AL^\beta}{K^{1-\alpha}}$$

which falls as K increases in value.

When there are more than two inputs in a production function, the same principles still apply. For example, if

$$Q = AX_1^a X_2^b X_3^c X_4^d$$

where X_1, X_2, X_3 and X_4 are inputs, then the marginal product of input X_3 will be

$$\frac{\partial Q}{\partial X_3} = cAX_1^a X_2^b X_3^{c-1} X_4^d = \frac{cAX_1^a X_2^b X_4^d}{X_3^{1-c}}$$

which decreases as X_3 increases, *ceteris paribus*.

We can also see that for a production function in the usual Cobb–Douglas format the marginal product functions will continuously decline towards zero and will never 'bottom out' for finite values of L; i.e. they will never reach a minimum point where the slope is zero. If, for example,

$$Q = 25K^{0.4} L^{0.5}$$

$$\text{MP}_\text{L} = \frac{\partial Q}{\partial L} = 12.5K^{0.4} L^{-0.5}$$

The first-order condition for a minimum is

$$\frac{12.5K^{0.4}}{L^{0.5}} = 0$$

This is satisfied only if $K = 0$ and hence $Q = 0$, or if L becomes infinitely large. Since, for finite values of L, MP_L will still remain positive however large L becomes, this means that on the isoquant map for a two-input Cob–Douglas production function the isoquants will never 'bend back'; i.e. there will not be an uneconomic region.

Other possible formats for production functions are possible though. For example, if

$$Q = 4.6K^2 + 3.5L^2 - 0.012K^3 L^3$$

then MP_L will first rise and then fall since

$$MP_L = \frac{\partial Q}{\partial L} = 7L - 0.036K^3L^2$$

The slope of the MP_L function will change from a positive to a negative value as L increases since

$$slope = \frac{\partial MP_L}{\partial L} = 7 - 0.072K^3L$$

The actual value and position of this MP_L function will depend on the value that the other input K takes.

Example 10.4

If $q = 20x^{0.6}y^{0.2}z^{0.3}$, find the rate of change of q with respect to x, y and z.

Solution

Although there are now three independent variables instead of two, the same rules still apply, this time with two variables treated as constants. Therefore, holding y and z constant

$$\frac{\partial q}{\partial x} = 12x^{-0.4}y^{0.2}z^{0.3}$$

Similarly, holding x and z constant

$$\frac{\partial q}{\partial y} = 4x^{0.6}y^{-0.8}z^{0.3}$$

and holding x and y constant

$$\frac{\partial q}{\partial z} = 6x^{0.6}y^{0.2}z^{-0.7}$$

To avoid making mistakes when partially differentiating a function with several variables, it may help if you write in the variables that do not change first and then differentiate. In the above example, when differentiating with respect to x for instance, this would mean first writing in $y^{0.2}z^{0.3}$ as y and z are held constant.

When a function has a large number of variables, a shorthand notation for the partial derivative is usually used. For example, for the function $f = f(x_1, x_2, \ldots, x_n)$ one can write

$$f_1 \text{ instead of } \frac{\partial f}{\partial x_1}, \qquad f_2 \text{ instead of } \frac{\partial f}{\partial x_2}, \text{ etc.}$$

Example 10.5

Find f_j where j is any input number for the production function

$$f(x_1, x_2, \ldots, x_n) = \sum_{i=1}^{n} 6x_i^{0.5}$$

Solution

This function is a summation of several terms. Only one term, the jth, will contain x_j. If one is differentiating with respect to x_j then all other terms are treated as constants and disappear. Therefore, one only has to differentiate the term $6x_j^{0.5}$ with respect to x_j, giving

$$f_j = 3x_j^{-0.5}$$

This shorthand notation can also be used to express second-order partial derivatives. For example,

$$f_{11} = \frac{\partial^2 f}{\partial x_1^2}$$

Uses of second-order partial derivatives will be explained in Section 10.3.

Test Yourself, Exercise 10.1

1. Find $\partial y/\partial x$ and $\partial y/\partial z$ when

   (a) $y = 6 + 3x + 16z + 4x^2 + 2z^2$
   (b) $y = 14x^3 z^2$
   (c) $y = 9 + 4xz - 3x^{-2}z^3$

2. Show that the law of diminishing marginal productivity holds for the production function $Q = 12K^{0.4}L^{0.4}$. Will the MP$_L$ schedule take the shape shown in Figure 10.1?

3. Derive formulae for the marginal products of the three inputs in the production function $Q = 40K^{0.3}L^{0.3}R^{0.4}$.

4. Use partial differentiation to explain why the production function

   $$Q = 0.4K + 0.7L$$

   does not obey the law of diminishing marginal productivity.

5. If $Q = 18K^{0.3}L^{0.2}R^{0.5}$, will the marginal products of any of the three inputs K, L and R become negative?

6. Derive a formula for the partial derivative Q_j, where j is an input number, for the production function

   $$Q(x_1, x_2, \ldots, x_n) = \sum_{i=1}^{n} 4x_i^{0.3}$$

10.2   Further applications of partial differentiation

Partial differentiation is basically a mathematical application of the assumption of *ceteris paribus* (i.e. other things being held equal) which is frequently used in economic analysis. Because the economy is a complex system to understand, economists often look at the effect of changes in one variable assuming all other influencing factors remain unchanged. When the relationship between the different economic variables can be expressed in a mathematical format, then the analysis of the effect of changes in one variable can be discovered via partial differentiation. We have already seen how partial differentiation can be applied to production functions and here we shall examine a few other applications.

Elasticity

In a market the quantity demanded, q, depends on several factors. These may include the price of the good (p), average consumer income (m), the price of a complement (p_c), the price of a substitute (p_s) and population (n). This relationship can be expressed as the demand function

$$q = f(p, m, p_c, p_s, n)$$

In introductory economics courses, price elasticity of demand is usually defined as

$$e = (-1)\frac{\text{percentage change in quantity demanded}}{\text{percentage change in price}}$$

This definition implicitly assumes *ceteris paribus*, even though there may be no mention of other factors that influence demand. The same implicit assumption is made in the more precise measure of point elasticity of demand with respect to price:

$$e = (1)\frac{p}{q}\frac{1}{dp/dq}$$

Recognizing that quantity demanded depends on factors other than price, then point elasticity of demand with respect to price can be more accurately redefined as

$$e = (1)\frac{p}{q}\frac{1}{\partial p/\partial q} = (-1)\frac{p}{q}\frac{\partial q}{\partial p}$$

Note that we have employed the *inverse function rule* here. This states that, for any function $y = f(x)$, then

$$\frac{dx}{dy} = \frac{1}{dy/dx}$$

as long as $dy/dx \neq 0$. This rule can also be used for partial derivatives and so

$$\frac{1}{\partial p/\partial q} = \frac{\partial q}{\partial p}$$

Point elasticity (with respect to own price) can now be determined for specific demand functions that include other explanatory variables.

Example 10.6

For the demand function

$$q = 35 - 0.4p + 0.15m - 0.25p_c + 0.12p_s + 0.003n$$

where the terms are as defined above, what is price elasticity of demand when $p = 24$?

Solution

We know the value of p and we can easily derive the partial derivative $\partial q / \partial p = -0.4$.
Substituting these values into the elasticity formula

$$e = (-1)\frac{p}{q}\frac{\partial q}{\partial p}$$

$$= (-1)\frac{24}{35 - 0.4(24) + 0.15m - 0.25p_c + 0.12p_s + 0.003n}(-0.4)$$

$$= \frac{9.6}{25.4 + 0.15m - 0.25p_c + 0.12p_s + 0.003n}$$

The actual value of elasticity cannot be calculated until specific values for m, p_c, p_s and n
are given. Thus this example shows that the value of point elasticity of demand with respect
to price will depend on the values of other factors that affect demand and thus determine the
position on the demand schedule.

Other measures of elasticity will also depend on the values of the different variables in the
demand function. For example, the basic definition of income elasticity of demand is

$$e_m = \frac{\text{percentage change in quantity demanded}}{\text{percentage change in income}}$$

If we assume an infinitesimally small change in income and recognize that all other factors
influencing demand are being held constant then income elasticity of demand can be defined as

$$e_m = \frac{\dfrac{\Delta q}{q}}{\dfrac{\Delta m}{m}} = \frac{m}{q}\frac{\Delta q}{\Delta m} = \frac{m}{q}\frac{\partial q}{\partial m}$$

Thus, for the demand function in Example 10.6 above, income elasticity of demand will be

$$e_m = \frac{m}{q}\frac{\partial q}{\partial m} = \frac{m}{q}(0.15)$$

If the value of m is given as 30, say, then

$$e_m = \frac{30}{35 - 0.4p + 0.15(30) - 0.25p_c + 0.12p_s + 0.003n}(0.15)$$

$$= \frac{4.5}{39.5 - 0.4p - 0.25p_c + 0.12p_s + 0.003n}$$

Thus the value of income elasticity of demand will depend on the value of the other factors
influencing demand as well as the level of income itself.

Consumer utility functions

The general form of a consumer's utility function is

$$U = U(x_1, x_2, \ldots, x_n)$$

where $x_1, x_2, \ldots, x_n$ represent the amounts of the different goods consumed.

Unlike output in a production function, one cannot actually measure utility and this theoretical concept is only of use in making general predictions about the behaviour of large numbers of consumers, as you should learn in your economics course. Modern economic theory assumes that utility is an 'ordinal concept', meaning that different combinations of goods can be ranked in order of preference but utility itself cannot be quantified in any way. However, economists also work with the concept of 'cardinal' utility where it is assumed that, hypothetically at least, each individual can quantify and compare different levels of their own utility. It is this cardinal utility concept which is used here.

If we assume that only the two goods A and B are consumed, then the utility function will take the form

$$U = U(A, B)$$

Marginal utility is defined as the rate of change of total utility with respect to the increase in consumption of one good. Therefore the marginal utility functions for goods A and B, respectively, will be

$$\text{MU}_\text{A} = \frac{\partial U}{\partial A} \quad \text{and} \quad \text{MU}_\text{B} = \frac{\partial U}{\partial B}$$

Three important principles of utility theory are:

(i) The law of diminishing marginal utility says that if, *ceteris paribus*, the quantity consumed of any one good is increased, then eventually its marginal utility will decline.

(ii) A consumer will consume a good up to the point where its marginal utility is zero if it is a free good, or if a fixed payment is made regardless of the quantity consumed, e.g. water rates.

(iii) A consumer maximizes satisfaction when each good is consumed up to the point where an extra pound spent on one good will derive the same utility as an extra pound spent on any other good.

Some applications of the first two principles are given in the following examples. We shall return to principle (iii) in Chapter 11, when we study constrained optimization.

Example 10.7

Find out whether the law of diminishing marginal utility holds for both goods A and B in the following utility functions:

(i) $U = A^{0.6} B^{0.8}$
(ii) $U = 85AB - 1.6A^2 B^2$
(iii) $U = 0.2A^{-1} B^{-1} + 5AB$

Solutions

(i) For the utility function $U = A^{0.6} B^{0.8}$ partial differentiation yields the marginal utility functions

$$\text{MU}_\text{A} = \frac{\partial U}{\partial A} = 0.6 A^{-0.4} B^{0.8} \quad \text{and} \quad \text{MU}_\text{B} = \frac{\partial U}{\partial B} = 0.8 A^{0.6} B^{-0.2}$$

Thus MU_A falls as A increases (when B is held constant) and MU_B falls as B increases (when A is held constant). As both marginal utility functions decline, the law of diminishing marginal utility holds.

(ii) For the utility function $U = 85AB - 1.6A^2B^2$ the marginal utility functions will be

$$MU_A = \frac{\partial U}{\partial A} = 85B - 3.2AB^2$$

$$MU_B = \frac{\partial U}{\partial B} = 85A - 3.2A^2B$$

Both MU_A and MU_B will be downward-sloping straight lines given that the quantity of the other good is held constant. Therefore the law of diminishing marginal utility holds.

(iii) When $U = 0.2A^{-1}B^{-1} + 5AB$ then

$$MU_A = \frac{\partial U}{\partial A} = -0.2A^{-2}B^{-1} + 5B$$

$$MU_B = \frac{\partial U}{\partial B} = -0.2A^{-1}B^{-2} + 5A$$

As A increases, the term $0.2A^{-2}B^{-1}$ gets smaller. As this term is subtracted from $5B$, which will be constant as B remains unchanged, this means that MU_A rises. Similarly, MU_B will rise as B increases. Therefore the law of diminishing marginal utility does not hold for this function.

Example 10.8

Given the following utility functions, how much of A will be consumed if it is a free good? If necessary give answers in terms of the fixed amount of B.

(i) $U = 96A + 35B - 0.8A^2 - 0.3B^2$
(ii) $U = 72AB - 0.6A^2B^2$
(iii) $U = A^{0.3}B^{0.4}$

Solutions

In each case we need to try to find the value of A where MU_A is zero. (The law of diminishing marginal utility holds for all three functions.) Consumers will not consume extra units of A which have negative marginal utility and hence decrease total utility.

(i) For utility function $U = 96A + 35B - 0.8A^2 - 0.3B^2$ marginal utility of A is zero when

$$MU_A = \frac{\partial U}{\partial A} = 96 - 1.6A = 0$$

$$96 = 1.6A$$

$$60 = A$$

Thus 60 units of A are consumed if A is free, regardless of the amount of B consumed.

(ii) When $U = 72AB - 0.6A^2B^2$ then MU_A is zero when

$$\frac{\partial U}{\partial A} = 72B - 1.2AB^2 = 0$$

$$72B = 1.2AB^2$$

$$60B^{-1} = A$$

Thus the amount of A consumed if it is free will depend inversely on the amount B consumed.

(iii) When $U = A^{0.3}B^{0.4}$ then the marginal utility of A will be

$$MU_A = \frac{\partial U}{\partial A} = 0.3A^{-0.7}B^{0.4}$$

This marginal utility function will decline continuously but, for any non-zero value of B, MU_A will not equal zero unless the amount of A consumed becomes infinitely large. Therefore no finite solution can be found.

The Keynesian multiplier

If a government sector and foreign trade are introduced then the basic Keynesian macroeconomic model becomes the accounting identity

$$Y = C + I + G + X - M \tag{1}$$

and the functional relationships of the consumption function

$$C = cY_d \tag{2}$$

where c is the marginal propensity to consume, plus

$$M = mY_d \tag{3}$$

where M is imports and m is the marginal propensity to import, and

$$Y_d = (1-t)Y \tag{4}$$

where Y_d is disposable income and t is the tax rate.

Investment I, government expenditure G and exports X are exogenously determined and c, m and t are given parameters. Substituting (2), (3) and (4) into (1) we get

$$Y = c(1-t)Y + I + G + X - m(1-t)Y$$

$$Y[1 - c(1-t) + m(1-t)] = I + G + X$$

$$Y = \frac{I + G + X}{1 - c(1-t) + m(1-t)} = \frac{I + G + X}{1 - (c-m)(1-t)} \tag{5}$$

In the basic Keynesian model without G and X, the investment multiplier is simply dY/dI. However, in this extended model one also has to assume that G and X are constant in order to derive the investment multiplier. Thus the investment multiplier is found by partially differentiating (5) with respect to I, which gives

$$\frac{\partial Y}{\partial I} = \frac{1}{1 - (c - m)(1 - t)}$$

You should also be able to see that the government expenditure and export multipliers will also take this format as

$$\frac{\partial Y}{\partial I} = \frac{\partial Y}{\partial G} = \frac{\partial Y}{\partial X} = \frac{1}{1 - (c - m)(1 - t)}$$

Example 10.9

In a Keynesian macroeconomic system, the following relationships and values hold:

$$Y = C + I + G + X - M$$

$$C = 0.8Y_d \quad M = 0.2Y_d \quad Y_d = (1 - t)Y$$

$$t = 0.2 \qquad G = 400 \qquad I = 300 \qquad X = 288$$

What is the equilibrium level of Y? What increase in G would be necessary to increase Y to 2,500? If this increased expenditure takes place, what will happen to

(i) the government's budget surplus/deficit, and
(ii) the balance of payments?

Solution

First we derive the relationship between C and Y. Thus

$$C = 0.8Y_d = 0.8(1 - t)Y = 0.8(1 - 0.2)Y \tag{1}$$

Next we substitute (1) and the other functional relationships and given values into the accounting identity to find the equilibrium Y. Thus

$$Y = C + I + G + X - M$$
$$= 0.8(1 - 0.2)Y + 300 + 400 + 288 - 0.2(1 - 0.2)Y$$
$$= 0.64Y + 988 - 0.16Y$$
$$(1 - 0.48)Y = 988$$
$$Y = \frac{988}{0.52} = 1,900$$

At this equilibrium level of Y the total amount of tax raised will be

$$tY = 0.2(1,900) = 380$$

Thus budget deficit, which is the excess of government expenditure over the amount of tax raised, will be

$$G - tY = 400 - 380 = 20$$

The amount spent on imports will be

$$M = 0.2Y_d = 0.2(0.8Y) = 0.16 \times 1{,}900 = 304$$

and so the balance of payments will be

$$X - M = 288 - 304 = -16$$

i.e. a deficit of 16.

The government expenditure multiplier is

$$\frac{\partial Y}{\partial G} = \frac{1}{1 - (c - m)(1 - t)} = \frac{1}{1 - (0.8 - 0.2)(1 - 0.2)}$$

$$= \frac{1}{1 - (0.6)(0.8)} = \frac{1}{1 - 0.48} = \frac{1}{0.52} \tag{2}$$

As equilibrium Y is 1,900, the increase in Y required to get to the target level of 2,500 is

$$\Delta Y = 2{,}500 - 1{,}900 = 600 \tag{3}$$

Given that the impact of the multiplier on Y will always be equal to

$$\Delta G \frac{\partial Y}{\partial G} = \Delta Y \tag{4}$$

where ΔG is the change in government expenditure, then substituting (2) and (3) into (4) gives

$$\Delta G \frac{1}{0.52} = 600$$

$$\Delta G = 600(0.52) = 312$$

This is the increase in G required to raise Y to 2,500.

At the new level of national income, the amount of tax raised will be

$$tY = 0.2(2{,}500) = 500$$

The new government expenditure level including the 312 increase will be

$$400 + 312 = 712$$

Therefore, the budget deficit will be

$$G - tY = 712 - 500 = 212$$

i.e. there is an increase of 192 in the deficit.

The new level of imports will be

$$M = 0.2(0.8)2{,}500 = 400$$

and so the new balance of payments figure will be

$$X - M = 288 - 400 = -112$$

i.e. the deficit increases by 96.

Cost and revenue functions

Some firms produce several different products. When common production facilities are used the costs of the individual products will be related and this will be reflected in the total cost schedules. The marginal cost schedules of the individual products can then be derived by partial differentiation.

Example 10.10

A firm produces two goods, with output levels q_1 and q_2, and faces the total cost function

$$TC = 45 + 125q_1 + 84q_2 - 6q_1^2 q_2^2 + 0.8q_1^3 + 1.2q_2^3$$

What are the two relevant marginal cost functions?

Solution

Marginal cost is the rate of change of TC with respect to output. Therefore

$$MC_1 = \frac{\partial TC}{\partial q_1} = 125 - 12q_1 q_2^2 + 2.4q_1^2$$

$$MC_2 = \frac{\partial TC}{\partial q_2} = 84 - 12q_1^2 q_2 + 3.6q_2^2$$

These marginal cost schedules show that the level of marginal cost for one good will depend on the amount of the other good that is produced.

Some firms may produce different goods which compete with each other in the market place, or are complements. This means that the price of one good will influence the quantity demanded of the other goods sold by the same firm. Marginal revenue for one good will therefore be the partial derivative of total revenue with respect to the output level of that particular good, assuming that the price of the other goods are fixed.

Example 10.11

A firm produces goods A and B which are complements. Derive marginal revenue functions for the two goods if the relevant demand schedules are

$$q_A = 850 - 12.5p_A - 3.8p_B$$
$$q_B = 936 - 4.8p_A - 24p_B$$

Solution

Marginal revenue is usually expressed as a function of quantity. Therefore, in order to derive total and marginal revenue functions, the demand functions are first rearranged to get price as a function of quantity. Thus, for good A

$$q_A = 850 - 12.5p_A - 3.8p_B$$

$$12.5p_A = 850 - 3.8p_B - q_A$$

$$p_A = \frac{850 - 3.8p_B - q_A}{12.5}$$

$$TR_A = p_A q_A$$

$$= \left(\frac{850 - 3.8p_B - q_A}{12.5} \right) q_A$$

$$= \frac{850q_A - 3.8p_B q_A - q_A^2}{12.5}$$

$$MR_A = \frac{\partial TR}{\partial q_A}$$

$$= \frac{850 - 3.8p_B - 2q_A}{12.5}$$

$$= 68 - 0.304p_B - 0.16q_A \tag{1}$$

Similarly, for good B

$$q_B = 936 - 4.8p_A - 24p_B$$

$$24p_B = 936 - 4.8p_A - q_B$$

$$p_B = 39 - 0.2p_A - \frac{q_B}{24}$$

$$TR_B = p_B q_B$$

$$= \left(39 - 0.2p_A - \frac{q_B}{24} \right) q_B$$

$$= 39q_B - 0.2p_A q_B - \frac{q_B^2}{24}$$

$$MR_B = \frac{\partial TR}{\partial q_B}$$

$$= 39 - 0.2p_A - \frac{q_B}{12} \tag{2}$$

The marginal revenue functions (1) and (2) for MR_A and MR_B confirm that, because the demand functions for the two goods are interrelated, the marginal revenue function for one good will depend on the price level of the other good.

Test Yourself, Exercise 10.2

1. The demand function for a good is

    $$q = 56.6 - 0.25p - 0.03m + 0.45p_s + 0.6n$$

    where q is the quantity demanded per week, p is the price per unit, m is the average weekly income, p_s is the price of a competing good and n is the population in millions. Given values are $p = 65$, $m = 350$, $p_s = 60$ and $n = 24$.

    (a) Calculate the price elasticity of demand.
    (b) Find out what would happen to (a) if n rose to 26.
    (c) Explain why this is an inferior good.
    (d) If producers of the competing product and the manufacturer of this good both increased their prices by the same percentage, what would happen to the quantity demanded (of the original good), assuming that the proportional price change is small and relevant elasticity measures do not alter significantly.

2. Do the following utility functions obey the law of diminishing marginal utility?

    (a) $U = 5A + 8B + 2.2A^2B^2 - 0.3A^3B^3$
    (b) $U = 24A^{0.8}B^{1.2}$
    (c) $U = 6A^{0.7}B^{0.8}$

3. An individual consumes two goods and has the utility function $U = 2A^{0.4}B^{0.4}$, where A and B represent the quantities of the two goods consumed. Will she ever consume either good up to the point where its marginal utility is zero?

4. In a Keynesian macroeconomic model of an economy, using the usual terminology,

    $$Y = C + I + G + X - M \quad Y_d = (1 - t)Y \quad C = 0.75Y_d$$

    $$M = 0.25Y_d \quad I = 820 \quad G = 960 \quad t = 0.3 \quad X = 650$$

    What will be the equilibrium value of Y? Use the export multiplier to find out what will happen to the balance of payments if exports exogenously increase by 100.

5. A multiplant firm faces the total cost schedule

    $$TC = 850 + 18q_1 + 25q_2 + 0.6q_1^2q_2 + 1.2q_1q_2^2$$

    where q_1 and q_2 are output levels in its two plants. What marginal cost schedule does it face if output in plant 2 is expanded while output in plant 1 is kept unchanged?

6. In a closed economy (i.e. one with no foreign trade) the following relationships hold:

    $$C = 0.6Y_d \quad Y_d = (1 - t)Y \quad Y = C + I + G$$

    $$I = 120 \quad t = 0.25 \quad G = 210$$

where C is consumer expenditure, Y_d is disposable income, Y is national income, I is investment, t is the tax rate and G is government expenditure. What is the marginal propensity to consume out of Y? What is the value of the government expenditure multiplier? How much does government expenditure need to be increased to achieve a national income of 700?

10.3 Second-order partial derivatives

Second-order partial derivatives are found by differentiating the first-order partial derivatives of a function.

When a function has two independent variables there will be four second-order partial derivatives. Take, for example, the production function

$$Q = 25K^{0.4}L^{0.3}$$

There are two first-order partial derivatives

$$\frac{\partial Q}{\partial K} = 10K^{-0.6}L^{0.3} \qquad \frac{\partial Q}{\partial L} = 7.5K^{0.4}L^{-0.7}$$

These represent the marginal product functions for K and L. Differentiating these functions a second time we get

$$\frac{\partial^2 Q}{\partial K^2} = -6K^{-1.6}L^{0.3} \qquad \frac{\partial^2 Q}{\partial L^2} = -5.25K^{0.4}L^{-1.7}$$

These second-order partial derivatives represent the rate of change of the marginal product functions. In this example we can see that the slope of MP_L (i.e. $\partial^2 Q/\partial L^2$) will always be negative (assuming positive values of K and L) and as L increases, *ceteris paribus*, the absolute value of this slope diminishes.

We can also find the rate of change of $\partial Q/\partial K$ with respect to changes in L and the rate of change of $\partial Q/\partial L$ with respect to K. These will be

$$\frac{\partial^2 Q}{\partial K \partial L} = 3K^{-0.6}L^{-0.7} \qquad \frac{\partial^2 Q}{\partial L \partial K} = 3K^{-0.6}L^{-0.7}$$

and are known as 'cross partial derivatives'. They show how the rate of change of Q with respect to one input alters when the other input changes. In this example, the cross partial derivative $\partial^2 Q/\partial L \partial K$ tells us that the rate of change of MP_L with respect to changes in K will be positive and will fall in value as K increases.

You will also have noted in this example that

$$\frac{\partial^2 Q}{\partial K \partial L} = \frac{\partial^2 Q}{\partial L \partial K}$$

In fact, matched pairs of cross partial derivatives will always be equal to each other.

Thus, for any continuous two-variable function $y = f(x, z)$, there will be four second-order partial derivatives:

(i) $\dfrac{\partial^2 y}{\partial x^2}$ (ii) $\dfrac{\partial^2 y}{\partial z^2}$ (iii) $\dfrac{\partial^2 y}{\partial x \partial z}$ (iv) $\dfrac{\partial^2 y}{\partial z \partial x}$

with the cross partial derivatives (iii) and (iv) always being equal, i.e.

$$\frac{\partial^2 y}{\partial x \partial z} = \frac{\partial^2 y}{\partial z \partial x}$$

Example 10.12

Derive the four second-order partial derivatives for the production function

$$Q = 6K + 0.3K^2L + 1.2L^2$$

and interpret their meaning.

Solution

The two first-order partial derivatives are

$$\frac{\partial Q}{\partial K} = 6 + 0.6KL \qquad \frac{\partial Q}{\partial L} = 0.3K^2 + 2.4L$$

and these represent the marginal product functions MP_K and MP_L.

   The four second-order partial derivatives are as follows:

(i) $\dfrac{\partial^2 Q}{\partial K^2} = 0.6L$

This represents the slope of the MP_K function. It tells us that the MP_K function will have a constant slope along its length (i.e. it is linear) for any given value of L, but an increase in L will cause an increase in this slope

(ii) $\dfrac{\partial^2 Q}{\partial L^2} = 2.4$

This represents the slope of the MP_L function and tells us that MP_L is a straight line with slope 2.4. This slope does not depend on the value of K.

(iii) $\dfrac{\partial^2 Q}{\partial K \partial L} = 0.6K$

This tells us that MP_K increases if L is increased. The rate at which MP_K rises as L is increased will depend on the value of K.

(iv) $\dfrac{\partial^2 Q}{\partial L \partial K} = 0.6K$

This tells us that MP_L will increase if K is increased and that the rate of this increase will depend on the value of K. Thus, although the slope of the MP_L schedule will always be 2.4, from (ii) above, its actual position will depend on the amount of K used.

   Some other applications of second-order partial derivatives are given below.

Example 10.13

A firm sells two competing products whose demand schedules are

$$q_1 = 120 - 0.8p_1 + 0.5p_2 \qquad q_2 = 160 + 0.4p_1 - 12p_2$$

How will the price of good 2 affect the marginal revenue of good 1?

Solution

To find the total revenue function for good 1 (TR_1) in terms of q_1 we first need to derive the inverse demand function $p_1 = f(q_1)$. Thus, given

$$\begin{aligned}
q_1 &= 120 - 0.8p_1 + 0.5p_2 \\
0.8p_1 &= 120 + 0.5p_2 - q_1 \\
p_1 &= 150 + 0.625p_2 - 1.25q_1 \\
TR_1 &= p_1 q_1 \\
&= (150 + 0.625p_2 - 1.25q_1)q_1 \\
&= 150q_1 + 0.625p_2 q_1 - 1.25q_1^2
\end{aligned}$$

Thus

$$MR_1 = \frac{\partial TR_1}{\partial q_1} = 150 + 0.625p_2 - 2.5q_1$$

This marginal revenue function will have a constant slope of -2.5 regardless of the value of p_2 or the amount of q_1 sold.

The effect of a change in p_2 on MR_1 is shown by the cross partial derivative

$$\frac{\partial TR_1}{\partial q_1 \partial p_2} = 0.625$$

Thus an increase in p_2 of one unit will cause an increase in the marginal revenue from good 1 of 0.625, i.e. although the slope of the MR_1 schedule remains constant at -2.5, its position shifts upward if p_2 rises. (Note that in order to answer this question, we have formulated the total revenue for good 1 as a function of one price and one quantity, i.e. $TR_1 = f(q_1, p_2)$.)

Example 10.14

A firm operates with the production function $Q = 820K^{0.3}L^{0.2}$ and can buy inputs K and L at £65 and £40 respectively per unit. If it can sell its output at a fixed price of £12 per unit, what is the relationship between increases in L and total profit? Will a change in K affect the extra profit derived from marginal increases in L?

Solution

$$TR = PQ = 12(820K^{0.3}L^{0.2})$$

$$TC = P_K K + P_L L = 65K + 40L$$

Therefore profit will be

$$\pi = TR - TC$$

$$= 12(820K^{0.3}L^{0.2}) - (65K + 40L)$$

$$= 9{,}840K^{0.3}L^{0.2} - 65K - 40L$$

The effect of an increase in L on profit is shown by the first-order partial derivative:

$$\frac{\partial \pi}{\partial L} = 1{,}968K^{0.3}L^{-0.8} - 40 \tag{1}$$

This effect will be positive as long as

$$1{,}968K^{0.3}L^{-0.8} > 40$$

However, if L is continually increased while K is held constant, the value of the term $1{,}968K^{0.3}L^{-0.8}$ will eventually fall below 40 and so $\partial \pi / \partial L$ will become negative.

To determine the effect of a change in K on the marginal profit function with respect to L, we need to differentiate (1) with respect to K, giving

$$\frac{\partial^2 \pi}{\partial L \partial K} = 0.3(1{,}968K^{-0.7}L^{-0.8}) = 590.4K^{-0.7}L^{-0.8}$$

This cross partial derivative will be positive as long as K and L are positive. This is what we would expect and so an increase in K will have a positive effect on the extra profit generated by marginal increases in L. The magnitude of this impact will depend on the values of K and L.

Second-order and cross partial derivatives can also be derived for functions with three or more independent variables. For a function with three independent variables, such as $y = f(w, x, z)$ there will be the three second-order partial derivatives

$$\frac{\partial^2 y}{\partial w^2} \qquad\qquad \frac{\partial^2 y}{\partial x^2} \qquad\qquad \frac{\partial^2 y}{\partial z^2}$$

plus the six cross partial derivatives

$$\frac{\partial^2 y}{\partial w \partial x} = \frac{\partial^2 y}{\partial x \partial w} \qquad \frac{\partial^2 y}{\partial x \partial z} = \frac{\partial^2 y}{\partial z \partial x} \qquad \frac{\partial^2 y}{\partial w \partial z} = \frac{\partial^2 y}{\partial z \partial w}$$

These are arranged in pairs because, as with the two-variable case, cross partial derivatives will be equal if the two stages of differentiation involve the same two variables.

Example 10.15

For the production function $Q = 32K^{0.5}L^{0.25}R^{0.4}$ derive all the second-order and cross partial derivatives and show that the cross partial derivatives with respect to each possible pair of independent variables will be equal to each other.

Solution

The three first-order partial derivatives will be

$$\frac{\partial Q}{\partial K} = 16K^{-0.5}L^{0.25}R^{0.4} \qquad \frac{\partial Q}{\partial L} = 8K^{0.5}L^{-0.75}R^{0.4}$$

$$\frac{\partial Q}{\partial R} = 12.8K^{0.5}L^{0.25}R^{-0.6}$$

The second-order partial derivatives will be

$$\frac{\partial^2 Q}{\partial K^2} = -8K^{-1.5}L^{0.25}R^{0.4} \qquad \frac{\partial^2 Q}{\partial L^2} = -6K^{0.5}L^{-1.75}R^{0.4}$$

$$\frac{\partial^2 Q}{\partial R^2} = -7.68K^{0.5}L^{0.25}R^{-1.6}$$

plus the six cross partial derivatives:

$$\frac{\partial^2 Q}{\partial K \partial L} = 4K^{-0.5}L^{-0.75}R^{0.4} = \frac{\partial^2 Q}{\partial L \partial K}$$

$$\frac{\partial^2 Q}{\partial L \partial R} = 3.2K^{0.5}L^{-0.75}R^{-0.6} = \frac{\partial^2 Q}{\partial R \partial L}$$

$$\frac{\partial^2 Q}{\partial R \partial K} = 6.4K^{-0.5}L^{0.25}R^{-0.6} = \frac{\partial^2 Q}{\partial K \partial R}$$

Second-order derivatives for multi-variable functions are needed to check second-order conditions for optimization, as explained in the next section.

Test Yourself, Exercise 10.3

1. For the production function $Q = 8K^{0.6}L^{0.5}$ derive a function for the slope of the marginal product of L. What effect will a marginal increase in K have upon this MP_L function?

2. Derive all the second-order and cross partial derivatives for the production function $Q = 35KL + 1.4LK^2 + 3.2L^2$ and interpret their meaning.

3. A firm operates three plants with the joint total cost function

$$TC = 58 + 18q_1 + 9q_2q_3 + 0.004q_1^2q_3^2 + 1.2q_1q_2q_3$$

Find all the second-order partial derivatives for TC and demonstrate that the cross partial derivatives can be arranged in three equal pairs.

10.4   Unconstrained optimization: functions with two variables

For the two variable function $y = f(x, z)$ to be at a maximum or at a minimum, the first-order conditions which must be met are

$$\frac{\partial y}{\partial x} = 0 \quad \text{and} \quad \frac{\partial y}{\partial z} = 0$$

These are similar to the first-order conditions for optimization of a single variable function that were explained in Chapter 9. To be at a maximum or minimum, the function must be at a stationary point with respect to changes in both variables.

The second-order conditions and the reasons for them were relatively easy to explain in the case of a function of one independent variable. However, when two or more independent variables are involved the rationale for all the second-order conditions is not quite so straightforward. We shall therefore just state these second-order conditions here and give a brief intuitive explanation for the two-variable case before looking at some applications. The second-order conditions for the optimization of multi-variable functions with more than two variables are explained in Chapter 15 using matrix algebra.

For the optimization of two variable functions there are two sets of second-order conditions. For any function $y = f(x, z)$.

(1) $\dfrac{\partial^2 y}{\partial x^2} < 0$ and $\dfrac{\partial^2 y}{\partial z^2} < 0$ for a maximum

$\dfrac{\partial^2 y}{\partial x^2} > 0$ and $\dfrac{\partial^2 y}{\partial z^2} > 0$ for a minimum

These are similar to the second-order conditions for the optimization of a single variable function. The rate of change of a function (i.e. its slope) must be decreasing at a stationary point for that point to be a maximum and it must be increasing for a stationary point to be a minimum. The difference here is that these conditions must hold with respect to changes in both independent variables.

(2)   The other second-order condition is

$$\left(\frac{\partial^2 y}{\partial x^2}\right)\left(\frac{\partial^2 y}{\partial z^2}\right) > \left(\frac{\partial^2 y}{\partial x \partial z}\right)^2$$

This must hold at both maximum *and* minimum stationary points.

To get an idea of the reason for this condition, imagine a three-dimensional model with x and z being measured on the two axes of a graph and y being measured by the height above the flat surface on which the x and z axes are drawn. For a point to be the peak of the y 'hill' then, as well as the slope being zero at this point, one needs to ensure that, whichever direction one moves, the height will fall and the slope will become steeper. Similarly, for a point to be the minimum of a y 'trough' then, as well as the slope being zero, one needs to ensure that the height will rise and the slope will become steeper whichever direction one moves in. As moves can be made in directions other than those parallel to the two axes, it can be mathematically proved that the condition

$$\left(\frac{\partial^2 y}{\partial x^2}\right)\left(\frac{\partial^2 y}{\partial z^2}\right) > \left(\frac{\partial^2 y}{\partial x \partial z}\right)^2$$

satisfies these requirements as long as the other second-order conditions for a maximum or minimum also hold.

Note also that all the above conditions refer to the requirements for *local* maximum or minimum values of a function, which may or may not be *global* maxima or minima. Refer back to Chapter 9 if you cannot remember the difference between these two concepts.

Let us now look at some applications of these rules for the unconstrained optimization of a function with two independent variables.

Example 10.16

A firm produces two products which are sold in two separate markets with the demand schedules

$$p_1 = 600 - 0.3q_1 \qquad p_2 = 500 - 0.2q_2$$

Production costs are related and the firm faces the total cost function

$$TC = 16 + 1.2q_1 + 1.5q_2 + 0.2q_1q_2$$

If the firm wishes to maximize total profits, how much of each product should it sell? What will the maximum profit level be?

Solution

The total revenue is

$$TR = TR_1 + TR_2$$
$$= p_1q_1 + p_2q_2$$
$$= (600 - 0.3q_1)q_1 + (500 - 0.2q_2)q_2$$
$$= 600q_1 - 0.3q_1^2 + 500q_2 - 0.2q_2^2$$

Therefore profit is

$$\pi = TR - TC$$
$$= 600q_1 - 0.3q_1^2 + 500q_2 - 0.2q_2^2 - (16 + 1.2q_1 + 1.5q_2 + 0.2q_1q_2)$$
$$= 600q_1 - 0.3q_1^2 + 500q_2 - 0.2q_2^2 - 16 - 1.2q_1 - 1.5q_2 - 0.2q_1q_2$$
$$= -16 + 598.8q_1 - 0.3q_1^2 + 498.5q_2 - 0.2q_2^2 - 0.2q_1q_2$$

First-order conditions for maximization of this profit function are

$$\frac{\partial \pi}{\partial q_1} = 598.8 - 0.6q_1 - 0.2q_2 = 0 \tag{1}$$

and

$$\frac{\partial \pi}{\partial q_2} = 498.5 - 0.4q_2 - 0.2q_1 = 0 \tag{2}$$

Simultaneous equations (1) and (2) can now be solved to find the optimal values of q_1 and q_2.

Multiplying (2) by 3 $1,495.5 - 1.2q_2 - 0.6q_1 = 0$
Rearranging (1) $598.8 - 0.2q_2 - 0.6q_1 = 0$

Subtracting gives $896.7 - q_2 = 0$
Giving the optimal value $896.7 = q_2$

Substituting this value for q_2 into (1)

$$598.8 - 0.6q_1 - 0.2(896.7) = 0$$

$$598.8 - 179.34 = 0.6q_1$$

$$419.46 = 0.6q_1$$

$$699.1 = q_1$$

Checking second-order conditions by differentiating (1) and (2) again:

$$\frac{\partial^2 \pi}{\partial q_1^2} = -0.6 < 0 \qquad \frac{\partial^2 \pi}{\partial q_2^2} = -0.4 < 0$$

This satisfies one set of second-order conditions for a maximum.
The cross partial derivative will be

$$\frac{\partial^2 \pi}{\partial q_1 \partial q_2} = -0.2$$

Therefore

$$\left(\frac{\partial^2 \pi}{\partial q_1^2}\right)\left(\frac{\partial^2 \pi}{\partial q_2^2}\right) = (-0.6)(-0.4) = 0.24 > 0.04 = (-0.2)^2 = \left(\frac{\partial^2 \pi}{\partial q_1 \partial q_2}\right)^2$$

and so the remaining second-order condition for a maximum is satisfied.
The actual profit is found by substituting the optimum values $q_1 = 699.1$ and $q_2 = 896.7$. into the profit function. Thus

$$\pi = -16 + 598.8q_1 - 0.3q_1^2 + 498.5q_2 - 0.2q_2^2 - 0.2q_1q_2$$

$$= -16 + 598.8(699.1) - 0.3(699.1)^2 + 498.5(896.7) - 0.2(896.7)^2$$

$$- 0.2(699.1)(896.7)$$

$$= £432,797.02$$

Example 10.17

A firm sells two products which are partial substitutes for each other. If the price of one product increases then the demand for the other substitute product rises. The prices of the two products (in £) are p_1 and p_2 and their respective demand functions are

$$q_1 = 517 - 3.5p_1 + 0.8p_2 \qquad q_2 = 770 - 4.4p_2 + 1.4p_1$$

What price should the firm charge for each product to maximize its total sales revenue?

Solution

For this problem it is more convenient to express total revenue as a function of price rather than quantity. Thus

$$\begin{aligned}
TR &= TR_1 + TR_2 = p_1 q_1 + p_2 q_2 \\
&= p_1(517 - 3.5p_1 + 0.8p_2) + p_2(770 - 4.4p_2 + 1.4p_1) \\
&= 517p_1 - 3.5p_1^2 + 0.8p_1 p_2 + 770p_2 - 4.4p_2^2 + 1.4p_1 p_2 \\
&= 517p_1 - 3.5p_1^2 + 770p_2 - 4.4p_2^2 + 2.2p_1 p_2
\end{aligned}$$

First-order conditions for a maximum are

$$\frac{\partial TR}{\partial p_1} = 517 - 7p_1 + 2.2p_2 = 0 \tag{1}$$

and

$$\frac{\partial TR}{\partial p_2} = 770 - 8.8p_2 + 2.2p_1 = 0 \tag{2}$$

$$\begin{aligned}
\text{Multiplying (1) by 4} && 2{,}068 - 28p_1 + 8.8p_2 &= 0 \\
\text{Rearranging and adding (2)} && \underline{770 + 2.2p_1 - 8.8p_2} &= 0 \\
&& 2{,}838 - 25.8p_1 &= 0 \\
&& 2{,}838 &= 25.8p_1 \\
&& 110 &= p_1
\end{aligned}$$

Substituting this value of p_1 into (1)

$$\begin{aligned}
517 - 7(110) + 2.2p_2 &= 0 \\
2.2p_2 &= 253 \\
p_2 &= 115
\end{aligned}$$

Checking second-order conditions:

$$\frac{\partial^2 TR}{\partial p_1^2} = -7 < 0 \qquad \frac{\partial^2 TR}{\partial p_2^2} = -8.8 < 0 \qquad \frac{\partial^2 TR}{\partial p_1 \partial p_2} = 2.2$$

$$\left(\frac{\partial^2 \mathrm{TR}}{\partial q_1^2}\right)\left(\frac{\partial^2 \mathrm{TR}}{\partial q_2^2}\right) = (-7)(-8.8) = 61.6 > 4.84 = (2.2)^2 = \left(\frac{\partial^2 \mathrm{TR}}{\partial q_1 \partial q_2}\right)^2$$

Therefore all second-order conditions for a maximum value of total revenue are satisfied when $p_1 = \pounds 110$ and $p_2 = \pounds 115$.

Example 10.18

A multiplant monopoly operates two plants whose total cost schedules are

$$\mathrm{TC}_1 = 8.5 + 0.03q_1^2 \qquad \mathrm{TC}_2 = 5.2 + 0.04q_2^2$$

If it faces the demand schedule

$$p = 60 - 0.04q$$

where $q = q_1 + q_2$, how much should it produce in each plant in order to maximize profits?

Solution

The total revenue is

$$\mathrm{TR} = pq = (60 - 0.04q)q = 60q - 0.04q^2$$

Substituting $(q_1 + q_2)$ for q gives

$$\mathrm{TR} = 60(q_1 + q_2) - 0.04(q_1 + q_2)^2$$
$$= 60q_1 + 60q_2 - 0.04q_1^2 - 0.08q_1q_2 - 0.04q_2^2$$

Thus, subtracting the two total cost schedules, profit is

$$\pi = \mathrm{TR} - \mathrm{TC}_1 - \mathrm{TC}_2$$
$$= 60q_1 + 60q_2 - 0.04q_1^2 - 0.08q_1q_2 - 0.04q_2^2 - 8.5 - 0.03q_1^2 - 5.2 - 0.04q_2^2$$
$$= -13.7 + 60q_1 + 60q_2 - 0.07q_1^2 - 0.08q_2^2 - 0.08q_1q_2$$

First-order conditions for a maximum value of π require

$$\frac{\partial \pi}{\partial q_1} = 60 - 0.14q_1 - 0.08q_2 = 0 \tag{1}$$

and

$$\frac{\partial \pi}{\partial q_2} = 60 - 0.16q_2 - 0.08q_1 = 0 \tag{2}$$

Multiplying (1) by 2 $\qquad\qquad\qquad 120 - 0.28q_1 - 0.16q_2 = 0$

Rearranging and subtracting (2) $\quad \underline{60 - 0.08q_1 - 0.16q_2 = 0}$

$$60 - 0.2q_1 = 0$$
$$60 = 0.2q_1$$
$$300 = q_1$$

Substituting this value of q_1 into (1)

$$60 - 0.14(300) - 0.08q_2 = 0$$
$$18 = 0.08q_2$$
$$225 = q_2$$

Checking second-order conditions:

$$\frac{\partial^2 \pi}{\partial q_1^2} = -0.14 < 0 \qquad\qquad \frac{\partial^2 \pi}{\partial q_2^2} = -0.16 < 0 \qquad\qquad \frac{\partial^2 \pi}{\partial q_1 \partial q_2} = -0.08$$

$$\left(\frac{\partial^2 \pi}{\partial q_1^2}\right)\left(\frac{\partial^2 \pi}{\partial q_2^2}\right) = (-0.14)(-0.16) = 0.0224 > 0.0064 = (-0.08)^2 = \left(\frac{\partial^2 \pi}{\partial q_1 \partial q_2}\right)^2$$

Therefore all second-order conditions are satisfied for profit maximization when $q_1 = 300$ and $q_2 = 225$.

We can also check that the total profit is positive for these output levels. Total output is

$$q = q_1 + q_2 = 300 + 225 = 525$$

Substituting this value into the demand schedule

$$p = 60 - 0.04q = 60 - 0.04(525) = 39$$

Therefore

$$\text{TR} = pq = 39(525) = 20{,}475$$
$$\text{TC} = \text{TC}_1 + \text{TC}_2$$
$$= [8.5 + 0.03(300)^2] + [5.2 + 0.04(225)^2]$$
$$= 2{,}708.5 + 2{,}030.2 = 4{,}738.7$$

$$\pi = \text{TR} - \text{TC} = 20{,}475 - 4{,}738.7 = £15{,}736.30$$

Note that this method could also be used to solve the multiplant monopoly problems in Chapter 5 that only involved linear functions. The unconstrained optimization method used here is, however, a more general method that can be used for both linear and non-linear functions.

Example 10.19

A firm sells its output in a perfectly competitive market at a fixed price of £200 per unit. It buys the two inputs K and L at prices of £42 per unit and £5 per unit respectively, and faces the production function

$$q = 3.1K^{0.3}L^{0.25}$$

What combination of K and L should it use to maximize profit?

Solution

$$TR = pq = 200(3.1K^{0.3}L^{0.25}) = 620K^{0.3}L^{0.25}$$

$$TC = 42K + 5L$$

Therefore the profit function the firm wishes to maximize is

$$\pi = TR - TC = 620K^{0.3}L^{0.25} - 42K - 5L$$

First-order conditions for a maximum require

$$\frac{\partial \pi}{\partial K} = 186K^{-0.7}L^{0.25} - 42 = 0 \qquad \frac{\partial \pi}{\partial L} = 155K^{0.3}L^{-0.75} - 5 = 0$$

giving

$$186L^{0.25} = 42K^{0.7} \qquad \text{and} \qquad 155K^{0.3} = 5L^{0.75}$$

$$L^{0.25} = \frac{42}{186}K^{0.7} \qquad (1) \qquad 31K^{0.3} = L^{0.75} \qquad (2)$$

Taking (1) to the power of 3

$$L^{0.75} = \left(\frac{42}{186}K^{0.7}\right)^3 = \frac{42^3}{186^3}K^{2.1} \qquad (3)$$

Setting (3) equal to (2)

$$\frac{42^3}{186^3}K^{2.1} = 31K^{0.3}$$

$$K^{1.8} = \frac{31(186)^3}{42^3} = 2,692.481$$

$$K = 80.471179 \qquad (4)$$

Substituting (4) into (1)

$$L^{0.25} = \frac{42}{186}(80.471179)^{0.7} = 4.8717455$$

$$L = (L^{0.25})^4 = (4.8717455)^4 = 563.29822$$

Therefore, first-order conditions suggest that the optimum values are $L = 563.3$ and $K = 80.47$ (to 2 dp).

Checking second-order conditions:

$$\frac{\partial^2 \pi}{\partial K^2} = (-0.7)186K^{-1.7}L^{0.25}$$

$$= -130.2(80.47)^{-1.7}(563.3)^{0.25}$$

$$= -0.3653576 < 0$$

$$\frac{\partial^2 \pi}{\partial L^2} = (-0.75)155K^{0.3}L^{-1.75}$$

$$= -116.25(80.47)^{0.3}(563.3)^{-1.75}$$

$$= -0.0066572 < 0$$

$$\frac{\partial^2 \pi}{\partial K \partial L} = (0.25)186K^{-0.7}L^{-0.75}$$

$$= 46.5(80.47)^{-0.7}(563.3)^{-0.75}$$

$$= 0.0186404$$

$$\left(\frac{\partial^2 \pi}{\partial K^2}\right)\left(\frac{\partial^2 \pi}{\partial L^2}\right) = (-0.3653576)(-0.0066572) = 0.0024323$$

$$\left(\frac{\partial^2 \pi}{\partial K \partial L}\right)^2 = (-0.0186404)^2 = 0.0003475$$

Therefore

$$\frac{\partial^2 \pi}{\partial K^2}\frac{\partial^2 \pi}{\partial L^2} > \left(\frac{\partial^2 \pi}{\partial K \partial L}\right)^2$$

and so all second-order conditions for maximum profit are satisfied when $K = 80.47$ and $L = 563.3$.

The actual profit will be

$$\pi = 620K^{0.3}L^{0.25} - 42K - 5L$$

$$= 620(80.47)^{0.3}(563.3)^{0.25} - 42(80.47) - 5(563.3)$$

$$= 11,265.924 - 3,379.74 - 2,816.5$$

$$= £5,069.68$$

Note that in this problem, and other similar ones in this section, the indices in the Cobb–Douglas production function add up to less than unity, giving decreasing returns to scale and hence rising average and marginal (long-run) cost schedules. If there were increasing returns to scale and the average and marginal cost schedules continued to fall, a firm facing a fixed price would wish to expand output indefinitely and so no profit-maximizing solution would be found by this method.

Example 10.20

A multiplant monopoly operates two plants whose total cost schedules are

$$TC_1 = 36 + 0.003q_1^3 \qquad TC_2 = 45 + 0.005q_2^3$$

If its total output is sold in a market where the demand schedule is $p = 320 - 0.1q$, where $q = q_1 + q_2$, how much should it produce in each plant to maximize total profits?

Solution

The total revenue is

$$TR = pq = (320 - 0.1q)q = 320q - 0.1q^2$$

Substituting $q_1 + q_2 = q$ gives

$$TR = 320(q_1 + q_2) - 0.1(q_1 + q_2)^2$$
$$= 320q_1 + 320q_2 - 0.1(q_1^2 + 2q_1q_2 + q_2^2)$$
$$= 320q_1 + 320q_2 - 0.1q_1^2 - 0.2q_1q_2 - 0.1q_2^2$$

Thus profit will be

$$\pi = TR - TC = TR - TC_1 - TC_2$$
$$= (320q_1 + 320q_2 - 0.1q_1^2 - 0.2q_1q_2 - 0.1q_2^2) - (36 + 0.003q_1^3) - (45 + 0.005q_2^3)$$
$$= 320q_1 + 320q_2 - 0.1q_1^2 - 0.2q_1q_2 - 0.1q_2^2 - 36 - 0.003q_1^3 - 45 - 0.005q_2^3$$

First-order conditions for a maximum require

$$\frac{\partial \pi}{\partial q_1} = 320 - 0.2q_1 - 0.2q_2 - 0.009q_1^2 = 0 \tag{1}$$

and

$$\frac{\partial \pi}{\partial q_2} = 320 - 0.2q_1 - 0.2q_2 - 0.015q_2^2 = 0 \tag{2}$$

Subtracting (2) from (1)

$$-0.009q_1^2 + 0.015q_2^2 = 0$$

$$q_2^2 = \left(\frac{0.009}{0.015}\right)q_1^2 = 0.6q_1^2$$

$$q_2 = \sqrt{0.6q_1^2} = 0.07746q_1 \tag{3}$$

Substituting (3) for q_2 in (1)

$$320 - 0.2q_1 - 0.2(0.7746q_1) - 0.009q_1^2 = 0$$

$$320 - 0.2q_1 - 0.15492q_1 - 0.009q_1^2 = 0$$

$$0 = 0.009q_1^2 + 0.35492q_1 - 320 \qquad (4)$$

Using the quadratic formula to solve (4)

$$q_1 = \frac{-b \pm \sqrt{b^2 - 4ac}}{2a} = \frac{-0.35492 \pm \sqrt{(0.35492)^2 - 4(0.009)(-320)}}{0.018}$$

$$= \frac{-0.35492 \pm \sqrt{11.64598}}{0.018}$$

$$= \frac{-0.35492 \pm 3.412619}{0.018}$$

Disregarding the negative solution, this gives plant 1 output

$$q_1 = \frac{3.057699}{0.018}$$

$$= 169.87216 = 169.87 \quad \text{(to 2 dp)}$$

Substituting this value for q_1 into (3)

$$q_2 = 0.7746(169.87216) = 131.58 \quad \text{(to 2 dp)}$$

Checking second-order conditions:

$$\frac{\partial^2 \pi}{\partial q_1^2} = -0.2 - 0.018q_1 = -0.2 - 0.018(169.87) = -3.25766 < 0$$

$$\frac{\partial^2 \pi}{\partial q_2^2} = -0.2 - 0.03q_2 = -0.2 - 0.03(131.58) = -4.1474 < 0$$

$$\frac{\partial^2 \pi}{\partial q_1 \partial q_2} = -0.2$$

Thus, using the shorthand notation for the above second-order derivatives,

$$(\pi_{11})(\pi_{22}) = (-3.25766)(-4.1474) = 13.51 > 0.04 = (-0.2)^2 = (\pi_{12})^2$$

Therefore all second-order conditions for a maximum value of profit are satisfied when $q_1 = 169.87$ and $q_2 = 131.58$.

When a function involves more than two independent variables the second-order conditions for a maximum or minimum become even more complex and matrix algebra is needed to check them. However, until we get to Chapter 15, for economic problems involving three or more independent variables, we shall just consider how the first-order conditions can be used to determine optimum values. From the way these problems are constructed it will be obvious whether or not a maximum or a minimum value is being sought, and it will be assumed that second-order conditions are satisfied for the values that meet the first-order conditions.

Example 10.21

A firm operates with the production function

$$Q = 95K^{0.3}L^{0.2}R^{0.25}$$

and buys the three inputs K, L and R at prices of £30, £16 and £12 respectively per unit. If it can sell its output at a fixed price of £4 a unit, what is the maximum profit it can make? (Assume that second-order conditions for a maximum are met at stationary points.)

Solution

$$\pi = \text{TR} - \text{TC} = PQ - (P_K K + P_L L + P_R R)$$
$$= 4(95K^{0.3}L^{0.2}R^{0.25}) - (30K + 16L + 12R)$$
$$= 380K^{0.3}L^{0.2}R^{0.25} - 30K - 16L - 12R$$

First-order conditions for a maximum are

$$\frac{\partial \pi}{\partial K} = 114K^{-0.7}L^{0.2}R^{0.25} - 30 = 0 \tag{1}$$

$$\frac{\partial \pi}{\partial L} = 76K^{0.3}L^{-0.8}R^{0.25} - 16 = 0 \tag{2}$$

$$\frac{\partial \pi}{\partial R} = 95K^{0.3}L^{0.2}R^{-0.75} - 12 = 0 \tag{3}$$

From (1)

$$114L^{0.2}R^{0.25} = 30K^{0.7}$$

$$R^{0.25} = \frac{30K^{0.7}}{114L^{0.2}} \tag{4}$$

Substituting (4) into (2)

$$76K^{0.3}L^{-0.8}\left[\frac{30K^{0.7}}{114L^{0.2}}\right] = 16$$

$$76K^{0.3}(30K^{0.7}) = 16L^{0.8}(114L^{0.2})$$

$$2{,}280K = 1{,}824L$$

$$K = 0.8L \tag{5}$$

Substituting (5) into (4)

$$R^{0.25} = \frac{30(0.8L)^{0.7}}{114L^{0.2}} = \frac{30(0.8)^{0.7}L^{0.5}}{114} \tag{6}$$

Taking each side of (6) to the power of 3

$$R^{0.75} = \frac{27,000(0.8)^{2.1}L^{1.5}}{114^3}$$

Inverting

$$R^{-0.75} = \frac{114^3}{27,000(0.8)^{2.1}L^{1.5}} \tag{7}$$

Substituting (7) and (5) into (3)

$$95K^{0.3}L^{0.2}R^{-0.75} - 12 = 0$$

$$\frac{95(0.8L)^{0.3}L^{0.2}(114)^3}{27,000(0.8)^{2.1}L^{1.5}} = 12$$

$$\frac{95(0.8)^{0.3}L^{0.3}L^{0.2}(114)^3}{(0.8)^{2.1}L^{1.5}} = 324,000$$

$$\frac{95(114)^3}{(0.8)^{1.8}L} = 324,000$$

$$\frac{95(114)^3}{324,000(0.8)^{1.8}} = L$$

$$649.12924 = L \tag{8}$$

Substituting (8) into (5)

$$K = 0.8(649.12924) = 519.3034 \tag{9}$$

Substituting (8) into (6)

$$R^{0.25} = \frac{30(0.8)^{0.7}(649.12924)^{0.5}}{114}$$

$$R = \frac{30^4(0.8)^{2.8}(649.12924)^2}{114^4}$$

$$= 1,081.882$$

It is assumed that the second-order conditions for a maximum are met when K, L and R take these values.

The maximum profit level will therefore be (taking quantities to 1dp)

$$\pi = 3800(519.3)^{0.3}(649.1)^{0.2}(1,081.9)^{0.25}$$

$$- 30(519.3) - 16(649.1) - 12(1,081.9)$$

$$= 51,929.98 - 15,579 - 10,385.6 - 12,982.8$$

$$= £12,982.58$$

Test Yourself, Exercise 10.4

(Ensure that you check that second-order conditions are satisfied for these uncon-strained optimization problems.)

1. A firm produces two products which are sold in separate markets with the demand schedules

$$p_1 = 210 - 0.4q_1^2 \qquad p_2 = 491 - 6q_2$$

Production costs are related and the firm's total cost schedule is

$$TC = 32 + 0.8q_1^2 + 0.7q_2^2 + 0.1q_1q_2$$

How much should the firm sell in each market in order to maximize total profits?

2. A company produces two competing products whose demand schedules are

$$q_1 = 219 - 1.8p_1 + 0.5p_2 \qquad q_2 = 303 - 2.1p_2 + 0.8p_1$$

What price should it charge in the two markets to maximize total sales revenue?

3. A price-discriminating monopoly sells in two separable markets with demand schedules

$$p_1 = 215 - 0.012q_1 \qquad p_2 = 324 - 0.023q_2$$

and faces the total cost schedule $TC = 4{,}200 + 0.3q^2$, where $q = q_1 + q_2$. What should it sell in each market to maximize total profit? (*Note that negative quantities are not allowed, as was explained in Chapter 5.*)

4. A monopoly sells its output in two separable markets with the demand schedules

$$p_1 = 20 - \frac{q_1}{6} \qquad p_2 = 13.75 - \frac{q_2}{8}$$

If it faces the total cost schedule $TC = 74 + 2.26q + 0.01q^2$ where $q = q_1 + q_2$, what is the maximum profit it can make?

5. A multiplant monopoly operates two plants whose cost schedules are

$$TC_1 = 2.4 + 0.015q_1^2 \qquad TC_2 = 3.5 + 0.012q_2^2$$

and sells its total output in a market where $p = 32 - 0.02q$.
   How much should it produce in each plant to maximize total profits?

6. A firm operates two plants with the total cost schedules

$$TC_1 = 62 + 0.00018q_1^3 \qquad TC_2 = 48 + 0.00014q_2^3$$

and faces the demand schedule $p = 2{,}360 - 0.15q$.
   To maximize profits, how much should it produce in each plant?

7. A firm faces the production function $Q = 0.8K^{0.4}L^{0.3}$. It sells its output at a fixed price of £450 a unit and can buy the inputs K and L at £15 per unit and £8 per unit respectively. What input mix will maximize profit?

8. A firm selling in a perfectly competitive market where the ruling price is £40 can buy inputs K and L at prices per unit of £20 and £6 respectively. If it operates with the production function $Q = 21K^{0.4}L^{0.2}$, what is the maximum profit it can make?

9. A firm faces the production function $Q = 2.4K^{0.6}L^{0.2}$, where K costs £25 per unit and L costs £9 per unit, and sells Q at a fixed price of £82 per unit. Explain why it cannot make a profit of more than £20,000, no matter how efficiently it plans its input mix.

10. A firm can buy inputs K and L at £32 per unit and £20 per unit respectively and sell its output at a fixed price of £5 per unit. How should it organize production to ensure maximum profit if it faces the production function $Q = 82K^{0.5}L^{0.3}$?

10.5   Total differentials and total derivatives

(Note that the mathematical methods developed in this section are mainly used for the proofs of different economic theories rather than for direct numerical applications. These proofs may be omitted if your course does not include these areas of economics as they are not essential in order to understand the following chapters.)

In Chapter 8, when the concept of differentiation was introduced, you learned that the derivative dy/dx measured the rate of change of y with respect to x for infinitesimally small changes in x and y. For any non-linear function $y = f(x)$, the value of dy/dx will alter if x and y alter. It is therefore not possible to predict the effect of a given increase in x on y with complete accuracy. However, for a very small change (Δx) in x, we can say that it will be approximately true that the resulting change in y will be

$$\Delta y = \frac{dy}{dx}\Delta x$$

The closer the function $y = f(x)$ is to a straight line, the more accurate will be the prediction, as the following example demonstrates.

Example 10.22

For the functions below assume that the value of x increases from 10 to 11. Predict the effect on y using the derivative dy/dx evaluated at the first value of x and check the answer against the new value of the function.

(i) $y = 2x$ (ii) $y = 2x^2$ (iii) $y = 2x^3$

Solution

In all cases the change in x is $\Delta x = 11 - 10 = 1$.

(i) $y = 2x$ $\dfrac{dy}{dx} = 2$

Therefore, predicted change in y is

$$\Delta y = \frac{dy}{dx}\Delta x = 1 \times 2 = 2$$

The actual values are $y = 2(10) = 20$ when $x = 10$

$y = 2(11) = 22$ when $x = 11$

Thus actual change is $22 - 20 = 2$ (accuracy of prediction 100%)

(ii) $y = 2x^2$ $\dfrac{dy}{dx} = 4x = 4(10) = 40$

Therefore, predicted change in y is

$$\Delta y = \frac{dy}{dx}\Delta x = 40 \times 1 = 40$$

The actual values are $y = 2(10)^3 = 200$ when $x = 10$

$y = 2(11)^2 = 242$ when $x = 11$

Thus actual change is $242 - 200 = 42$ (accuracy of prediction 95%)

(iii) $y = 2x^3$ $\dfrac{dy}{dx} = 6x^2 = 6(10)^2 = 600$

Therefore, predicted change in y is

$$\Delta y = \frac{dy}{dx}\Delta x = 600 \times 1 = 600$$

The actual values are $y = 2(10)^3 = 2,000$ when $x = 10$

$y = 2(11)^3 = 2,662$ when $x = 11$

Thus actual change is $2,662 - 2,000 = 662$ (accuracy of prediction 91%).

The above method of predicting approximate actual changes in a variable can itself be useful for practical purposes. However, in economic theory this mathematical method is taken a stage further and helps yield some important results.

Total differentials

If the changes in variables x and y become infinitesimally small then even for non-linear functions

$$\Delta y = \frac{dy}{dx}\Delta x$$

These infinitesimally small changes in x and y are known as 'differentials'. When y is a function of more than one independent variable, e.g. $y = f(x, z)$, and there are infinitesimally small changes in all variables, then the total effect will be

$$\Delta y = \frac{\partial y}{\partial x}\Delta x + \frac{\partial y}{\partial z}\Delta z$$

This is known as the 'total differential' as it shows the total effect on y of changes in all independent variables.

It is usual to write dy, dx, dz etc. to represent infinitesimally small changes instead of $\Delta y, \Delta x, \Delta z$, which usually represent small, but finite, changes. Thus

$$dy = \frac{\partial y}{\partial x}dx + \frac{\partial y}{\partial z}dz$$

Example 10.23

What is the total differential of $y = 6x^2 + 8z^2 - 0.3xz$?

Solution

The total differential is

$$dy = \frac{\partial y}{\partial x}dx + \frac{\partial y}{\partial z}dz$$
$$= (12x - 0.3z)dx + (16z - 0.3x)dz$$

We can now demonstrate some examples of how the concept of a total differential can be used in economics.

In production theory, the slope of an isoquant represents the marginal rate of technical substitution (MRTS) between two inputs. The use of the total differential can help demonstrate that the MRTS will equal the ratio of the marginal products of the two inputs.

In introductory economics texts the MRTS of K for L (usually written as $MRTS_{KL}$) is usually defined as the amount of K that would be needed to compensate for the loss of one unit of L so that the production level remains unchanged. This is only an approximate measure though and more accuracy can be obtained when the $MRTS_{KL}$ is defined at a point on an isoquant. For infinitesimally small changes in K and L the $MRTS_{KL}$ measures the rate at which K needs to be substituted for L to keep output unchanged, i.e. it is equal to the negative of the slope of the isoquant at the point corresponding to the given values of K and L, when K is measured on the vertical axis and L on the horizontal axis.

For any given output level, K is effectively a function of L (and vice versa) and so, moving along an isoquant,

$$MRTS_{KL} = -\frac{dK}{dL} \tag{1}$$

For the production function $Q = f(K, L)$, the total differential is

$$dQ = \frac{\partial Q}{\partial K}dK + \frac{\partial Q}{\partial L}dL$$

If we are looking at a movement along the same isoquant then output is unchanged and so dQ is zero and thus

$$\frac{\partial Q}{\partial K}dK + \frac{\partial Q}{\partial L}dL = 0$$

$$\frac{\partial Q}{\partial K}dK = -\frac{\partial Q}{\partial L}dL$$

$$-\frac{dK}{dL} = \frac{\dfrac{\partial Q}{\partial L}}{\dfrac{\partial Q}{\partial K}} \tag{2}$$

We already know that $\partial Q/\partial L$ and $\partial Q/\partial K$ represent the marginal products of K and L. Therefore, from (1) and (2) above,

$$\mathrm{MRTS_{KL}} = \frac{\mathrm{MP_L}}{\mathrm{MP_K}}$$

Euler's theorem

Another use of the total differential is to prove *Euler's theorem* and demonstrate the conditions for the 'exhaustion of the total product'. This relates to the marginal productivity theory of factor pricing and the normative idea of what might be considered a 'fair wage', which was debated for many years by political economists.

Consider a firm that uses several different inputs. Each will contribute a different amount to total production. One suggestion for what might be considered a 'fair wage' was that each input, including labour, should be paid the 'value of its marginal product' (VMP). This is defined, for any input i, as marginal product (MP_i) multiplied by the price that the finished good is sold at (P_Q), i.e.

$$\mathrm{VMP}_i = P_Q \mathrm{MP}_i$$

Any such suggestion is, of course, a normative concept and the value judgements on which it is based can be questioned. However, what we are concerned with here is whether it is even *possible* to pay each input the value of its marginal product. If it is not possible, then it would not be a practical idea to set this as an objective even if it seemed a 'fair' principle.

Before looking at Euler's theorem we can illustrate how the conditions for product exhaustion can be derived for a Cobb–Douglas production function with two inputs. This example also shows how the product price is irrelevant to the product exhaustion question and it is the properties of the production function that matter.

Assume that a firm sells its output Q at a given price P_Q and that $Q = AK^\alpha L^\beta$ where A, α and β are constants. If each input was paid a price equal to the value of its marginal product then the prices of the two inputs K and L would be

$$P_K = VMP_K = P_Q \times \mathrm{MP_K} = P_Q \frac{\partial Q}{\partial K}$$

$$P_L = VMP_L = P_Q \times \mathrm{MP_L} = P_Q \frac{\partial Q}{\partial L}$$

The total expenditure on inputs would therefore be

$$\mathrm{TC} = KP_K + LP_L$$

$$= K \left(P_Q \frac{\partial Q}{\partial K} \right) + L \left(P_Q \frac{\partial Q}{\partial L} \right)$$

$$= P_Q \left(K \frac{\partial Q}{\partial K} + L \frac{\partial Q}{\partial L} \right) \tag{1}$$

Total revenue from the sale of the firm's output will be

$$\mathrm{TR} = P_Q Q$$

Total expenditure on inputs (which are paid the value of their marginal product) will equal total revenue when TR = TC. Therefore

$$P_Q Q = P_Q \left(K \frac{\partial Q}{\partial K} + L \frac{\partial Q}{\partial L} \right)$$

Cancelling P_Q, this gives

$$Q = K \frac{\partial Q}{\partial K} + L \frac{\partial Q}{\partial L} \tag{2}$$

Thus the *conditions of product exhaustion* are based on the physical properties of the production function. If (2) holds then the product is exhausted. If it does not hold then there will be either not enough revenue or a surplus.

For the Cobb–Douglas production function $Q = AK^\alpha L^\beta$ *we* know that

$$\frac{\partial Q}{\partial K} = \alpha A K^{\alpha-1} L^\beta \qquad \frac{\partial Q}{\partial L} = \beta A K^\alpha L^{\beta-1}$$

Substituting these values into (2), this gives

$$Q = K(\alpha A K^{\alpha-1} L^\beta) + L(\beta A K^\alpha L^{\beta-1})$$
$$= \alpha A K^\alpha L^\beta + \beta A K^\alpha L^\beta$$
$$= \alpha Q + \beta Q$$
$$Q = Q(\alpha + \beta) \tag{3}$$

The condition required for (3) to hold is that $\alpha + \beta = 1$. This means that product exhaustion occurs for a Cobb–Douglas production function when there are constant returns to scale.

We can also see from (3) and (1) that:

(i) when there are decreasing returns to scale and $\alpha + \beta < 1$, then

$$\text{TC} = P_Q(\alpha + \beta)Q < P_Q Q = \text{TR}$$

and so there will be a surplus left over if all inputs are paid their VMP, and

(ii) when there are increasing returns to scale and $\alpha + \beta > 1$, then

$$\text{TC} = P_Q(\alpha + \beta)Q > P_Q Q = \text{TR}$$

and so there will not be enough revenue to pay each input its VMP.

Euler's theorem also applies to the case of a general production function

$$Q = f(x_1, x_2, \ldots, x_n)$$

The previous example showed that the price will always cancel in the TR and TC formulae and what we are interested in is whether or not

$$Q = x_1 \frac{\partial Q}{\partial x_1} + x_2 \frac{\partial Q}{\partial x_2} + \cdots + x_n \frac{\partial Q}{\partial x_n} \tag{1}$$

Using the notation

$$f_1 = \frac{\partial Q}{\partial x_1}, f_2 = \frac{\partial Q}{\partial x_2}, \ldots \text{etc.}$$

the total differential of this production function will be

$$dQ = f_1 dx_1 + f_2 dx_2 + \cdots + f_n dx_n \tag{2}$$

Assume that all inputs are increased by the same proportion λ. Thus

$$\frac{dx_i}{x_i} = \lambda \quad \text{for all } i$$

and so $dx_i = \lambda x_i$ $\qquad\qquad$ (3)

Substituting (3) into (2) gives

$$dQ = f_1 \lambda x_1 + f_2 \lambda x_2 + \cdots + f_n \lambda x_n$$
$$= \lambda(f_1 dx_1 + f_2 dx_2 + \cdots + f_n dx_n)$$
$$\frac{dQ}{\lambda} = f_1 x_1 + f_2 x_2 + \cdots + f_n x_n \tag{4}$$

Multiplying top and bottom of the left-hand side of (4) by Q gives

$$\left(\frac{1}{\lambda}\frac{dQ}{Q}\right) Q = f_1 x_1 + f_2 x_2 + \cdots + f_n x_n$$

Thus, product exhaustion will only hold if

$$\frac{1}{\lambda}\frac{dQ}{Q} = 1$$

If this result does hold it means that output increases by the same proportion as the inputs,

$$\frac{dQ}{Q} = \lambda$$

i.e. there are constant returns to scale.

If there are decreasing returns to scale, output increases by a smaller proportion than the inputs. Therefore,

$$\frac{dQ}{Q} < \lambda$$

and so

$$\frac{1}{\lambda}\frac{dQ}{Q} < 1$$

This means that

$$f_1 x_1 + f_2 x_2 + \cdots + f_n x_n < Q$$

so that if each input is paid the value of its marginal product there will be some surplus left over.

Similarly, if there are increasing returns to scale then

$$\frac{dQ}{Q} > \lambda$$

Therefore,

$$\frac{1}{\lambda}\frac{dQ}{Q} > 1$$

and

$$f_1 x_1 + f_2 x_2 + \cdots + f_n x_n > Q$$

which means that the total cost of paying each input the value of its marginal product will sum to more than the total revenue earned, i.e. it will not be possible.

To sum up, Euler's theorem proves that if each input is paid the value of its marginal product the total cost of the inputs will

(i)   equal total revenue if there are constant returns to scale;
(ii) be less than total revenue if there are decreasing returns to scale;
(iii) be greater than total revenue if there are increasing returns to scale.

Example 10.24

Is it possible for a firm to pay each input the value of its marginal product if it operates with the production function $Q = 14K^{0.6}L^{0.8}$?

Solution

If each input is paid its VMP then the price of input K will be

$$P_K = \text{VMP}_K = P\text{MP}_K = P\frac{\partial Q}{\partial K} = P(8.4K^{-0.4}L^{0.8})$$

where P is the price of the final product. For input L,

$$P_L = \text{VMP}_L = P\text{MP}_L = P\frac{\partial Q}{\partial L} = P(11.2K^{0.6}L^{-0.2})$$

The total cost of inputs will therefore be

$$\begin{aligned}
\text{TC} &= P_K K + P_L L \\
&= P(8.4K^{-0.4}L^{0.8})K + P(11.2K^{0.6}L^{-0.2})L \\
&= P(8.4K^{0.6}L^{0.8}) + P(11.2K^{0.6}L^{0.8}) \\
&= P(8.4K^{0.6}L^{0.8} + 11.2K^{0.6}L^{0.8}) \\
&= P(19.6K^{0.6}L^{0.8})
\end{aligned}$$

The total revenue from selling the product will be

$$\text{TR} = PQ = P(14K^{0.6}L^{0.8})$$

Therefore,

$$\frac{TC}{TR} = \frac{P(19.6K^{0.6}L^{0.8})}{P(14K^{0.6}L^{0.8})} = \frac{19.6}{14} = 1.4$$

Thus the total revenue is not enough to pay each input the value of its marginal product. This checks out with the predictions of Euler's theorem, given that there are increasing returns to scale for this production function.

Total derivatives

In partial differentiation it is assumed that one variable changes while all other independent variables are held constant. However, in some instances there may be a connection between the independent variables and so this *ceteris paribus* assumption will not apply. For example, in a production function the amount of one input used may affect the amount of another input that can be used with it. From the total differential of a function we can derive a total derivative which can cope with this additional effect.

Assume $y = f(x, z)$ and also that $x = g(z)$.
Thus any change in z will affect y:

(a) directly via the function $f(x, z)$, and
(b) indirectly by changing x via the function $g(z)$, which in turn will affect y via the function $f(x, z)$.

The total differential of $y = f(x, z)$ is

$$dy = \frac{\partial y}{\partial x}dx + \frac{\partial y}{\partial z}dz$$

Dividing through by dz gives

$$\frac{dy}{dz} = \frac{\partial y}{\partial x}\frac{dx}{dz} + \frac{\partial y}{\partial z}$$

The first term shows the indirect effect of z, via its impact on x, and the second term shows the direct effect.

Example 10.25

If $Q = 25K^{0.4}L^{0.5}$ and $K = 0.8L^2$ what is the total effect of a change in L on Q? Identify the direct and indirect effects.

Solution
The total differential is

$$dQ = \frac{\partial Q}{\partial K}dK + \frac{\partial Q}{\partial L}dL$$

The total derivative with respect to L will be

$$\frac{dQ}{dL} = \frac{\partial Q}{\partial K}\frac{dK}{dL} + \frac{\partial Q}{\partial L} \tag{1}$$

From the functions given in the question we can derive

$$\frac{\partial Q}{\partial K} = 10K^{-0.6}L^{0.5} \qquad \frac{\partial Q}{\partial L} = 12.5K^{0.4}L^{-0.5} \qquad \frac{dK}{dL} = 1.6L$$

Substituting these derivatives into (1), we get

$$\frac{dQ}{dL} = (10K^{-0.6}L^{0.5})1.6L + 12.5K^{0.4}L^{-0.5}$$

$$= 16K^{-0.6}L^{1.5} + 12.5K^{0.4}L^{-0.5}$$

The first term shows the indirect effect of changes in L on Q and the second term shows the direct effect.

Test Yourself, Exercise 10.5

1. Derive the total differentials of the following production functions:

   (a) $Q = 20K^{0.6}L^{0.4}$
   (b) $Q = 48K^{0.3}L^{0.2}R^{0.4}$
   (c) $Q = 6K^{0.8} + 5L^{0.7} + 0.8K^2L^2$

2. If each input is paid the value of its marginal product, will this exhaust a firm's total revenue if the relevant production function is

   (a) $Q = 4K + 1.5L$?
   (b) $Q = 8K^{0.4}L^{0.3}$?
   (c) $Q = 3.5K^{0.25}L^{0.35}R^{0.3}$?

3. If $y = 40x^{0.4}z^{0.3}$ and $x = 5z^{0.25}$, find the total effect of a change in z on y.

4. A consumer spends all her income on the two goods A and B. The quantity of good A bought is determined by the demand function $Q_A = f(P_A, P_B, M)$ where P_A and P_B are the prices of the two goods and M is real income. A change in the price of A will also affect real income M via the function $M = g(P_A, P_B, £M)$ where $£M$ is money income. Derive an expression for the total effect of a change in P_A on Q_A.

11 Constrained optimization

Learning objectives

After completing this chapter students should be able to:
- Solve constrained optimization problems by the substitution method.
- Use the Lagrange method to set up and solve constrained maximization and constrained minimization problems.
- Apply the Lagrange method to resource allocation problems in economics.

11.1 Constrained optimization and resource allocation

Chapters 9 and 10 dealt with the optimization of functions without any constraints imposed. However, in economics we often come across resource allocation problems that involve the optimization of some variable subject to certain limitations. For example, a firm may try to maximize output subject to a budget constraint for expenditure on inputs, or it may wish to minimize costs subject to a specified output being produced. We have already seen in Chapter 5 how constrained optimization problems with linear constraints and objective functions can be tackled using linear programming. This chapter now explains how problems involving the constrained optimization of non-linear functions can be tackled, using partial differentiation.

We shall consider two methods:

 (i) constrained optimization by substitution, and
(ii) the Lagrange multiplier method.

The Lagrange multiplier method can be used for most types of constrained optimization problems. The substitution method is mainly suitable for problems where a function with only two variables is maximized or minimized subject to one constraint. We shall consider this simpler substitution method first.

11.2 Constrained optimization by substitution

Consider the example of a firm that wishes to maximize output $Q = f(K, L)$, with a fixed budget M for purchasing inputs K and L at set prices P_K and P_L. This problem is illustrated in Figure 11.1. The firm needs to find the combination of K and L that will allow it to reach

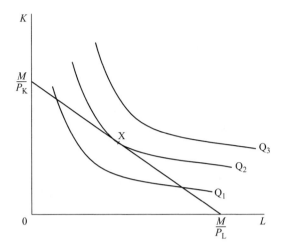

Figure 11.1

the optimum point X which is on the highest possible isoquant within the budget constraint with intercepts M/P_K and M/P_L.

To determine a solution for this type of economic resource allocation problem we have to reformulate it as a mathematical constrained optimization problem. The following examples suggest ways in which this can be done.

Example 11.1

A firm faces the production function $Q = 12K^{0.4}L^{0.4}$ and can buy the inputs K and L at prices per unit of £40 and £5 respectively. If it has a budget of £800 what combination of K and L should it use in order to produce the maximum possible output?

Solution

The problem is to maximize the function $Q = 12K^{0.4}L^{0.4}$ subject to the budget constraint

$$40K + 5L = 800 \tag{1}$$

(In all problems in this chapter, it is assumed that each constraint 'bites'; e.g. all the budget is used in this example.)

The theory of the firm tells us that a firm is optimally allocating a fixed budget if the last £1 spent on each input adds the same amount to output, i.e. marginal product over price should be equal for all inputs. This optimization condition can be written as

$$\frac{MP_K}{P_K} = \frac{MP_L}{P_L} \tag{2}$$

The marginal products can be determined by partial differentiation:

$$\text{MP}_K = \frac{\partial Q}{\partial K} = 4.8K^{-0.6}L^{0.4} \tag{3}$$

$$\text{MP}_L = \frac{\partial Q}{\partial L} = 4.8K^{0.4}L^{-0.6} \tag{4}$$

Substituting (3) and (4) and the given prices for P_K and P_L into (2)

$$\frac{4.8K^{-0.6}L^{0.4}}{40} = \frac{4.8K^{0.4}L^{-0.6}}{5}$$

Dividing both sides by 4.8 and multiplying by 40 gives

$$K^{-0.6}L^{0.4} = 8K^{0.4}L^{-0.6}$$

Multiplying both sides by $K^{0.6}L^{0.6}$ gives

$$L = 8K \tag{5}$$

Substituting (5) for L into the budget constraint (1) gives

$$40K + 5(8K) = 800$$
$$40K + 40K = 800$$
$$80K = 800$$

Thus the optimal value of K is

$$K = 10$$

and, from (5), the optimal value of L is

$$L = 80$$

Note that although this method allows us to derive optimum values of K and L that satisfy condition (2) above, it does not provide a check on whether this is a unique solution, i.e. there is no second-order condition check. However, it may be assumed that in all the problems in this section the objective function is maximized (or minimized depending on the question) when the basic economic rules for an optimum are satisfied.

The above method is not the only way of tackling this problem by substitution. An alternative approach, explained below, is to encapsulate the constraint within the function to be maximized, and then maximize this new objective function.

Example 11.1 (reworked)

Solution

From the budget constraint

$$40K + 5L = 800$$
$$5L = 800 - 40K \tag{1}$$
$$L = 160 - 8K \tag{2}$$

Substituting (2) into the objective function $Q = 12K^{0.4}L^{0.4}$ gives

$$Q = 12K^{0.4}(160 - 8K)^{0.4} \tag{3}$$

We are now faced with the unconstrained optimization problem of finding the value of K that maximizes the function (3) which has the budget constraint (1) 'built in' to it by substitution. This requires us to set d $Q/dK = 0$. However, it is not straightforward to differentiate the function in (3), and we must wait until further topics in calculus have been covered before proceeding with this solution (see Chapter 12, Example 12.9).

To make sure that you understand the basic substitution method, we shall use it to tackle another constrained maximization problem.

Example 11.2

A firm faces the production function $Q = 20K^{0.4}L^{0.6}$. It can buy inputs K and L for £400 a unit and £200 a unit respectively. What combination of L and K should be used to maximize output if its input budget is constrained to £6,000?

Solution

$$\text{MP}_\text{L} = \frac{\partial Q}{\partial L} = 12K^{0.4}L^{-0.4} \qquad \text{MP}_\text{K} = \frac{\partial Q}{\partial K} = 8K^{-0.6}L^{0.6}$$

Optimal input mix requires

$$\frac{\text{MP}_\text{L}}{P_\text{L}} = \frac{\text{MP}_\text{K}}{P_\text{K}}$$

Therefore

$$\frac{12K^{0.4}L^{-0.4}}{200} = \frac{8K^{-0.6}L^{0.6}}{400}$$

Cross multiplying gives

$$4,800K = 1,600L$$

$$3K = L$$

Substituting this result into the budget constraint

$$200L + 400K = 6,000$$

gives

$$200(3K) + 400K = 6,000$$
$$600K + 400K = 6,000$$
$$1,000K = 6,000$$
$$K = 6$$

Therefore

$$L = 3K = 18$$

The examples of constrained optimization considered so far have only involved output maximization when a firm faces a Cobb–Douglas production function, but the same technique can also be applied to other forms of production functions.

Example 11.3

A firm faces the production function

$$Q = 120L + 200K - L^2 - 2K^2$$

for positive values of Q. It can buy L at £5 a unit and K at £8 a unit and has a budget of £70. What is the maximum output it can produce?

Solution

$$\text{MP}_\text{L} = \frac{\partial Q}{\partial L} = 120 - 2L \qquad \text{MP}_\text{K} = \frac{\partial Q}{\partial K} = 200 - 4K$$

For optimal input combination

$$\frac{\text{MP}_\text{L}}{P_\text{L}} = \frac{\text{MP}_\text{K}}{P_\text{K}}$$

Therefore, substituting MP_K and MP_L and the given input prices

$$\frac{120 - 2L}{5} = \frac{200 - 4K}{8}$$

$$8(120 - 2L) = 5(200 - 4K)$$

$$960 - 16L = 1,000 - 20K$$

$$20K = 40 + 16L$$

$$K = 2 + 0.8L \tag{1}$$

Substituting (1) into the budget constraint

$$5L + 8K = 70$$

gives

$$5L + 8(2 + 0.8L) = 70$$

$$5L + 16 + 6.4L = 70$$

$$11.4L = 54$$

$$L = 4.74 \quad \text{(to 2 dp)}$$

Substituting this result into (1)

$$K = 2 + 0.8(4.74) = 5.79$$

Therefore maximum output is

$$Q = 120L + 200K - L^2 - 2K^2$$
$$= 120(4.74) + 200(5.79) - (4.74)^2 - 2(5.79)^2$$
$$= 1,637.28$$

This technique can also be applied to consumer theory, where utility is maximized subject to a budget constraint.

Example 11.4

The utility a consumer derives from consuming the two goods A and B can be assumed to be determined by the utility function $U = 40A^{0.25}B^{0.5}$. If A costs £4 a unit and B costs £10 a unit and the consumer's income is £600, what combination of A and B will maximize utility?

Solution
The marginal utility of A is

$$\text{MU}_A = \frac{\partial U}{\partial A} = 10A^{-0.75}B^{0.5}$$

The marginal utility of B is

$$\text{MU}_B = \frac{\partial U}{\partial B} = 20A^{0.25}B^{-0.5}$$

Consumer theory tells us that total utility will be maximized when the utility derived from the last pound spent on each good is equal to the utility derived from the last pound spent on any other good. This optimization rule can be expressed as

$$\frac{\text{MU}_A}{P_A} = \frac{\text{MU}_B}{P_B}$$

Therefore, substituting the above MU functions and the given prices of £4 and £10, this condition becomes

$$\frac{10A^{-0.75}B^{0.5}}{4} = \frac{20A^{0.25}B^{-0.5}}{10}$$
$$100B = 80A$$
$$B = 0.8A \tag{1}$$

Substituting (1) for B in the budget constraint

$$4A + 10B = 600$$

gives

$$A + 10(0.8A) = 600$$
$$4A + 8A = 600$$
$$12A = 600$$
$$A = 50$$

Thus from (1)

$$B = 0.8(50) = 40$$

The substitution method can also be used for **constrained minimization** problems. If output is given and a firm is required to minimize the cost of this output, then one variable can be eliminated from the production function before it is substituted into the cost function which is to be minimized.

Example 11.5

A firm operates with the production function $Q = 4K^{0.6}L^{0.4}$ and buys inputs K and L at prices per unit of £40 and £15 respectively. What is the cheapest way of producing 600 units of output?

Solution

The output constraint is

$$600 = 4K^{0.6}L^{0.4}$$

Therefore

$$\frac{150}{K^{0.6}} = L^{0.4}$$

$$\left(\frac{150}{K^{0.6}}\right)^{2.5} = L$$

$$\frac{275{,}567.6}{K^{1.5}} = L \tag{1}$$

The total cost of inputs, which is to be minimized, is

$$\text{TC} = 40K + 15L \tag{2}$$

Substituting (1) into (2) gives

$$\text{TC} = 40K + 15(275{,}567.6)K^{-1.5}$$

Differentiating and setting equal to zero to find a stationary point

$$\frac{dTC}{dK} = 40 - 22.5(275,567.6)K^{-2.5} = 0 \qquad (3)$$

$$40 = \frac{22.5(275,567.6)}{K^{2.5}}$$

$$K^{2.5} = \frac{22.5(275,567.6)}{40} = 155,006.78$$

$$K = 119.16268$$

Substituting this value into (1) gives

$$L = \frac{275,567.6}{(119.1628)^{1.5}} = 211.84478$$

This time we can check the second-order condition for minimization. Differentiating (3) again gives

$$\frac{d^2TC}{dK^2} = (2.5)22.5(275,567.6)K^{-3.5} > 0 \text{ for any } K > 0$$

This confirms that these values minimize TC. We can also check that these values give 600 when substituted back into the production function.

$$Q = 4K^{0.6}L^{0.4} = 4(119.16268)^{0.6}(211.84478)^{0.4} = 600$$

Thus cost minimization is achieved when $K = 119.16$ and $L = 212.84$ (to 2 dp) and so total production costs will be

$$TC = 40(119.16) + 15(211.84) = £7,944$$

Test Yourself, Exercise 11.1

1. If a firm has a budget of £378 what combination of K and L will maximize output given the production function $Q = 40K^{0.6}L^{0.3}$ and prices for K and L of £20 per unit and £6 per unit respectively?

2. A firm faces the production function $Q = 6K^{0.4}L^{0.5}$. If it can buy input K at £32 a unit and input L at £8 a unit, what combination of L and K should it use to maximize production if it is constrained by a fixed budget of £36,000?

3. A consumer spends all her income of £120 on the two goods A and B. Good A costs £10 a unit and good B costs £15. What combination of A and B will she purchase if her utility function is $U = 4A^{0.5}B^{0.5}$?

4. If a firm faces the production function $Q = 4K^{0.5}L^{0.5}$, what is the maximum output it can produce for a budget of £200? The prices of K and L are given as £4 per unit and £2 per unit respectively.

5. Make up your own constrained optimization problem for an objective function with two independent variables and solve it using the substitution method.

6. A firm faces the production function $Q = 2K^{0.2}L^{0.6}$ and can buy L at £240 a unit and K at £4 a unit.

    (a) If it has a budget of £16,000 what combination of K and L should it use to maximize output?
    (b) If it is given a target output of 40 units of Q what combination of K and L should it use to minimize the cost of this output?

7. A firm has a budget of £1,140 and can buy inputs K and L at £3 and £8 respectively a unit. Its output is determined by the production function

    $$Q = 6K + 20L - 0.025K^2 - 0.05L^2$$

    for positive values of Q. What is the maximum output it can produce?

8. A firm operates with the production function $Q = 30K^{0.4}L^{0.2}$ and buys inputs K and L at £12 per unit and £5 per unit respectively. What is the cheapest way of producing 750 units of output? (Work to nearest whole units of K and L.)

11.3   The Lagrange multiplier: constrained maximization with two variables

The best way to explain how to use the Lagrange multiplier is with an example and so we shall work through the problem in Example 11.1 from the last section using the Lagrange multiplier method.

The firm is trying to maximize output $Q = 12K^{0.4}L^{0.4}$ subject to the budget constraint $40K + 5L = 800$. The first step is to rearrange the budget constraint so that zero appears on one side of the equality sign. Therefore

$$0 = 800 - 40K - 5L \tag{1}$$

We then write the 'Lagrange equation' or 'Lagrangian' in the form

$$G = \text{(function to be optimized)} + \lambda\text{(constraint)}$$

where G is just the value of the Lagrangian function and λ is known as the 'Lagrange multiplier'. (Do not worry about where these terms come from or what their actual values are. They are just introduced to help the analysis. Note also that in other texts a 'curly L' is often used to represent the Lagrange function. This can confuse students because economics problems frequently involve labour, represented by L, as one of the variables in the function to be optimized. This text therefore uses the notation 'G' to avoid this confusion. However, if you are already accustomed to using the 'curly L' you can, of course, continue to use it when answering problems yourself. What matters is whether you understand the analysis, not what symbols you use.)

In this problem the Lagrange function is thus

$$G = 12K^{0.4}L^{0.4} + \lambda(800 - 40K - 5L) \tag{2}$$

Next, derive the partial derivatives of G with respect to K, L and λ and set them equal to zero, i.e. find the stationary points of G that satisfy the first-order conditions for a maximum.

$$\frac{\partial G}{\partial K} = 4.8K^{-0.6}L^{0.4} - 40\lambda = 0 \tag{3}$$

$$\frac{\partial G}{\partial L} = 4.8K^{0.4}L^{-0.6} - 5\lambda = 0 \tag{4}$$

$$\frac{\partial G}{\partial \lambda} = 800 - 40K - 5L = 0 \tag{5}$$

You will note that (5) is the same as the budget constraint (1). We now have a set of three linear simultaneous equations in three unknowns to solve for K and L. The Lagrange multiplier λ can be eliminated as, from (3),

$$0.12K^{-0.6}L^{0.4} = \lambda$$

and from (4)

$$0.96K^{0.4}L^{-0.6} = \lambda$$

Therefore

$$0.12K^{-0.6}L^{0.4} = 0.96K^{0.4}L^{-0.6}$$

Multiplying both sides by $K^{0.6}L^{0.6}$,

$$0.12L = 0.96K$$

$$L = 8K \tag{6}$$

Substituting (6) into (5),

$$800 - 40K - 5(8K) = 0$$

$$800 = 80K$$

$$10 = K$$

Substituting back into (5),

$$800 - 40(10) - 5L = 0$$

$$400 = 5L$$

$$80 = L$$

These are the same values of K and L as those obtained by the substitution method. Thus, the values of K and L that satisfy the first-order conditions for a maximum value of the Lagrangian function G are the values that will maximize output subject to the given budget constraint. We shall just accept this result without going into the proof of why this is so.

Strictly speaking we should now check the second-order conditions in the above problem to be sure that we actually have a maximum rather than a minimum. These, however, are rather complex, involving an examination of the function at and near the stationary points found, and are discussed in the next section. For the time being you can assume that once the stationary points of a Lagrangian function have been found the second-order conditions for a maximum will automatically be met. Some more examples are worked through so that you can become familiar with this method.

Example 11.6

A firm can buy two inputs K and L at £18 per unit and £8 per unit respectively and faces the production function $Q = 24K^{0.6}L^{0.3}$. What is the maximum output it can produce for a budget of £50,000? (Work to nearest whole units of K, L and Q.)

Solution

The budget constraint is $50,000 - 18K - 8L = 0$ and the function to be maximized is $Q = 24K^{0.6}L^{0.3}$. The Lagrangian for this problem is therefore

$$G = 24K^{0.6}L^{0.3} + \lambda(50,000 - 18K - 8L)$$

Partially differentiating to find the stationary points of G gives

$$\frac{\partial G}{\partial K} = 14.4K^{-0.4}L^{0.3} - 18\lambda = 0$$

$$\frac{14.4L^{0.3}}{18K^{0.4}} = \lambda \qquad (1)$$

$$\frac{\partial G}{\partial L} = 7.2K^{0.6}L^{-0.7} - 8\lambda = 0$$

$$\frac{7.2K^{0.6}}{8L^{0.7}} = \lambda \qquad (2)$$

$$\frac{\partial G}{\partial \lambda} = 50,000 - 18K - 8L = 0 \qquad (3)$$

Setting (1) equal to (2) to eliminate λ

$$\frac{14.4L^{0.3}}{18K^{0.4}} = \frac{7.2K^{0.6}}{8L^{0.7}}$$

$$115.2L = 129.6K$$

$$L = 1.125K \qquad (4)$$

Substituting (4) into (3)

$$50,000 - 18K - 8(1.125K) = 0$$

$$50,000 - 18K - 9K = 0$$

$$50,000 = 27K$$

$$1,851.8519 = K \qquad (5)$$

Substituting (5) into (4)

$$L = 1.125(1,851.8519) = 2,083.3334$$

Thus, to the nearest whole unit, optimum values of K and L are 1,852 and 2,083 respectively.

We can check that when these whole values of K and L are used the total cost will be

$$TC = 18K + 8L = 18(1,852) + 8(2,083) = 33,336 + 16,664 = £50,000$$

and so the budget constraint is satisfied. The actual maximum output level will be

$$Q = 24K^{0.6}L^{0.3} = 24(1,852)^{0.6}(2,083)^{0.3} = 21,697 \text{ units}$$

Although the same mathematical method can be used for various economic applications, you must learn to use your knowledge of economics to set up the mathematical problem in the first place. The example below demonstrates another application of the Lagrange method.

Example 11.7

A consumer has the utility function $U = 40A^{0.5}B^{0.5}$. The prices of the two goods A and B are initially £20 and £5 per unit respectively, and the consumer's income is £600. The price of A then falls to £10. Work out the income and substitution effects of this price change on the amount of A consumed using Hicks's method and say whether A and B are normal or inferior goods.

Solution

To help solve this problem the relevant budget schedules and indifference curves are illustrated in Figure 11.2, although the indifference curves are not accurately drawn to scale. The original optimum is at X. The price fall for A causes the budget line to become flatter and swing round, giving a new equilibrium at Y.

Hicks's method for splitting the total change in A into its income and substitution effects requires one to draw a 'ghost' budget line parallel to the new budget line (reflecting the new

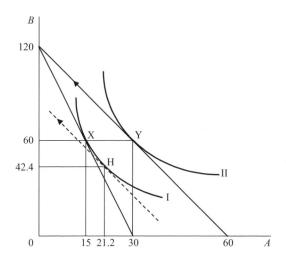

Figure 11.2

relative prices) but tangential to the original indifference curve. This is shown by the broken line tangential to indifference curve I at H. From X to H is the substitution effect and from H to Y is the income effect of the price change. This problem requires us to find the corresponding values of A and B for the three tangency points X, Y and H and then to comment on the direction of these changes.

The original equilibrium is the combination of A and B that maximizes the utility function $U = 40A^{0.5}B^{0.5}$ subject to the budget constraint $600 = 20A + 5B$. These values of A and B can be found by deriving the stationary points of the Lagrange function

$$G = 40A^{0.5}B^{0.5} + \lambda(600 - 20A - 5B)$$

Thus

$$\frac{\partial G}{\partial A} = 20A^{-0.5}B^{0.5} - 20\lambda = 0 \quad \text{giving} \quad A^{-0.5}B^{0.5} = \lambda \tag{1}$$

$$\frac{\partial G}{\partial B} = 20A^{0.5}B^{-0.5} - 5\lambda = 0 \quad \text{giving} \quad 4A^{0.5}B^{-0.5} = \lambda \tag{2}$$

$$\frac{\partial G}{\partial \lambda} = 600 - 20A - 5B = 0 \tag{3}$$

Setting (1) equal to (2)

$$A^{-0.5}B^{0.5} = 4A^{0.5}B^{-0.5}$$

$$B = 4A \tag{4}$$

Substituting (4) into (3)

$$600 - 20A - 5(4A) = 0$$

$$600 = 40A$$

$$15 = A$$

Substituting this value into (4)

$$B = 4(15) = 60$$

Thus, $A = 15$ and $B = 60$ at the original equilibrium at X.

When the price of A falls to 10, the budget constraint becomes

$$600 = 10A + 5B$$

and so the new Lagrange function is

$$G = 40A^{0.5}B^{0.5} + \lambda(600 - 10A - 5B)$$

New stationary points will be where

$$\frac{\partial G}{\partial A} = 20A^{-0.5}B^{0.5} - 10\lambda = 0 \quad \text{giving} \quad 2A^{-0.5}B^{0.5} = \lambda \tag{5}$$

$$\frac{\partial G}{\partial B} = 20A^{0.5}B^{-0.5} - 5\lambda = 0 \quad \text{giving} \quad 4A^{0.5}B^{-0.5} = \lambda \tag{6}$$

$$\frac{\partial G}{\partial \lambda} = 600 - 10A - 5B = 0 \tag{7}$$

Setting (5) equal to (6)

$$2A^{-0.5}B^{0.5} = 4A^{0.5}B^{-0.5}$$

$$B = 2A \tag{8}$$

Substituting (8) into (7)

$$600 - 10A - 5(2A) = 0$$

$$600 = 20A$$

$$30 = A$$

Substituting this value into (8) gives

$$B = 2(30) = 60.$$

Thus, the total effect of the price change is to increase consumption of A from 15 to 30 units and leave consumption of B unchanged at 60.

There are several ways of finding the values of A and B that correspond to point H. We know that H is on the same indifference curve as point X, and therefore the utility function will take the same value at both points. We can find the value of utility at X where $A = 15$ and $B = 60$. This will be

$$U = 40A^{0.5}B^{0.5} = 40(15)^{0.5}(60)^{0.5} = 40(900)^{0.5} = 40(30) = 1,200$$

Thus, at any point on the indifference curve I

$$40A^{0.5}B^{0.5} = 1,200$$

$$B^{0.5} = 30A^{-0.5}$$

$$B = 900A^{-1} \tag{9}$$

The slope of indifference curve I will therefore be

$$\frac{\mathrm{d}B}{\mathrm{d}A} = -900A^{-2} \tag{10}$$

At point X, the indifference curve I is tangential to the new budget line whose slope will be

$$\frac{-P_A}{P_B} = \frac{-10}{5} = -2 \tag{11}$$

Therefore, from (10) and (11)

$$-900A^{-2} = -2$$

$$450 = A^2$$

$$21.2132 = A$$

Substituting this value into (9)

$$B = 900(21.2132)^{-1} = 42.4264$$

Thus the substitution effect of A's price fall, from X to H, increases consumption of A from 15 to 21.2 units and decreases consumption of B from 60 to 42.4 units. This effect is negative (i.e. quantity rises when price falls) in line with standard consumer theory.

The income effect, from H to Y, increases consumption of A from 21.2 to 30 units and also increases consumption of B from 42.4 back to its original 60 unit level. As both income effects are positive, both A and B must be normal goods.

Test Yourself, Exercise 11.2

Use the Lagrange method to answer questions 1, 2, 3, 4, 6(a) and 7 from Test Yourself, Exercise 11.1.

11.4   The Lagrange multiplier: second-order conditions

Inasmuch as it involves setting the first derivatives of the objective function equal to zero, the Lagrange method of solving constrained optimization problems is similar to the method of solving unconstrained optimization problems involving functions of several variables that was explained in Chapter 10. However, one cannot simply apply the same set of second-order conditions to check for a maximum or minimum because of the special role that the Lagrange multiplier takes. The mathematics required to prove why this is so, and to explain what additional second-order conditions are necessary for a Lagrangian function to be a maximum or minimum, becomes rather complex. We shall therefore just look at an intuitive explanation of what these conditions involve here. The use of matrix algebra to check second-order conditions in constrained optimization problems will then be explained later in Chapter 15.

First, we shall consider the conditions for a maximum. If we assume that a function has two independent variables then both the function to be maximized, $f(A, B)$, and the constraint could take on several possible forms, as illustrated in Figure 11.3. These diagrams are all constructed on the same basis as isoquant maps. Thus the lines I, II and III represent different levels of the objective function with its value increasing as one moves away from the origin.

In Figure 11.3(a), the objective function is convex to the origin and the constraint CD is linear. Maximization of the objective function occurs at the tangency point T.

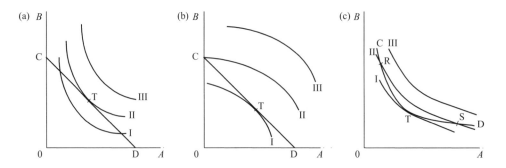

Figure 11.3

In Figure 11.3(b), the objective function is concave to the origin and so it is maximized subject to the linear constraint CD at the corner point C. Thus the tangency point T does *not* determine the maximum value.

In Figure 11.3(c), the objective function is convex to the origin but the constraint CD is non-linear and more sharply curved than the objective function. Thus the tangency point T is *not* the maximum value. Higher values of the objective function can be found at points R and S, for example.

From the above examples we can see that, for the two-variable case, a linear constraint and an objective function that is convex to the origin will ensure maximization at the point of tangency. In other cases, tangency may not ensure maximization. If a problem involves the maximization of production subject to a linear budget constraint this means that output is maximized where the slope of the isoquant is equal to the slope of the budget constraint. We have already seen in Chapter 10 how a Cobb–Douglas production function in the form $Q = AK^{\alpha}L^{\beta}$ will correspond to a set of isoquants which continually decline and become flatter as L is increased, i.e. are convex to the origin. Thus in any constrained optimization problem where one is attempting to maximize a production function in the Cobb–Douglas format subject to a linear budget constraint, the input combination that satisfies the first-order conditions will be a maximum.

If the Lagrangian represents other concepts with similar shaped functions and constraints, such as utility, the same conditions apply. In all such cases, one assumes that the independent variables in the objective function must take positive values and so any negative mathematical solutions can be disregarded.

Although we cannot illustrate functions with more than two variables diagrammatically, the same basic principles apply when one is attempting to maximize a function with three or more variables subject to a linear constraint. Thus any Cobb–Douglas production function with more than two inputs will be at a maximum, subject to a specified linear budget constraint, when the first-order conditions for optimization of the relevant Lagrange equation are met. For the purpose of answering the problems in this Chapter, and for most constrained maximization problems that you will encounter in a first-year economics course, it can be assumed that the stationary points of the Lagrange function will satisfy the second-order conditions for a maximum.

The properties of the objective function and constraint that guarantee that a Lagrange function is minimized when first-order conditions are met are the reverse of the properties required for a maximum, i.e. the objective function must be linear and the constraint must be convex to the origin. Thus if one is required to find values of K and L that minimize a budget function in the form

$$TC = P_K K + P_L L$$

subject to a given output Q^* being produced via the production function $Q = AK^{\alpha}L^{\beta}$, then the corresponding Lagrange function is

$$G = P_K K + P_L L + \lambda(Q^* - AK^{\alpha}L^{\beta})$$

and the values of K and L which satisfy the first-order conditions

$$\frac{\partial G}{\partial K} = 0 \qquad \frac{\partial G}{\partial L} = 0$$

will also satisfy second-order conditions for a minimum.

If you refer back to Figure 11.3(a) you can see the rationale for this rule in the two-input case. If input prices are given and one is trying to minimize the cost of the output represented by isoquant II, then one needs to find the budget constraint with slope equal to the negative of the price ratio which is nearest the origin and still goes through this isoquant. The linear objective function and the constraint convex to the origin guarantee that this will be at the tangency point T. Some examples of minimization problems that use this rule are given in the next section.

To conclude this section, let us reiterate what we have learned about second-order conditions and Lagrangians. When the objective function is in the Cobb–Douglas format and the constraint is linear, then the second-order conditions for a maximum are met at the stationary points of the Lagrange function. This rule is reversed for minimization.

11.5   Constrained minimization using the Lagrange multiplier

As was explained in Section 11.4, the same principles used to construct a Lagrange function for a constrained maximization problem are used to construct a Lagrange function for a constrained minimization problem. The difference is that the components of the function are reversed, as is shown in the following examples. In all these cases the constraints and objective function take formats which guarantee that second-order conditions for a minimum are met.

Example 11.8

A firm operates with the production function $Q = 4K^{0.6}L^{0.5}$ and can buy K at £15 a unit and L at £8 a unit. What input combination will minimize the cost of producing 200 units of output?

Solution

The output constraint is $200 = 4K^{0.6}L^{0.5}$ and the objective function to be minimized is the total cost function TC $= 15K + 8L$. The corresponding Lagrangian function is therefore

$$G = 15K + 8L + \lambda(200 - 4K^{0.6}L^{0.5})$$

Partially differentiating G and setting equal to zero, first-order conditions require

$$\frac{\partial G}{\partial K} = 15 - \lambda 2.4K^{-0.4}L^{0.5} = 0 \quad \text{giving} \quad \frac{15K^{0.4}}{2.4L^{0.5}} = \lambda \tag{1}$$

$$\frac{\partial G}{\partial L} = 8 - \lambda 2K^{0.6}L^{-0.5} = 0 \quad \text{giving} \quad \frac{4L^{0.5}}{K^{0.6}} = \lambda \tag{2}$$

$$\frac{\partial G}{\partial \lambda} = 200 - 4K^{0.6}L^{0.5} = 0 \tag{3}$$

Setting (1) equal to (2) to eliminate λ

$$\frac{15K^{0.4}}{2.4L^{0.5}} = \frac{4L^{0.5}}{K^{0.6}}$$

$$15K = 9.6L$$

$$1.5625K = L \tag{4}$$

Substituting (4) into (3)

$$200 - 4K^{0.6}(1.5625K)^{0.5} = 0$$

$$200 = 4K^{0.6}(1.5625)^{0.5}K^{0.5}$$

$$\frac{200}{4(1.5625)^{0.5}} = K^{1.1}$$

$$40 = K^{1.1}$$

$$K = \sqrt[1.1]{40} = 28.603434$$

Substituting this value into (4)

$$1.5625(28.603434) = L$$

$$44.692866 = L$$

Thus the optimal input combination is 28.6 units of K plus 44.7 units of L (to 1 dp). We can check that these input values correspond to the given output level by substituting them back into the production function. Thus

$$Q = 4K^{0.6}L^{0.5} = 4(28.6)^{0.6}(44.7)^{0.5} = 200$$

which is correct, allowing for rounding error. The actual cost entailed will be

$$TC = 15K + 8L = 15(28.6) + 8(44.7) = \pounds786.60$$

Example 11.9

The prices of inputs K and L are given as £12 per unit and £3 per unit respectively, and a firm operates with the production function $Q = 25K^{0.5}L^{0.5}$.

(i) What is the minimum cost of producing 1,250 units of output?
(ii) Demonstrate that the maximum output that can be produced for this budget will be the 1,250 units specified in (i) above.

Solution

This question essentially asks us to demonstrate that the constrained maximization and minimization methods give consistent answers.

(i) The output constraint is that

$$1,250 = 25K^{0.5}L^{0.5}$$

The objective function to be minimized is the cost function

$$TC = 12K + 3L$$

The corresponding Lagrange function is therefore

$$G = 12K + 3L + \lambda(1{,}250 - 25K^{0.5}L^{0.5})$$

First-order conditions require

$$\frac{\partial G}{\partial K} = 12 - \lambda 12.5K^{-0.5}L^{0.5} = 0 \quad \text{giving} \quad \frac{12K^{0.5}}{12.5L^{0.5}} = \lambda \tag{1}$$

$$\frac{\partial G}{\partial L} = 3 - \lambda 12.5K^{0.5}L^{-0.5} = 0 \quad \text{giving} \quad \frac{3L^{0.5}}{12.5K^{0.5}} = \lambda \tag{2}$$

$$\frac{\partial G}{\partial \lambda} = 1{,}250 - 25K^{0.5}L^{0.5} = 0 \tag{3}$$

Setting (1) equal to (2)

$$\frac{12K^{0.5}}{12.5L^{0.5}} = \frac{3L^{0.5}}{12.5K^{0.5}}$$

$$4K = L \tag{4}$$

Substituting (4) into (3)

$$1{,}250 - 25K^{0.5}(4K)^{0.5} = 0$$
$$1{,}250 = 25K^{0.5}(4)^{0.5}K^{0.5}$$
$$1{,}250 = 50K$$
$$25 = K$$

Substituting this value into (4)

$$4(25) = L$$
$$100 = L$$

When these optimum values of K and L are used the actual minimum cost will be

$$TC = 12K + 3L = 12(25) + 3(100) = 300 + 300 = £600$$

(ii) This part of the question requires us to find the values of K and L that will maximize output subject to a budget of £600, i.e. the answer to (i) above. The objective function to be maximized is therefore $Q = 25K^{0.5}L^{0.5}$ and the constraint is $12K + 3L = 600$. The corresponding Lagrange equation is thus

$$G = 25K^{0.5}L^{0.5} + \lambda(600 - 12K - 3L)$$

First-order conditions require

$$\frac{\partial G}{\partial K} = 12.5K^{-0.5}L^{0.5} - 12\lambda = 0 \quad \text{giving} \quad \frac{12.5L^{0.5}}{12K^{0.5}} = \lambda \tag{5}$$

$$\frac{\partial G}{\partial L} = 12.5K^{0.5}L^{-0.5} - 3\lambda = 0 \quad \text{giving} \quad \frac{12.5K^{0.5}}{3L^{0.5}} = \lambda \tag{6}$$

$$\frac{\partial G}{\partial \lambda} = 600 - 12K - 3L = 0 \tag{7}$$

Setting (5) equal to (6)

$$\frac{12.5L^{0.5}}{12K^{0.5}} = \frac{12.5K^{0.5}}{3L^{0.5}}$$

$$3L = 12K$$

$$L = 4K \tag{8}$$

Substituting (8) into (7)

$$600 - 12K - 3(4K) = 0$$

$$600 - 12K - 12K = 0$$

$$600 = 24K$$

$$25 = K$$

Substituting this value into (8) $L = 4(25) = 100$

These are the same optimum values of K and L that were found in part (i) above. The actual output produced by 25 of K plus 100 of L will be

$$Q = 25K^{0.5}L^{0.5} = 25(25)^{0.5}(100)^{0.5} = 1,250$$

which checks out with the amount specified in the question.

Although most of the examples of constrained optimization presented in this chapter are concerned with a firm's output and costs, or a consumer's utility level and income, the Lagrange method can be applied to other areas of economics. For instance, in environmental economics one may wish to find the cheapest way of securing a given level of environmental cleanliness.

Example 11.10

Assume that there are two sources of pollution into a lake. The local water authority can clean up the discharges and reduce pollution levels from these sources but there are, of course, costs involved. The damage effects of each pollution source are measured on a 'pollution scale'. The lower the pollution level the greater the cost of achieving it, as is shown by the cost schedules for cleaning up the two pollution sources:

$$Z_1 = 478 - 2C_1^{0.5} \quad \text{and} \quad Z_2 = 600 - 3C_2^{0.5}$$

where Z_1 and Z_2 are pollution levels and C_1 and C_2 are expenditure levels (in £000s) on reducing pollution.

To secure an acceptable level of water purity in the lake the water authority's objective is to reduce the total pollution level to 1,000 by the cheapest method. How can it do this?

Solution

This can be formulated as a constrained optimization problem where the constraint is the total amount of pollution $Z_1 + Z_2 = 1000$ and the objective function to be minimized is the cost of pollution control $TC = C_1 + C_2$. Thus the Lagrange function is

$$G = C_1 + C_2 + \lambda(1,000 - Z_1 - Z_2)$$

Substituting in the cost functions for Z_1 and Z_2, this becomes

$$G = C_1 + C_2 + \lambda[1,000 - (478 - 2C_1^{0.5}) - (600 - 3C_2^{0.5})]$$
$$G = C_1 + C_2 + \lambda(-78 + 2C_1^{0.5} + 3C_2^{0.5})$$

First-order conditions require

$$\frac{\partial G}{\partial C_1} = 1 + \lambda C_1^{-0.5} = 0 \qquad \text{giving} \quad \lambda = -C_1^{0.5} \tag{1}$$

$$\frac{\partial G}{\partial C_2} = 1 + \lambda 1.5 C_2^{-0.5} = 0 \qquad \text{giving} \quad \lambda = \frac{-C_2^{0.5}}{1.5} \tag{2}$$

$$\frac{\partial G}{\partial \lambda} = -78 + 2C_1^{0.5} + 3C_2^{0.5} = 0 \tag{3}$$

Equating (1) and (2)

$$-C_1^{0.5} = \frac{-C_2^{0.5}}{1.5}$$
$$1.5C_1^{0.5} = C_2^{0.5} \tag{4}$$

Substituting (4) into (3)

$$-78 + 2C_1^{0.5} + 3(1.5C_1^{0.5}) = 0$$
$$2C_1^{0.5} + 4.5C_1^{0.5} = 78$$
$$6.5C_1^{0.5} = 78$$
$$C_1^{0.5} = 12$$
$$C_1 = 144 \tag{5}$$

Substituting (5) into (4)

$$C_2^{0.5} = 1.5(12) = 18$$
$$C_2 = 324$$

We can use these optimum pollution control expenditure amounts to check the total pollution level:

$$Z_1 + Z_2 = [(478 - 2C_1^{0.5}) + (600 - 3C_2^{0.5})]$$
$$= 478 - 2(12) + 600 - 3(18)$$
$$= 1,000$$

which is the required level. Thus the water authority should spend £144,000 on reducing the first pollution source and £324,000 on reducing the second source.

Test Yourself, Exercise 11.3

1. Use the Lagrange multiplier to answer Questions 6(b) and 8 from Test Yourself, Exercise 11.1.
2. What is the cheapest way of producing 850 units of output if a firm operates with the production function $Q = 30K^{0.5}L^{0.5}$ and can buy input K at £75 a unit and L at £40 a unit?
3. Two pollution sources can be cleaned up if money is spent on them according to the functions $Z_1 = 780 - 12C_1^{0.5}$ and $Z_2 = 600 - 8C_2^{0.5}$ where Z_1 and Z_2 are the pollution levels from the two sources and C_1 and C_2 are expenditure levels (in £000s) on pollution reduction. What is the cheapest way of reducing the total pollution level from 1,380, which is the level it would be without any controls, to 1,000?
4. A firm buys inputs K and L at £70 a unit and £30 a unit respectively and faces the production function $Q = 40K^{0.5}L^{0.5}$. What is the cheapest way it can produce an output of 500 units?

11.6 Constrained optimization with more than two variables

The same procedures that were used for two-variable problems are also used for applying the Lagrange method to constrained optimization problems with three or more variables. The only difference is that one has a more complex set of simultaneous equations to solve for the optimum values that satisfy the first-order conditions. Although some of these sets of equations may initially look rather awkward to work with, they can usually be greatly simplified and solutions can be found by basic algebra, as the following examples show. As with the two-variable problems, it is assumed that second-order conditions for a maximum (or minimum) are satisfied at stationary points of the Lagrange function in the problems set out here.

Example 11.11

A firm has a budget of £300 to spend on the three inputs x, y and z whose prices per unit are £4, £1 and £6 respectively. What combination of x, y and z should it employ to maximize output if it faces the production function $Q = 24x^{0.3}y^{0.2}z^{0.3}$?

Solution

The budget constraint is

$$300 - 4x - y - 6z = 0$$

and the objective function to be maximized is

$$Q = 24x^{0.3}y^{0.2}z^{0.3}$$

Thus the Lagrange function is

$$G = 24x^{0.3}y^{0.2}z^{0.3} + \lambda(300 - 4x - y - 6z)$$

Differentiating with respect to each variable and setting equal to zero gives

$$\frac{\partial G}{\partial x} = 7.2x^{-0.7}y^{0.2}z^{0.3} - 4\lambda = 0 \qquad \lambda = 1.8x^{-0.7}y^{0.2}z^{0.3} \qquad (1)$$

$$\frac{\partial G}{\partial y} = 4.8x^{0.3}y^{-0.8}z^{0.3} - \lambda = 0 \qquad \lambda = 4.8x^{0.3}y^{-0.8}z^{0.3} \qquad (2)$$

$$\frac{\partial G}{\partial z} = 7.2x^{0.3}y^{0.2}z^{-0.7} - 6\lambda = 0 \qquad \lambda = 1.2x^{0.3}y^{0.2}z^{-0.7} \qquad (3)$$

$$\frac{\partial G}{\partial \lambda} = 300 - 4x - y - 6z = 0 \qquad (4)$$

A simultaneous three-linear-equation system in the three unknowns x, y and z can now be set up if λ is eliminated. There are several ways in which this can be done. In the method used below we set (1) and then (3) equal to (2) to eliminate x and z and then substitute into (4) to solve for y. Whichever method is used, the point of the exercise is to arrive at functions for any two of the unknown variables in terms of the remaining third variable.

Thus, setting (1) equal to (2)

$$1.8x^{-0.7}y^{0.2}z^{0.3} = 4.8x^{0.3}y^{-0.8}z^{0.3}$$

Multiplying both sides by $x^{0.7}y^{0.8}$ and dividing by $z^{0.3}$ gives

$$1.8y = 4.8x$$
$$0.375y = x \qquad (5)$$

We have now eliminated z and obtained a function for x in terms of y. Next we need to eliminate x and obtain a function for z in terms of y. To do this we set (2) equal to (3), giving

$$4.8x^{0.3}y^{-0.8}z^{0.3} = 1.2x^{0.3}y^{0.2}z^{-0.7}$$

Multiplying through by $z^{0.7}y^{0.8}$ and dividing by $x^{0.3}$ gives

$$4.8z = 1.2y$$
$$z = 0.25y \qquad (6)$$

Substituting (5) and (6) into the budget constraint (4)

$$300 - 4(0.375y) - y - 6(0.25y) = 0$$
$$300 - 1.5y - y - 1.5y = 0$$
$$300 = 4y$$
$$75 = y$$

Therefore, from (5)

$$x = 0.375(75) = 28.125$$

and from (6)

$$z = 0.25(75) = 18.75$$

If these optimal values of x, y and z are used then the maximum output will be

$$Q = 24x^{0.3}y^{0.2}z^{0.3} = 24(28.125)^{0.3}(75)^{0.2}(18.75)^{0.3} = 373.1 \text{ units.}$$

Example 11.12

A firm uses the three inputs K, L and R to manufacture good Q and faces the production function

$$Q = 50K^{0.4}L^{0.2}R^{0.2}$$

It has a budget of £24,000 and can buy K, L and R at £80, £12 and £10 respectively per unit. What combination of inputs will maximize its output?

Solution

The objective function to be maximized is $Q = 50K^{0.4}L^{0.2}R^{0.2}$ and the budget constraint is

$$24,000 - 80K - 12L - 10R = 0$$

Thus the Lagrange equation is

$$G = 50K^{0.4}L^{0.2}R^{0.2} + \lambda(24,000 - 80K - 12L - 10R)$$

Differentiating

$$\frac{\partial G}{\partial K} = 20K^{-0.6}L^{0.2}R^{0.2} - 80\lambda = 0 \qquad \lambda = 0.25K^{-0.6}L^{0.2}R^{0.2} \qquad (1)$$

$$\frac{\partial G}{\partial L} = 10K^{0.4}L^{-0.8}R^{0.2} - 12\lambda = 0 \qquad \lambda = \frac{10}{12}K^{0.4}L^{-0.8}R^{0.2} \qquad (2)$$

$$\frac{\partial G}{\partial R} = 10K^{0.4}L^{0.2}R^{-0.8} - 10\lambda = 0 \qquad \lambda = K^{0.4}L^{0.2}R^{-0.8} \qquad (3)$$

$$\frac{\partial G}{\partial \lambda} = 24,000 - 80K - 12L - 10R = 0 \qquad (4)$$

Equating (1) and (2) to eliminate R

$$0.25K^{-0.6}L^{0.2}R^{0.2} = \frac{10}{12}K^{0.4}L^{-0.8}R^{0.2}$$

$$3L = 10K$$

$$0.3L = K \tag{5}$$

Equating (2) and (3) to eliminate K and get R in terms of L

$$\frac{10}{12}K^{0.4}L^{-0.8}R^{0.2} = K^{0.4}L^{0.2}R^{-0.8}$$

$$10R = 12L$$

$$R = 1.2L \tag{6}$$

Substituting (5) and (6) into (4)

$$24{,}000 - 80(0.3L) - 12L - 10(1.2L) = 0$$

$$24{,}000 = 24L + 12L + 12L$$

$$24{,}000 = 48L$$

$$500 = L$$

Substituting this value for L into (5) and (6)

$$K = 0.3(500) = 150 \qquad R = 1.2(500) = 600$$

Using these optimal values for K, L and R, the firm's maximum output will be

$$Q = 50K^{0.4}L^{0.2}R^{0.2} = 50(150)^{0.4}(500)^{0.2}(600)^{0.2} = 4{,}622 \text{ units}$$

Example 11.13

A firm buys the four inputs K, L, R and M at per-unit prices of £50, £30, £25 and £20 respectively and operates with the production function

$$Q = 160K^{0.3}L^{0.25}R^{0.2}M^{0.25}$$

What is the maximum output it can make for a total cost of £30,000?

Solution

The relevant Lagrange function is

$$G = 160K^{0.3}L^{0.25}R^{0.2}M^{0.25} + \lambda(30{,}000 - 50K - 30L - 25R - 20M)$$

Differentiating to find stationary points, setting equal to zero and then equating to λ

$$\frac{\partial G}{\partial K} = 48K^{-0.7}L^{0.25}R^{0.2}M^{0.25} - 50\lambda = 0 \qquad \lambda = \frac{48L^{0.25}R^{0.2}M^{0.25}}{50K^{0.7}} \qquad (1)$$

$$\frac{\partial G}{\partial L} = 40K^{0.3}L^{-0.75}R^{0.2}M^{0.25} - 30\lambda = 0 \qquad \lambda = \frac{4K^{0.3}R^{0.2}M^{0.25}}{3L^{0.75}} \qquad (2)$$

$$\frac{\partial G}{\partial R} = 32K^{0.3}L^{0.25}R^{-0.8}M^{0.25} - 25\lambda = 0 \qquad \lambda = \frac{32K^{0.3}L^{0.25}M^{0.25}}{25R^{0.8}} \qquad (3)$$

$$\frac{\partial G}{\partial M} = 40K^{0.3}L^{0.25}R^{0.2}M^{-0.75} - 20\lambda = 0 \qquad \lambda = \frac{2K^{0.3}L^{0.25}R^{0.2}}{M^{0.75}} \qquad (4)$$

$$\frac{\partial G}{\partial \lambda} = 30{,}000 - 50K - 30L - 25R - 20M = 0 \qquad (5)$$

Equating (1) and (2)

$$\frac{48L^{0.25}R^{0.2}M^{0.25}}{50K^{0.7}} = \frac{4K^{0.3}R^{0.2}M^{0.25}}{3L^{0.75}}$$

Dividing through by $R^{0.2}M^{0.25}$ and cross multiplying

$$144L = 200K$$
$$0.72L = K \qquad (6)$$

Note that, because it is simpler to divide by 200 than 144, we have expressed K as a fraction of L rather than vice versa. Having done this we must now find R and M in terms of L and so (2) must be equated with (3) and (4) to ensure that L is not cancelled out in each set of equalities. Thus, equating (2) and (3)

$$\frac{4K^{0.3}R^{0.2}M^{0.25}}{3L^{0.75}} = \frac{32K^{0.3}L^{0.25}M^{0.25}}{25R^{0.8}}$$

Cancelling out $K^{0.3}M^{0.25}$ and cross multiplying

$$100R = 96L$$
$$R = 0.96L \qquad (7)$$

Equating (2) and (4)

$$\frac{4K^{0.3}R^{0.2}M^{0.25}}{3L^{0.75}} = \frac{2K^{0.3}L^{0.25}R^{0.2}}{M^{0.75}}$$

Cancelling $K^{0.3}R^{0.2}$ and cross multiplying

$$4M = 6L$$
$$M = 1.5L \qquad (8)$$

Substituting (6), (7) and (8) into (5)

$$30{,}000 - 50(0.72L) - 30L - 25(0.96L) - 20(1.5L) = 0$$
$$30{,}000 - 36L - 30L - 24L - 30L = 0$$
$$30{,}000 = 120L$$
$$250 = L$$

Substituting this value for L into (6), (7) and (8)

$$K = 0.72(250) = 180 \qquad R = 0.96(250) = 240 \qquad M = 1.5(250) = 375$$

Using these optimal values of L, K, M and R gives the maximum output level

$$Q = 160K^{0.3}L^{0.25}R^{0.2}M^{0.25}$$
$$= 160(180^{0.3})(250^{0.25})(240^{0.2})(375^{0.25}) = 39{,}786.6 \text{ units}$$

Example 11.14

A firm operates with the production function $Q = 20K^{0.5}L^{0.25}R^{0.4}$. The input prices per unit are £20 for K, £10 for L and £5 for R. What is the cheapest way of producing 1,200 units of output?

Solution

This time output is the constraint such that

$$20K^{0.5}L^{0.25}R^{0.4} = 1{,}200$$

and the objective function to be minimized is the cost function

$$\text{TC} = 20K + 10L + 5R$$

The corresponding Lagrange function is therefore

$$G = 20K + 10L + 5R + \lambda(1{,}200 - 20K^{0.5}L^{0.25}R^{0.4})$$

Differentiating to get stationary points

$$\frac{\partial G}{\partial K} = 20 - \lambda 10K^{-0.5}L^{0.25}R^{0.4} = 0 \qquad \lambda = \frac{2K^{0.5}}{L^{0.25}R^{0.4}} \tag{1}$$

$$\frac{\partial G}{\partial L} = 10 - \lambda 5K^{0.5}L^{-0.75}R^{0.4} = 0 \qquad \lambda = \frac{2L^{0.75}}{K^{0.5}R^{0.4}} \tag{2}$$

$$\frac{\partial G}{\partial R} = 5 - \lambda 8K^{0.5}L^{0.25}R^{-0.6} = 0 \qquad \lambda = \frac{5R^{0.6}}{8K^{0.5}L^{0.25}} \tag{3}$$

$$\frac{\partial G}{\partial \lambda} = 1{,}200 - 20K^{0.5}L^{0.25}R^{0.4} = 0 \tag{4}$$

Equating (1) and (2)

$$\frac{2K^{0.5}}{L^{0.25}R^{0.4}} = \frac{2L^{0.75}}{K^{0.5}R^{0.4}}$$

$$K = L \tag{5}$$

Equating (2) and (3)

$$\frac{2L^{0.75}}{K^{0.5}R^{0.4}} = \frac{5R^{0.6}}{8K^{0.5}L^{0.25}}$$

$$16L = 5R$$

$$3.2L = R \tag{6}$$

Substituting (5) and (6) into (4) to eliminate R and K

$$1{,}200 - 20(L)^{0.5}L^{0.25}(3.2L)^{0.4} = 0$$

$$1{,}200 - 20(3.2)^{0.4}L^{1.15} = 0$$

$$60 = 1.5924287L^{1.15}$$

$$37.678296 = L^{1.15}$$

$$23.47 = L$$

Substituting this value for L into (5) and (6) gives

$$K = 23.47 \qquad R = 3.2(23.47) = 75.1$$

Checking that these values do give the required 1,200 units of output:

$$Q = 20K^{0.5}L^{0.25}R^{0.4} = 20(23.47)^{0.5}(23.47)^{0.25}(75.1)^{0.4} = 1{,}200$$

The cheapest cost level for producing this output will therefore be

$$20K + 10L + 5R = 20(23.4) + 10(23.47) + 5(75.1) = £1{,}079.60$$

Example 11.15

A firm operates with the production function $Q = 45K^{0.4}L^{0.3}R^{0.3}$ and can buy input K at £80 a unit, L at £35 and R at £50. What is the cheapest way it can produce an output of 75,000 units?

Solution

The output constraint is $45K^{0.4}L^{0.3}R^{0.3} = 75{,}000$ and the objective function to be minimized is TC $= 80K + 35L + 50R$. The corresponding Lagrange function is thus

$$G = 80K + 35L + 50R + \lambda(75{,}000 - 45K^{0.4}L^{0.3}R^{0.3})$$

Differentiating to get first-order conditions for a minimum

$$\frac{\partial G}{\partial K} = 80 - \lambda 18 K^{-0.6} L^{0.3} R^{0.3} = 0 \qquad \lambda = \frac{80 K^{0.6}}{18 L^{0.3} R^{0.3}} \tag{1}$$

$$\frac{\partial G}{\partial L} = 35 - \lambda 13.5 K^{0.4} L^{-0.7} R^{0.3} = 0 \qquad \lambda = \frac{35 L^{0.7}}{13.5 K^{0.4} R^{0.3}} \tag{2}$$

$$\frac{\partial G}{\partial R} = 50 - \lambda 13.5 K^{0.4} L^{0.3} R^{-0.7} = 0 \qquad \lambda = \frac{50 R^{0.7}}{13.5 K^{0.4} L^{0.3}} \tag{3}$$

$$\frac{\partial G}{\partial \lambda} = 75,000 - 45 K^{0.4} L^{0.3} R^{0.3} = 0 \tag{4}$$

Equating (1) and (2)

$$\frac{80 K^{0.6}}{18 L^{0.3} R^{0.3}} = \frac{35 L^{0.7}}{13.5 K^{0.4} R^{0.3}}$$

$$1,080 K = 630 L$$

$$\frac{12 K}{7} = L \tag{5}$$

As we have L in terms of K we now need to use (1) and (3) to get R in terms of K. Thus equating (1) and (3)

$$\frac{80 K^{0.6}}{18 L^{0.3} R^{0.3}} = \frac{50 R^{0.7}}{13.5 K^{0.4} L^{0.3}}$$

$$1,080 K = 900 R$$

$$1.2 K = R \tag{6}$$

Substituting (5) and (6) into (4)

$$75,000 - 45 K^{0.4} \left(\frac{12 K}{7}\right)^{0.3} (1.2 K)^{0.3} = 0$$

$$75,000 - 45 \left(\frac{12}{7}\right)^{0.3} (1.2)^{0.3} K^{0.4} K^{0.3} K^{0.3} = 0$$

$$75,000 = 55.871697 K$$

$$1,342.3612 = K$$

Substituting this value into (5)

$$L = \frac{12}{7}(1,342.3612) = 2,301.1907$$

Substituting into (6)

$$R = 1.2(1,342.3612) = 1,610.8334$$

Thus, optimum values are

$$K = 1,342.4 \qquad L = 2,301.2 \qquad R = 1,610.8 \quad \text{(to 1 dp)}$$

Total expenditure on inputs will then be

$$80K + 35L + 50R = 80(1,342.4) + 35(2,301.2) + 50(1,610.8) = £268,474$$

Test Yourself, Exercise 11.4

1. A firm has a budget of £570 to spend on the three inputs x, y and z whose prices per unit are respectively £4, £6 and £3. What combination of x, y and z will maximize output given the production function $Q = 2x^{0.2}y^{0.3}z^{0.45}$?

2. A firm uses inputs K, L and R to manufacture good Q. It has a budget of £828 and its production function (for positive values of Q) is

$$Q = 20K + 16L + 12R - 0.2K^2 - 0.1L^2 - 0.3R^2$$

   If $P_K = £20$, $P_L = £10$ and $P_R = £6$, what is the maximum output it can produce? Assume that second-order conditions for a maximum are satisfied for the relevant Lagrangian.

3. What amounts of the inputs x, y and z should a firm use to maximize output if it faces the production function $Q = 2x^{0.4}y^{0.2}z^{0.6}$ and it has a budget of £600, given that the prices of x, y and z are respectively £4, £1 and £2 per unit?

4. A firm buys the inputs x, y and z for £5, £10 and £2 respectively per unit. If its production function is $Q = 60x^{0.2}y^{0.4}z^{0.5}$ how much can it produce for an outlay of £8,250?

5. Inputs K, L, R and M cost £10, £6, £15 and £3 respectively per unit. What is the cheapest way of producing an output of 900 units if a firm operates with the production function $Q = 20K^{0.4}L^{0.3}R^{0.2}M^{0.25}$?

6. Make up your own constrained optimization problem for an objective function with three variables and solve it.

7. A firm faces the production function $Q = 50K^{0.5}L^{0.2}R^{0.25}$ and is required to produce an output level of 1,913 units. What is the cheapest way of doing this if the per-unit costs of inputs K, L and R are £80, £24 and £45 respectively?

12 Further topics in calculus

Learning objectives

After completing this chapter students should be able to:

- Use the chain, product and quotient rules for differentiation.
- Choose the most appropriate method for differentiating different forms of functions.
- Check the second-order conditions for optimization of relevant economic functions using the quotient rule for differentiation.
- Integrate simple functions.
- Use integration to determine total cost and total revenue from marginal cost and marginal revenue functions.
- Understand how a definite integral relates to the area under a function and apply this concept to calculate consumer surplus.

12.1 Overview

In this chapter, some techniques are introduced that can be used to differentiate functions that are rather more complex than those encountered in Chapters 8, 9, 10 and 11. These are the chain rule, the product rule and the quotient rule. As you will see in the worked examples, it is often necessary to combine several of these methods to differentiate some functions. The concept of integration is also introduced.

12.2 The chain rule

The chain rule is used to differentiate 'functions within functions'. For example, if we have the function

$$y = f(z)$$

and we also know that there is a second functional relationship

$$z = g(x)$$

then we can write y as a function of x in the form

$$y = f[g(x)]$$

To differentiate y with respect to x in this type of function we use the chain rule which states that

$$\frac{dy}{dx} = \frac{dy}{dz}\frac{dz}{dx}$$

One economics example of a function within a function occurs in the marginal revenue productivity theory of the demand for labour, where a firm's total revenue depends on output which, in turn, depends on the amount of labour employed. An applied example is explained later. However, we shall first look at what is perhaps the most frequent use of the chain rule, which is to break down an awkward function artificially into two components in order to allow differentiation via the chain rule. Assume, for example, that you wish to find an expression for the slope of the non-linear demand function

$$p = (150 - 0.2q)^{0.5} \tag{1}$$

The basic rules for differentiation explained in Chapter 8 cannot cope with this sort of function. However, if we define a new function

$$z = 150 - 0.2q \tag{2}$$

then (1) above can be rewritten as

$$p = z^{0.5} \tag{3}$$

(Note that in both (1) and (3) the functions are assumed to hold for $p \geq 0$ only, i.e. negative roots are ignored.)

Differentiating (2) and (3) we get

$$\frac{dz}{dq} = -0.2 \qquad \frac{dp}{dz} = 0.5z^{-0.5}$$

Thus, using the chain rule and then substituting equation (2) back in for z, we get

$$\frac{dp}{dq} = \frac{dp}{dz}\frac{dz}{dq} = 0.5z^{-0.5}(-0.2) = \frac{-0.1}{z^{0.5}} = \frac{-0.1}{(150 - 0.2q)^{0.5}}$$

Some more examples of the use of the chain rule are set out below.

Example 12.1

The present value of a payment of £1 due in 8 years' time is given by the formula

$$PV = \frac{1}{(1+i)^8}$$

where i is the given interest rate. What is the rate of change of PV with respect to i?

Solution

If we let

$$z = 1 + i \qquad\qquad (1)$$

then we can write

$$\text{PV} = \frac{1}{z^8} = z^{-8} \qquad\qquad (2)$$

Differentiating (1) and (2) gives

$$\frac{dz}{di} = 1 \qquad \frac{d\text{PV}}{dz} = -8z^{-9}$$

Therefore, using the chain rule, the rate of change of PV with respect to i will be

$$\frac{d\text{PV}}{di} = \frac{d\text{PV}}{dz}\frac{dz}{di} = -8z^{-9} = \frac{-8}{(1+i)^9}$$

Example 12.2

If $y = (48 + 20x^{-1} + 4x + 0.3x^2)^4$, what is dy/dx?

Solution

Let

$$z = 48 + 20x^{-1} + 4x + 0.3x^2 \qquad\qquad (1)$$

and so

$$\frac{dz}{dx} = -20x^{-2} + 4 + 0.6x \qquad\qquad (2)$$

Substituting (1) into the function given in the question

$$y = z^4$$

and so

$$\frac{dy}{dz} = 4z^3 \qquad\qquad (3)$$

Therefore, using the chain rule and substituting (2) and (3)

$$\frac{dy}{dx} = \frac{dy}{dz}\frac{dz}{dx}$$

$$= 4z^3(-20x^{-2} + 4 + 0.6x)$$

$$= 4(48 + 20x^{-1} + 4x + 0.3x^2)^3(-20x^{-2} + 4 + 0.6x)$$

The marginal revenue productivity theory of the demand for labour

In the marginal revenue productivity theory of the demand for labour, the rule for profit maximization is to employ additional units of labour as long as the extra revenue generated by selling the extra output produced by an additional unit of labour exceeds the marginal cost of employing this additional unit of labour. This rule applies in the short run when inputs other than labour are assumed fixed.

The optimal amount of labour is employed when

$$\text{MRP}_\text{L} = \text{MC}_\text{L}$$

where MRP_L is the marginal revenue product of labour, defined as the additional revenue generated by an additional unit of labour, and MC_L is the marginal cost of an additional unit of labour. The MC_L is normally equal to the wage rate unless the firm is a monopsonist (sole buyer) in the labour market.

If all relevant functions are assumed to be continuous then the above definitions can be rewritten as

$$\text{MRP}_\text{L} = \frac{\text{dTR}}{\text{d}L} \qquad \text{MC}_\text{L} = \frac{\text{dTC}_\text{L}}{\text{d}L}$$

where TR is total sales revenue (i.e. pq) and TC_L is the total cost of labour. If a firm is a monopoly seller of a good, then we effectively have to deal with two functions in order to derive its MRP_L function since total revenue will depend on output, i.e. $\text{TR} = \text{f}(q)$, and output will depend on labour input, i.e. $q = \text{f}(L)$. Therefore, using the chain rule,

$$\text{MRP}_\text{L} = \frac{\text{dTR}}{\text{d}L} = \frac{\text{dTR}}{\text{d}q}\frac{\text{d}q}{\text{d}L} \tag{1}$$

We already know that

$$\frac{\text{dTR}}{\text{d}q} = \text{MR} \qquad \frac{\text{d}q}{\text{d}L} = \text{MP}_\text{L}$$

Therefore, substituting these into (1),

$$\text{MRP}_\text{L} = \text{MR} \times \text{MP}_\text{L}$$

This is the rule for determining the profit-maximizing amount of labour which you should encounter in your microeconomics course.

Example 12.3

A firm is a monopoly seller of good q and faces the demand schedule $p = 200 - 2q$, where p is the price in pounds, and the short-run production function $q = 4L^{0.5}$. If it can buy labour at a fixed wage of £8, how much L should be employed to maximize profit, assuming other inputs are fixed?

Solution

Using the chain rule we need to derive a formula for MRP_L in terms of L and then set it equal to £8, given that MC_L is fixed at this wage rate. As

$$MRP_L = \frac{dTR}{dL} = \frac{dTR}{dq}\frac{dq}{dL} \tag{1}$$

we need to find dTR/dq and dq/dL.

Given $p = 200 - 2q$, then

$$TR = pq = (200 - 2q)q = 200q - 2q^2$$

Therefore

$$\frac{dTR}{dq} = 200 - 4q \tag{2}$$

Given $q = 4L^{0.5}$, then the marginal product of labour will be

$$\frac{dq}{dL} = 2L^{-0.5} \tag{3}$$

Thus, substituting (2) and (3) into (1)

$$MRP_L = (200 - 4q)2L^{-0.5} = (400 - 8q)L^{-0.5}$$

As all units of L cost £8, setting this function for MRP_L equal to the wage rate we get

$$\frac{400 - 8q}{L^{0.5}} = 8$$
$$400 - 8q = 8L^{0.5} \tag{4}$$

Substituting the production function $q = 4L^{0.5}$ into (4), as we are trying to derive a formula in terms of L, gives

$$400 - 8(4L^{0.5}) = 8L^{0.5}$$
$$400 - 32L^{0.5} = 8L^{0.5}$$
$$400 = 40L^{0.5}$$
$$10 = L^{0.5}$$
$$100 = L$$

which is the optimal employment level.

In the example above the idea of a 'short-run production function' was used to simplify the analysis, where the input of capital (K) was implicitly assumed to be fixed. Now that you understand how an MRP_L function can be derived we can work with full production functions in the format $Q = f(K, L)$. The effect of one input increasing while the other is held constant can now be shown by the relevant *partial derivative*.

Thus

$$\mathrm{MP_L} = \frac{\partial Q}{\partial L}$$

The same chain rule can be used for partial derivatives, and full and partial derivatives can be combined, as in the following examples.

Example 12.4

A firm operates with the production function $q = 45K^{0.7}L^{0.4}$ and faces the demand function $p = 6{,}980 - 6q$. Derive its $\mathrm{MRP_L}$ function.

Solution

By definition $\mathrm{MRP_L} = \partial TR/\partial L$, where K is assumed fixed.
   We know that

$$\mathrm{TR} = pq = (6{,}980 - 6q)q = 6{,}980q - 6q^2$$

Therefore

$$\frac{d\mathrm{TR}}{dq} = 6{,}980 - 12q \tag{1}$$

From the production function $q = 45K^{0.7}L^{0.4}$ we can derive

$$\mathrm{MP_L} = \frac{\partial q}{\partial L} = 18K^{0.7}L^{-0.6} \tag{2}$$

Using the chain rule and substituting (1) and (2)

$$\mathrm{MRP_L} = \frac{\partial \mathrm{TR}}{\partial L} = \frac{d\mathrm{TR}}{dq}\frac{\partial q}{\partial L} = (6{,}980 - 12q)18K^{0.7}L^{-0.6} \tag{3}$$

As we wish to derive $\mathrm{MRP_L}$ as a function of L, we substitute the production function given in the question into (3) for q. Thus

$$\mathrm{MRP_L} = [6{,}980 - 12(45K^{0.7}L^{0.4})]18K^{0.7}L^{-0.6}$$

$$= 125{,}640K^{0.7}L^{-0.6} - 9{,}720K^{1.4}L^{-0.2}$$

Note that the value $\mathrm{MRP_L}$ will depend on the amount that K is fixed at, as well as the value of L.

Point elasticity of demand

The chain rule can help the calculation of point elasticity of demand for some non-linear demand functions.

Example 12.5

Find point elasticity of demand when $q = 10$ if $p = (120 - 2q)^{0.5}$.

Solution

Point elasticity is defined as

$$e = (-1)\frac{p}{q}\frac{1}{\left(\dfrac{dp}{dq}\right)} \tag{1}$$

Create a new variable $z = 120 - 2q$. Thus $p = z^{0.5}$ and so, by differentiating:

$$\frac{dz}{dq} = -2 \qquad \frac{dp}{dz} = 0.5z^{-0.5}$$

Therefore

$$\frac{dp}{dq} = \frac{dp}{dz}\frac{dz}{dq}$$

$$= 0.5z^{-0.5}(-2)$$

$$= 0.5(120 - 2q)^{-0.5}(-2)$$

$$= \frac{-1}{(120 - 2q)^{0.5}}$$

and so, inverting this result,

$$\frac{1}{dp/dq} = -(120 - 2q)^{0.5}$$

When $q = 10$, then from the original demand function price can be calculated as

$$p = (120 - 20)^{0.5} = 100^{0.5} = 10$$

Thus, substituting these results into formula (1), point elasticity will be

$$e = (-1)\frac{10}{10}(-1)(120 - 2q)^{0.5} = (120 - 20)^{0.5} = 100^{0.5} = 10$$

Sometimes it may be possible to simplify an expression in order to be able to differentiate it, but one may instead use the chain rule if it is more convenient. The same result will be obtained by both methods, of course.

Example 12.6

Differentiate the function $y = (6 + 4x)^2$.

Solution

(i) By multiplying out

$$y = (6 + 4x)^2 = 36 + 48x + 16x^2$$

Therefore

$$\frac{dy}{dx} = 48 + 32x$$

(ii) Using the chain rule, let $z = 6 + 4x$ so that $y = z^2$. Thus

$$\frac{dy}{dx} = \frac{dy}{dz}\frac{dz}{dx} = 2z \times 4 = 2(6 + 4x)4 = 48 + 32x$$

Test Yourself, Exercise 12.1

1. A firm operates in the short run with the production function $q = 2L^{0.5}$ and faces the demand schedule $p = 60 - 4q$ where p is price in pounds. If it can employ labour at a wage rate of £4 per hour, how much should it employ to maximize profit?

2. If a supply schedule is given by $p = (2 + 0.05q)^2$ show (a) by multiplying out, and (b) by using the chain rule, that its slope is 2.2 when q is 400.

3. The return R on a sum M invested at i per cent for 3 years is given by the formula

$$R = M(1 + i)^3$$

   What is the rate of change of R with respect to i?

4. If $y = (3 + 0.6x^2)^{0.5}$ what is dy/dx?

5. If a firm faces the total cost function $TC = (6 + x)^{0.5}$, what is its marginal cost function?

6. A firm operates with the production function $q = 0.4K^{0.5}L^{0.5}$ and sells its output in a market where it is a monopoly with the demand schedule $p = 60 - 2q$. If K is fixed at 25 units and the wage rate is £7 per unit of L, derive the MRP_L function and work out how much L the firm should employ to maximize profit.

7. A firm faces the demand schedule $p = 650 - 3q$ and the production function $q = 4K^{0.5}L^{0.5}$ and has to pay £8 per unit to buy L. If K is fixed at 4 units how much L should the firm use if it wishes to maximize profits?

8. If a firm operates with the total cost function $TC = 4 + 10(9 + q^2)^{0.5}$, what is its marginal cost when q is 4?

9. Given the production function $q = (6K^{0.5} + 0.5L^{0.5})^{0.3}$, find MP_L when K is 16 and L is 576.

12.3   The product rule

The product rule allows one to differentiate two functions which are multiplied together.

If $y = uv$ where u and v are both functions of x, then according to the product rule

$$\frac{dy}{dx} = u\frac{dv}{dx} + v\frac{du}{dx}$$

As with the chain rule, one may find it convenient to split a single awkward function into two artificial functions even if these functions do not have any particular economic meaning. The following examples show how this rule can be used.

Example 12.7

If $y = (7.5 + 0.2x^2)(4 + 8x^{-1})$, what is dy/dx?

Solution

This function could in fact be multiplied out and differentiated without using the product rule. However, let us first use the product rule and then we can compare the answers obtained by the two methods. They should, of course, be the same.

We are given the function

$$y = (7.5 + 0.2x^2)(4 + 8x^{-1})$$

so let

$$u = 7.5 + 0.2x^2 \qquad v = 4 + 8x^{-1}$$

Therefore

$$\frac{du}{dx} = 0.4x \qquad\qquad \frac{dv}{dx} = -8x^{-2}$$

Thus, using the product rule and substituting these results in, we get

$$\frac{dy}{dx} = u\frac{dv}{dx} + v\frac{du}{dx}$$

$$= (7.5 + 0.2x^2)(-8x^{-2}) + (4 + 8x^{-1})0.4x$$

$$= -60x^{-2} - 1.6 + 1.6x + 3.2$$

$$= 1.6 + 1.6x - 60x^{-2} \tag{1}$$

The alternative method of differentiation is to multiply out the original function. Thus

$$y = (7.5 + 0.2x^2)(4 + 8x^{-1}) = 30 + 60x^{-1} + 0.8x^2 + 1.6x$$

and so

$$\frac{dy}{dx} = -60x^{-2} + 1.6x + 1.6 \tag{2}$$

The answers (1) and (2) are the same, as we expected.

When it is not possible to multiply out the different components of a function then one must use the product rule to differentiate. One may also need to use the chain rule to help differentiate the different sub-functions.

Example 12.8

A firm faces the non-linear demand schedule $p = (650 - 0.25q)^{1.5}$. What output should it sell to maximize total revenue?

Solution

When the demand function in the question is substituted for p then

$$TR = pq = (650 - 0.25q)^{1.5}q$$

To differentiate TR using the product rule, first let

$$u = (650 - 0.25q)^{1.5} \qquad v = q$$

Thus, employing the chain rule

$$\frac{du}{dq} = 1.5(650 - 0.25q)^{0.5}(-0.25) = -0.375(650 - 0.25q)^{0.5}$$

and also

$$\frac{dv}{dq} = 1$$

Therefore, using the product rule

$$\frac{dTR}{dq} = u\frac{dv}{dq} + v\frac{du}{dq}$$

$$= (650 - 0.25q)^{1.5} + q(-0.375)(650 - 0.25q)^{0.5}$$

$$= (650 - 0.25q)^{0.5}(650 - 0.25q - 0.375q)$$

$$= (650 - 0.25q)^{0.5}(650 - 0.625q) \tag{1}$$

For a stationary point

$$\frac{dTR}{dq} = (650 - 0.25q)^{0.5}(650 - 0.625q) = 0$$

Therefore, either

$$650 - 0.25q = 0 \quad \text{or} \quad 650 - 0.625q = 0$$
$$2{,}600 = q \quad \text{or} \quad 1{,}040 = q$$

We now need to check which of these values of q satisfies the second-order condition for a maximum. (You should immediately be able to see why it will not be 2,600 by observing what happens when this quantity is substituted into the demand function.) To derive d^2TR/dq^2 we need to use the product rule again to differentiate dTR/dq. From (1) above

$$\frac{dTR}{dq} = (650 - 0.25q)^{0.5}(650 - 0.625q) = 0$$

Let

$$u = (650 - 0.25q)^{0.5} \quad \text{and} \quad v = 650 - 0.625q$$

giving

$$\frac{du}{dq} = 0.5(650 - 0.25q)^{-0.5}(-0.25) = -0.125(650 - 0.25q)^{-0.5}$$

using the chain rule and

$$\frac{dv}{dq} = -0.625$$

Therefore, employing the product rule

$$\frac{d^2TR}{dq^2} = u\frac{dv}{dq} + v\frac{du}{dq}$$

$$= (650 - 0.25q)^{0.5}(-0.625) + (650 - 0.625q)(-0.125)(650 - 0.25q)^{-0.5}$$

$$= \frac{(650 - 0.25q)(-0.625) + (650 - 0.625q)(-0.125)}{(650 - 0.25q)^{0.5}} \tag{2}$$

Substituting the value $q = 1{,}040$ into (2) gives

$$\frac{d^2TR}{dq^2} = \frac{(390)(-0.625) + 0}{390^{0.5}} = -12.34 < 0$$

Therefore, the second-order condition is met and TR is maximized when $q = 1{,}040$. We can double check that the other stationary point will not maximize TR by substituting the value $q = 2{,}600$ into (2) giving

$$\frac{d^2TR}{dq^2} = \frac{0 + (-975)(-0.125)}{0} \rightarrow +\infty$$

Therefore this second value for q obviously does not satisfy second-order conditions for a maximum.

Example 12.9

At what level of K is the function $Q = 12K^{0.4}(160 - 8K)^{0.4}$ at a maximum? (*This is Example 11.1 (reworked) which was not completed in the last chapter.*)

Solution

We need to differentiate the function $Q = 12K^{0.4}(160 - 8K)^{0.4}$ to check the first-order condition for a maximum. To use the product rule, let

$$u = 12K^{0.4} \quad \text{and} \quad v = (160 - 8K)^{0.4}$$

and so

$$\frac{du}{dK} = 4.8K^{-0.6} \quad \text{and} \quad \frac{dv}{dK} = 0.4(160 - 8K)^{-0.6}(-8)$$

$$= -3.2(160 - 8K)^{-0.6}$$

Therefore,

$$\frac{dQ}{dK} = 12K^{0.4}(-3.2)(160 - 8K)^{-0.6} + (160 - 8K)^{0.4}4.8K^{-0.6}$$

$$= \frac{-38.4K + (160 - 8K)4.8}{(160 - 8K)^{0.6}K^{0.6}}$$

$$= \frac{768 - 76.8K}{(160 - 8K)^{0.6}K^{0.6}} \tag{1}$$

Setting (1) equal to zero for a stationary point must mean

$$768 - 76.8K = 0$$

$$K = 10$$

As we have already left this example in mid-solution once already, it will not do any harm to leave it once again. Although the second-order condition could be worked out using the product rule it is more convenient to use the quotient rule in this case and so we shall continue this problem later, in Example 12.13.

Example 12.10

In a perfectly competitive market the demand schedule is $p = 120 - 0.5q^2$ and the supply schedule is $p = 20 + 2q^2$. If the government imposes a per-unit tax t on the good sold in this market, what level of t will maximize the government's tax yield?

Solution

With the tax the supply schedule shifts upwards by the amount of the tax and becomes

$$p = 20 + 2q^2 + t$$

In equilibrium, demand price equals supply price. Therefore

$$120 - 0.5q^2 = 20 + 2q^2 + t$$

$$100 - t = 2.5q^2$$

$$40 - 0.4t = q^2$$

$$(40 - 0.4t)^{0.5} = q \tag{1}$$

The government's tax yield (TY) is tq. Substituting (1) for q, this gives

$$TY = t(40 - 0.4t)^{0.5} \tag{2}$$

We need to set $dTY/dt = 0$ for the first-order condition for maximization of TY.
From (2) let

$$u = t \quad \text{and} \quad v = (40 - 0.4t)^{0.5}$$

giving

$$\frac{du}{dt} = 1 \qquad \frac{dv}{dt} = 0.5(40 - 0.4t)^{-0.5}(-0.4)$$

$$= -0.2(40 - 0.4t)^{-0.5}$$

Therefore, using the product rule

$$\frac{dTY}{dt} = t(-0.2)(40 - 0.4t)^{-0.5} + (40 - 0.4t)^{0.5}$$

$$= \frac{-0.2t + 40 - 0.4t}{(40 - 0.4t)^{0.5}}$$

$$= \frac{40 - 0.6t}{(40 - 0.4t)^{0.5}} = 0 \tag{3}$$

For finite values of t the first-order condition (3) will only hold when

$$40 - 0.6t = 0$$

$$66.67 = t$$

To check second-order conditions for this stationary point we need to find d^2TY/dt^2.
From (3)

$$\frac{dTY}{dt} = (40 - 0.6t)(40 - 0.4t)^{-0.5}$$

To differentiate using the product rule, let

$$u = 40 - 0.6t \quad \text{and} \quad v = (40 - 0.4t)^{-0.5}$$

giving

$$\frac{du}{dt} = -0.6 \qquad \frac{dv}{dt} = -0.5(40 - 0.4t)^{-1.5}(-0.4)$$

Therefore

$$\frac{d^2TY}{dt^2} = (40 - 0.6t)[0.2(40 - 0.4t)^{-1.5}] + (40 - 0.4t)^{-0.5}(-0.6) \tag{4}$$

When $t = 66.67$ then $40 - 0.6t = 0$ and so the first term in (4) disappears giving

$$\frac{d^2TY}{dt^2} = [40 - 0.4(66.67)]^{-0.5}(-0.6) = -0.1644 < 0$$

Therefore, the second-order condition for a maximum is satisfied when $t = 66.67$. Maximum tax revenue is raised when the per-unit tax is £66.67.

Test Yourself, Exercise 12.2

1. If $y = (6x + 7)^{0.5}(2.6x^2 - 1.9)$, what is dy/dx?
2. What output will maximize total revenue given the non-linear demand schedule $p = (60 - 2q)^{1.5}$?
3. Derive a function for the marginal product of L given the production function $Q = 85(0.5K^{0.8} + 3L^{0.5})^{0.6}$.
4. If $Q = 120K^{0.5}(250 - 0.5K)^{0.3}$ at what value of K will $dQ/dK = 0$? (That is, find the first-order condition for maximization of Q.)
5. In a perfectly competitive market the demand schedule is $p = 600 - 4q^{0.5}$ and the supply schedule is $p = 30 + 6q^{0.5}$. What level of a per-unit tax levied on the good sold in this market will maximize the government's tax yield?
6. Make up your own function involving the product of two sub-functions and then differentiate it using the product rule.
7. For the demand schedule $p = (60 - 0.1q)^{0.5}$:

   (a) derive an expression for the slope of the demand schedule;
   (b) demonstrate that this slope gets flatter as q increases from 0 to 600;
   (c) find the output at which total revenue is a maximum.

12.4 The quotient rule

The quotient rule allows one to differentiate two functions where one function is divided by the other function.

If $y = u/v$ where u and v are functions of x, then according to the quotient rule

$$\frac{dy}{dx} = \frac{v\dfrac{du}{dx} - u\dfrac{dv}{dx}}{v^2}$$

Example 12.11

What is $\dfrac{dy}{dx}$ if $y = \dfrac{4x^2}{8 + 0.2x}$?

Solution

Defining relevant sub-functions and differentiating them

$$u = 4x^2 \quad \text{and} \quad v = 8 + 0.2x$$

$$\frac{du}{dx} = 8x \qquad \frac{dv}{dx} = 0.2$$

Therefore, according to the quotient rule,

$$\frac{dy}{dx} = \frac{v\dfrac{du}{dx} - u\dfrac{dv}{dx}}{v^2}$$

$$= \frac{(8 + 0.2x)8x - 4x^2(0.2)}{(8 + 0.2x)^2}$$

$$= \frac{64x + 1.6x^2 - 0.8x^2}{(8 + 0.2x)^2}$$

$$= \frac{64x + 0.8x^2}{(8 + 0.2x)^2} \tag{1}$$

This solution could also have been found using the product rule, since any function in the form $y = u/v$ can be written as $y = uv^{-1}$. We can check this by reworking Example 12.11 and differentiating the function $y = 4x^2(8 + 0.2x)^{-1}$.

Defining relevant sub-functions and differentiating them

$$u = 4x^2 \qquad v = (8 + 0.2x)^{-1}$$

$$\frac{du}{dx} = 8x \qquad \frac{dv}{dx} = -0.2(8 + 0.2x)^{-2}$$

Thus, using the product rule

$$\frac{dy}{dx} = u\frac{dv}{dx} + v\frac{du}{dx}$$

$$= 4x^2[-0.2(8 + 0.2x)^{-2}] + (8 + 0.2x)^{-1}8x$$

$$= \frac{-0.8x^2 + (8 + 0.2x)8x}{(8 + 0.2x)^2}$$

$$= \frac{-0.8x^2 + 64x + 1.6x^2}{(8 + 0.2x)^2}$$

$$= \frac{64x + 0.8x^2}{(8 + 0.2x)^2} \tag{2}$$

The answers (1) and (2) are identical, as expected.

Whether one chooses to use the quotient rule or the product rule depends on the functions to be differentiated. Only practice will give you an idea of which will be the easier to use for specific examples.

Example 12.12

Derive a function for marginal revenue (in terms of q) if a monopoly faces the non-linear demand schedule $p = \dfrac{252}{(4 + q)^{0.5}}$.

Solution

$$TR = pq = \frac{252q}{(4+q)^{0.5}}$$

Defining

$$u = 252q \quad \text{and} \quad v = (4+q)^{0.5}$$

gives $\quad \dfrac{du}{dq} = 252 \qquad\qquad \dfrac{dv}{dq} = 0.5(4+q)^{-0.5}$

Therefore, using the quotient rule

$$
\begin{aligned}
MR = \frac{dTR}{dq} &= \frac{v\dfrac{du}{dq} - u\dfrac{dv}{dq}}{v^2} \\[2mm]
&= \frac{(4+q)^{0.5}252 - 252q(0.5)(4+q)^{-0.5}}{4+q} \\[2mm]
&= \frac{(4+q)252 - 126q}{(4+q)^{1.5}} \\[2mm]
&= \frac{1{,}008 + 126q}{(4+q)^{1.5}}
\end{aligned}
$$

Note that, in this example, MR only becomes zero when q becomes infinitely large. TR will therefore rise continually as q increases.

All three rules may be used in some problems. In particular, one may find it convenient to use the chain rule and the product rule to derive the first-order condition in an optimization problem and then use the quotient rule to check the second-order condition. If we return to the unfinished Example 12.9 we can now see how the quotient rule can be used to check the second-order condition.

Example 12.13

The objective is to find the value of K which maximizes $Q = 12K^{0.4}(160 - 8K)^{0.4}$. In Example 12.9, first-order conditions were satisfied when

$$\frac{dQ}{dK} = \frac{768 - 76.8K}{(160 - 8K)^{0.6}K^{0.6}}$$

which holds when $K = 10$.

To derive d^2Q/dK^2 let $u = 768 - 76.8K$ and $v = (160 - 8K)^{0.6}K^{0.6}$. Therefore,

$$\frac{du}{dK} = -76.8 \tag{1}$$

and, using the product rule,

$$\frac{dv}{dK} = (160 - 8K)^{0.6}0.6K^{-0.4} + K^{0.6}0.6(160 - 8K)^{-0.4}(-8)$$

$$= \frac{(160 - 8K)0.6 - 4.8K}{K^{0.4}(160 - 8K)^{0.4}}$$

$$= \frac{96 - 9.6K}{K^{0.4}(160 - 8K)^{0.4}} \qquad (2)$$

Therefore, using the quotient rule and substituting (1) and (2)

$$\frac{d^2Q}{dK^2} = \frac{(160 - 8K)^{0.6}K^{0.6}(-76.8) - (768 - 76.8K)\dfrac{96 - 9.6K}{K^{0.4}(160 - 8K)^{0.4}}}{(160 - 8K)^{1.2}K^{1.2}}$$

$$= \frac{(160 - 8K)K(-76.8) - 76.8(10 - K)9.6(10 - K)}{(160 - 8K)^{1.6}K^{1.6}}$$

At the stationary point when $K = 10$ several terms become zero, giving

$$\frac{d^2Q}{dK^2} = \frac{-76.8(800)}{(800)^{1.6}} < 0$$

Therefore, the second-order condition for a maximum is satisfied when $K = 10$.

Minimum average cost

In your introductory economics course you were probably given an intuitive geometrical explanation of why a marginal cost schedule cuts a U-shaped average cost curve at its minimum point. The quotient rule can now be used to prove this rule.

In the short run, with only one variable input, assume that total cost (TC) is a function of q. Thus, MC = dTC/dq (as explained in Chapter 8) and, by definition, AC = TC/q.

To differentiate AC using the quotient rule let

$$u = \text{TC} \quad \text{and} \qquad v = q$$

giving

$$\frac{du}{dq} = \frac{d\text{TC}}{dq} = \text{MC} \qquad \frac{dv}{dq} = 1$$

Therefore, using the quotient rule, first-order conditions for a minimum are

$$\frac{d\text{AC}}{dq} = \frac{q\text{MC} - \text{TC}}{q^2} = 0 \qquad (1)$$

$$q\text{MC} - \text{TC} = 0 \qquad (\text{or } q \rightarrow \infty, \text{ which we disregard})$$

$$\text{MC} = \frac{\text{TC}}{q} = \text{AC} \qquad (2)$$

Therefore, MC = AC when AC is at a stationary point.

To check second-order conditions we need to find d^2AC/dq^2. From (1) above we know that

$$\frac{dAC}{dq} = \frac{qMC - TC}{q^2}$$

Again use the quotient rule and let $u = qMC - TC$ and $v = q^2$ giving

$$\frac{du}{dq} = \left(q\frac{dMC}{dq} + MC\right) - MC = q\frac{dMC}{dq}$$

and

$$\frac{dv}{dq} = 2q$$

Therefore

$$\frac{d^2AC}{dq^2} = \frac{q^2\left(q\dfrac{dMC}{dq}\right) - (qMC - TC)2q}{q^4} \tag{3}$$

The first-order condition for a minimum is satisfied when $qMC = TC$, from (2) above. Substituting this result into (3) the second term in the numerator disappears and we get

$$\frac{d^2AC}{dq^2} = \frac{q^2\left(q\dfrac{dMC}{dq}\right)}{q^4}$$

$$= \frac{1}{q}\frac{dMC}{dq} > 0 \quad \text{when} \quad \frac{dMC}{dq} > 0$$

Therefore, the second-order condition for a minimum is satisfied when MC = AC and MC is rising. Thus, although MC may cut AC at another point when MC is falling, when MC is rising it cuts AC at its minimum point.

12.5 Individual labour supply

Not all of you will have encountered the theory of individual labour supply. Nevertheless you should now be able to understand the following example which shows how the utility-maximizing combination of work and leisure hours can be found when an individual's utility function, wage rate and maximum working day are specified.

Example 12.14

In the theory of individual labour supply it is assumed that an individual derives utility from both leisure (L) and income (I). Income is determined by hours of work (H) multiplied by the hourly wage rate (w), i.e. $I = wH$.

Assume that each day a total of 12 hours is available for an individual to split between leisure and work, the wage rate is given as £4 an hour and that the individual's utility function is $U = L^{0.5}I^{0.75}$. How will this individual balance leisure and income so as to maximize utility?

Solution

Given a maximum working day of 12 hours, then hours of work $H = 12 - L$.
   Therefore, given an hourly wage of £4, income earned will be

$$I = wH = w(12 - L) = 4(12 - L) = 48 - 4L \tag{1}$$

Substituting (1) into the utility function

$$U = L^{0.5}I^{0.75} = L^{0.5}(48 - 4L)^{0.75} \tag{2}$$

To differentiate U using the product rule let

$$u = L^{0.5} \quad \text{and} \quad v = (48 - 4L)^{0.75}$$

giving

$$\frac{du}{dL} = 0.5L^{-0.5} \qquad \frac{dv}{dL} = 0.75(48 - 4L)^{-0.25}(-4)$$

$$= -3(48 - 4L)^{-0.25}$$

Therefore

$$\frac{dU}{dL} = L^{0.5}[-3(48 - 4L)^{-0.25}] + (48 - 4L)^{0.75}(0.5L^{-0.5})$$

$$= \frac{-3L + (48 - 4L)0.5}{(48 - 4L)^{0.25}L^{0.5}}$$

$$= \frac{24 - 5L}{(48 - 4L)^{0.25}L^{0.5}} = 0 \tag{3}$$

for a stationary point. Therefore

$$24 - 5L = 0$$
$$24 = 5L$$
$$4.8 = L$$

and so

$$H = 12 - 4.8 = 7.2 \text{ hours}$$

To check the second-order condition we need to differentiate (3) again. Let

$$u = 24 - 5L \quad \text{and} \quad v = (48 - 4L)^{0.25}L^{0.5}$$

giving

$$\frac{du}{dL} = -5$$

and

$$\frac{dv}{dL} = (48 - 4L)^{0.25}0.5L^{-0.5} + L^{0.5}0.25(48 - 4L)^{-0.75}(-4)$$

$$= \frac{(48 - 4L)0.5 - L}{L^{0.5}(48 - 4L)^{0.75}}$$

$$= \frac{24 - 3L}{L^{0.5}(48 - 4L)^{0.75}}$$

Therefore, using the quotient rule,

$$\frac{d^2U}{dL^2} = \frac{(48 - 4L)^{0.25}L^{0.5}(-5) - (24 - 5L)[(24 - 3L)/L^{0.5}(48 - 4L)^{0.75}]}{(48 - 4L)^{0.5}L}$$

When $L = 4.8$ then $24 - 5L = 0$ and so the second part of the numerator disappears. Then, dividing through top and bottom by $(48 - 4L)^{0.25}L^{0.5}$ we get

$$\frac{d^2U}{dL^2} = \frac{-5}{(48 - 4L)^{0.25}L^{0.5}} = -0.985 < 0$$

and so the second-order condition for maximization of utility is satisfied when 7.2 hours are worked and 4.8 hours are taken as leisure.

Test Yourself, Exercise 12.3

1. If $y = \dfrac{(3x + 0.4x^2)}{(8 - 6x^{1.5})^{0.5}}$ what is $\dfrac{dy}{dx}$?
2. Derive a function for marginal revenue for the demand schedule

    $$p = \frac{720}{(25 + q)^{0.5}}$$

3. Using your answer from Test Yourself, Exercise 12.2.4, show that the second-order condition for a maximum value of the function $Q = 120^{0.5}(250 - 0.5K)^{0.3}$ is satisfied when K is 312.5 and evaluate d^2Q/dK^2.
4. For the demand schedule $p = (800 - 0.4q)^{0.5}$ find which value of q will maximize total revenue, using the quotient rule to check the second-order condition.
5. Assume that an individual can choose the number of hours per day that they work up to a maximum of 12 hours. This individual attempts to maximize the utility function $U = L^{0.4}I^{0.6}$ where L is defined as hours not worked out of the 12-hour maximum working day, and I is income, equal to hours worked (H) times the hourly wage rate of £15. What mix of leisure and work will be chosen?
6. Show that when a firm faces a U-shaped short-run average variable cost (AVC) schedule, its marginal cost schedule will always cut the AVC schedule at its minimum point when MC is rising.

12.6   Integration

Integrating a function means finding another function which, when it is differentiated, gives the first function. It is basically differentiation in reverse, and the rules for integration are the reverse of those for differentiation. Unlike differentiation, which we have seen to be very useful in optimization problems, the mathematical technique of integration is not as widely used in economics and so we shall only look at some of the basic ideas involved.

Assume that you wish to integrate the function

$$f'(x) = 12x + 24x^2$$

This means that you wish to find a function $y = f(x)$ such that

$$\frac{dy}{dx} = f'(x) = 12x + 24x^2$$

From your knowledge of differentiation you should be able to work out that if

$$y = 6x^2 + 8x^3$$

then

$$\frac{dy}{dx} = 12x + 24x^2$$

However, although this is one solution, the same derivative can be obtained from other functions. For example, if $y = 35 + 6x^2 + 8x^3$ then we also get

$$\frac{dy}{dx} = 12x + 24x^2$$

In fact, whatever constant term starts the function the same derivative will be obtained. Because constant numbers disappear when a function is differentiated, we cannot know what constant should appear in an integrated function unless further information is available. We therefore simply include a 'constant of integration' (C) in the integral.

The notation used for integration is

$$y = \int f'(x)dx$$

This means that y is the integral of the function $f'(x)$. The sign $\int$ is known as the integration sign. The 'dx' signifies that if y is differentiated with respect x the result will equal $f'(x)$. We can therefore write the integral of the above example as

$$y = \int (12x + 24x^2)dx = 6x^2 + 8x^3 + C \tag{4}$$

The general rule for the integration of individual terms in an expression is

$$\int ax^n dx = \frac{ax^{n+1}}{n+1} + C$$

where a and n are given parameters and $n \neq -1$. As this procedure is simply the reverse of the rule for differentiation you should have no problems in seeing how the answers below are derived.

Example 12.15

Find the following integrals: Solutions:

(i) $\int 30x^4 dx$ $y = 6x^5 + C$

(ii) $\int (24 + 7.2x)dx$ $y = 24x + 3.6x^2 + C$

(iii) $\int 0.5x^{-0.5} dx$ $y = x^{0.5} + C$

(iv) $\int (48x - 0.4x^{-1.4})dx$ $y = 24x^2 + x^{-0.4} + C$

(v) $\int (65 + 1.5x^{-2.5} + 1.5x^2)dx$ $y = 65x - x^{-1.5} + 0.5x^3 + C$

There are rules for integrating more complex functions, based on the chain, product and quotient rules for differentiation. However, they are awkward to use and will not be much use to you at present and so they are not covered in this text. The special case when $n = -1$, i.e. the integral $\int (1/x)dx$, will be dealt with in Chapter 14 when we cover exponential functions.

In earlier chapters we have seen how differentiation of total cost, total revenue and other functions gives the corresponding marginal function. For example, using the usual terminology

$$\frac{dTC}{dq} = MC \qquad \frac{dTR}{dq} = MR$$

Therefore the integration of the marginal function will give the corresponding total function, apart from the unknown constant.

Total cost functions can usually be split into fixed and variable components. The integral of marginal cost will give total variable costs plus a constant of integration which should equal Total Fixed Cost (TFC). For example, if we are given the information that total variable cost is

$$TVC = 25q - 6q^2 + 0.8q^3$$

and total fixed cost is

$$TFC = 10$$

then, by definition,

$$TC = TVC + TFC$$

$$= 10 + 25q - 6q^2 + 0.8q^3$$

Thus, marginal cost will be

$$MC = \frac{dTC}{dq} = 25 - 12q + 2.4q^2$$

On the other hand, if we were given the information that total fixed cost was 10 and that

$$MC = 25 - 12q + 2.4q^2$$

then we could find total variable cost by integration as

$$TVC = \int MC \, dq - TFC = 25q - 6q^2 + 0.8q^3$$

Thus

$$TC = TFC + TVC = 10 + 25q - 6q^2 + 0.8q^3$$

Example 12.16

If a firm spends £650 on fixed costs what is its total cost function if its marginal cost function is $MC = 82 - 16q + 1.8q^2$?

Solution

We know that for any cost function

$$\int MC \, dq = TC$$

Therefore

$$\int (82 - 16q + 1.8q^2)dq = TC$$

$$82q - 8q^2 + 0.6q^3 + TFC = TC$$

We are told that $TFC = 650$ and so

$$TC = 650 + 82q - 8q^2 + 0.6q^3$$

If one is given a firm's marginal revenue function then one can integrate this to find the total revenue function. For example, if

$$MR = 360 - 2.5q$$

$$TR = \int MR \, dq = 360q - 1.25q^2 + C$$

When q is zero, TR must also be zero. Thus $C = 0$ and so

$$\int MR \, dq = 360q - 1.25q^2 = TR \tag{1}$$

Example 12.17

If $MR = 520 - 3q^{0.5}$ what is the corresponding TR function?

Solution

$$TR = \int MR \, dq = \int (520 - 3q^{0.5})dq = 520q - 2q^{1.5}$$

Once the TR function corresponding to a given MR function has been derived then one simply has to divide this through by q to arrive at the demand function.

Example 12.18

What total revenue will a firm earn if it charges a price of £715 and its marginal revenue function is MR $= 960 - 0.15q^2$?

Solution

As we have established that the integral of this form of MR function will not have a constant of integration then

$$\text{TR} = \int \text{MR } dq = \int (960 - 0.15q^2)dq = 960q - 0.05q^3$$

In this example we need to use the TR function to find the price. Since TR $= pq$ then $p = \text{TR}/q$ and so

$$p = \frac{1}{q}(960q - 0.05q^3) = 960 - 0.05q^2$$
$$0.05q^2 = 960 - p$$
$$q^2 = 19{,}200 - 20p$$
$$q = (19{,}200 - 20p)^{0.5}$$

When $p = 715$ then

$$q = (19{,}200 - 14{,}300)^{0.5} = 4{,}900^{0.5} = 70$$

and so total revenue will be

$$\text{TR} = pq = 715(70) = £50{,}050$$

If both MC and MR functions are specified then one can use integration to work out what the actual profit is at any given output, provided that TFC is specified.

Example 12.19

If a firm faces the marginal cost schedule   MC $= 180 + 0.3q^2$

and the marginal revenue schedule   MR $= 540 - 0.6q^2$

and total fixed costs are £65, what is the maximum profit it can make? (Assume that the second-order condition for a maximum is met.)

Solution

Profit is maximized when MC = MR. Therefore,

$$180 + 0.3q^2 = 540 - 0.6q^2$$
$$0.9q^2 = 360$$
$$q^2 = 400$$
$$q = 20$$

To find the actual profit (π), we now integrate to get TR and TC and then subtract TC from TR.

$$\text{TR} = \int \text{MR} \, dq = \int (540 - 0.6q^2)dq = 540q - 0.2q^3$$

$$\text{TC} = \int \text{MC} \, dq + \text{TFC} = \int (180 + 0.3q^2)dq + 65 = 180q + 0.1q^3 + 65$$

$$\pi = \text{TR} - \text{TC}$$
$$= 540q - 0.2q^3 - (180q + 0.1q^3 + 65)$$
$$= 540q - 0.2q^3 - 180q - 0.1q^3 - 65$$
$$= 360q - 0.3q^3 - 65$$

Thus when $q = 20$ the maximum profit level is

$$\pi = 360(20) - 0.3(20)^3 - 65 = \pounds 4,735$$

Test Yourself, Exercise 12.4

1. Find the integrals for the following functions:

    (a) $25x$ (b) $5 + 1.2x + 0.15x^2$

    (c) $120x^4 - 60x^3$ (d) $42 - 18x^{-2}$

    (e) $90x^{0.5} - 44x^{-1.2}$

2. Find the total variable cost functions corresponding to the following marginal cost functions:

    (a) $\text{MC} = 4 + 0.1q$ (b) $\text{MC} = 42 - 18q + 6q^2$

    (c) $\text{MC} = 35 + 0.9q^2$ (d) $\text{MC} = 62 - 16q + 1.5q^2$

    (e) $\text{MC} = 185 - 24q + 1.2q^3$

12.7   Definite integrals

The integrals we have looked at so far are called 'indefinite integrals'. Another form of integral is the 'definite integral'. This is specified with two values of the independent variable (placed at the top and bottom of the integration sign) and is defined as the value of the integral at one value minus its value at the other value.

For example, the definite integral $\int_3^8 6x^2 dx$ is the value of this integral when x is 8 minus its value when x is 3. Thus, given that

$$\int 6x^2 dx = 2x^3 + C$$

then

$$\int_3^8 6x^2 dx = [2(8)^3 + C] - [2(3)^3 + C]$$

$$= 1024 + C - 54 - C = 970$$

In any definite integral the two constants of integration will always cancel out, as in the example above, and so they can be omitted.

The usual notation used when evaluating a definite integral is to write the relevant values outside a set of squared brackets which contains the integral of the given function. For example

$$\int_3^8 6x^2 dx = [2x^3]_3^8 = 2(8)^3 - 2(3)^3 = 1024 - 54 = 970$$

The same procedure is used for more complex functions.

Example 12.20

Evaluate the definite integral

$$\int_5^6 (6x^{0.5} - 3x^{-2} + 85x^4) dx$$

Solution

$$\int_5^6 (6x^{0.5} - 3x^{-2} + 85x^4) dx = [4x^{1.5} + 3x^{-1} + 17x^5]_5^6$$

$$= (58.787752 + 0.5 + 132,192)$$

$$- (44.72136 + 0.6 + 53,125)$$

$$= 132,251.29 - 53,170.32$$

$$= 79,080.97$$

An important feature of definite integrals is that they are equal to the area between a function and the horizontal axis that is between the two specified values of the independent variable. For example, assume that you wished to find the area between $x = 1$ and $x = 3$ under the

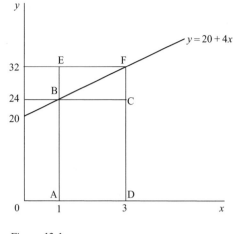

Figure 12.1

function $y = 20 + 4x$ which is illustrated in Figure 12.1, i.e. the area BFDA. This would be equal to the definite integral

$$\int_1^3 (20 + 4x)\mathrm{d}x = [20x + 2x^2]_1^3$$

$$= (60 + 18) - (20 + 2)$$

$$= 78 - 22 = 56$$

(Although the spacing on the x and y axes in Figure 12.1 differs, if the same unit of measurement is used for both x and y then this area is measured in abstract 'square units'.)

As this example uses a linear function, the answer obtained by integration can be checked using basic geometry.

$$\text{Area of rectangle ABCD} = 2 \times 24 = 48$$

$$\text{Area of triangle BFC} = \tfrac{1}{2} \text{ area EBCF} = \tfrac{1}{2}(2 \times 8) = \underline{\ 8\ }$$

$$\text{Total area BFDA} = \text{ABCD} + \text{BFC} = 56$$

Definite integrals of marginal cost functions

This concept of the definite integral has several applications in economics. To evaluate TVC from an MC function for a given value of output one simply substitutes the given quantity into the TVC function. This is the same as evaluating the definite integral between zero and the given quantity. For example, assume that you wished to find the value of TVC when $q = 8$ and you are given the function $MC = 7.5 + 0.3q^2$. This value would be

$$\int_0^8 (7.5 + 0.3q^2)\mathrm{d}q = [7.5q + 0.1q^3]_0^8 = 60 + 51.2 = 111.2$$

Therefore, TVC is equal to the area under the MC schedule between zero and the given quantity.

We can also see that the increase in TVC between two quantities will be equal to the area under the corresponding MC schedule between the given quantities. Assume that marginal cost is the function $MC = 20 + 4x$, where x is output and cost is in pounds (same format as Figure 12.1), and you wish to determine the increase in TVC when output is increased from 1 to 3 units. This will be the area BFDA which we have already found to be 56 'square units', or £56 when the function represents MC.

This must be so because this area represents the definite integral $[20x + 2x^2]_1^3$ which is the value of TVC when quantity is 3 minus its value when quantity is 1.

The definite integral of a function between two given quantities has been shown to be equal to the area under the function between the two quantities but above the horizontal axis. If a function takes negative values, i.e. it goes below the horizontal axis, then areas below the axis but above the function are treated as 'negative' areas.

Definite integrals of marginal revenue functions

This phenomenon is illustrated in Figure 12.2 which shows the linear demand schedule $p = 60 - 2q$ and the linear marginal revenue schedule $MR = 60 - 4q$. The corresponding total revenue schedule $TR = 60q - 2q^2$ is shown in the lower part of the diagram. Total revenue is at its maximum when

$$MR = 60 - 4q = 0$$

$$q = 15$$

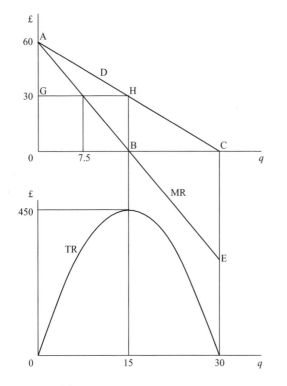

Figure 12.2

Therefore

$$p = 60 - 2(15) = 30$$

and so the maximum value of TR is $pq = 450$.

Given the linear demand and marginal revenue schedules we can see that TR rises from 0 to 450 when q increases from 0 to 15, and then falls back again to zero when q increases from 15 to 30. These changes in TR correspond to the values of the definite integrals over these quantity ranges and are represented by the area between the MR schedule and the quantity axis.

When q is 15, TR will be equal to the area 0AB which is

$$\int_0^{15} \text{MR } dq = \int_0^{15} (60 - 4q)dq = [60q - 2q^2]_0^{15} = 900 - 450 = 450$$

The change in TR when q increases from 15 to 30 will be the 'negative' area BCE which lies above the MR schedule and below the quantity axis. This will be equal to

$$\int_{15}^{30} \text{MR } dq = \int_{15}^{30} (60 - 4q)dq = [60q - 2q^2]_{15}^{30}$$
$$= (1,800 - 1,800) - (900 - 450) = -450$$

This checks with our initial assessment. Total revenue rises by 450 and then falls by the same amount.

Finally, let us see what happens when we look at the definite integral of the MR function over the entire output range 0–30. This will be

$$\int_0^{30} \text{MR } dq = \int_0^{30} (60 - 4q)dq = [60q - 2q^2]_0^{30} = 1,800 - 1,800 = 0$$

The negative area BCE has exactly cancelled out the positive area 0AB, giving zero TR when q is 30, which is correct.

Integration and consumer surplus

We can also use the demand schedule in Figure 12.2 to determine consumer surplus. This is defined as the area below a demand schedule but above the ruling price. This difference between what consumers are willing to pay for a good and what they actually have to pay is often used as a measure of welfare.

Returning to Figure 12.2, if price were zero then consumer surplus would be the entire area under the demand schedule, the triangle OAC. Geometrically this area can be calculated as

$$\tfrac{1}{2} \text{ (height} \times \text{base)} = \tfrac{1}{2}(60 \times 30) = 900$$

Using the definite integral of the demand function the area will be

$$\int_0^{30} (60 - 2q)dq = [60q - q^2]_0^{30} = 1,800 - 900 = 900$$

Both answers are, of course, the same.

If price is 30 then consumer surplus is the area AHG. The corresponding quantity is 15 and so the area AHB0 is equal to the definite integral

$$\int_0^{15} (60 - 2q)\mathrm{d}q = [60q - q^2]_0^{15} = 900 - 225 = 675$$

Thus consumer surplus = AHG = AHB0 − GHB0 = 675 − 450 = 225

The above examples of linear functions were used so that the integration method of finding areas under functions could be easily compared with the geometric solutions. The same principles can also be applied to a non-linear function.

Example 12.21

For the non-linear demand function $p = 1{,}800 - 0.6q^2$ and the corresponding marginal revenue function MR $= 1{,}800 - 1.8q^2$, use definite integrals to find:

(i) TR when q is 10;
(ii) the change in TR when q increases from 10 to 20;
(iii) consumer surplus when q is 10.

Solution

(i) TR when q is 10 will be

$$\int_0^{10} \text{MR } \mathrm{d}q = \int_0^{10} (1{,}800 - 1.8q^2)\mathrm{d}q$$

$$= [1{,}800q - 0.6q^3]_0^{10}$$

$$= 18{,}000 - 600 = £17{,}400$$

(ii) The change in TR when q increases from 10 to 20 will be

$$\int_{10}^{20} \text{MR } \mathrm{d}q = \int_{10}^{20} (1{,}800 - 1.8q^2)\mathrm{d}q$$

$$= [1{,}800q - 0.6q^3]_{10}^{20}$$

$$= (36{,}000 - 4{,}800) - (18{,}000 - 600) = £13{,}800$$

(iii) Consumer surplus when q is 10 will be the definite integral of the demand function minus total revenue actually spent by consumers. This integral will be

$$\int_0^{10} (1{,}800 - 0.6q^2)\mathrm{d}q = [1{,}800q - 0.2q^3]_0^{10} = 18{,}000 - 200 = £17{,}800$$

and

TR $= pq = 1{,}800q - 0.6q^3 = 18{,}000 - 600 = £17{,}400$

Therefore consumer surplus = £17,800 − £17,400 = £400

Test Yourself, Exercise 12.5

1. Given the non-linear demand schedule $p = 600 - 6q^{0.5}$ and the corresponding marginal revenue function $MR = 600 - 9q^{0.5}$, use definite integrals to find:

    (a) total revenue when q is 2,500;
    (b) the change in total revenue when q increases from 2,025 to 2,500;
    (c) consumer surplus when q is 2,500 and price is £300;
    (d) the change in consumer surplus when q increases from 2,025 to 2,500 owing to a price fall from £330 to £300.

2. If a firm faces the marginal cost function $MC = 40 - 18q + 4.5q^2$, what would be the increase in total cost if output were increased from 30 to 40 ?

3. Specify your own function representing a marginal concept in economics, find the indefinite integral and the definite integral over a specified range of values and interpret the meaning of your answers.

13 Dynamics and difference equations

Learning objectives

After completing this chapter students should be able to:

- Demonstrate how a time lag can affect the pattern of adjustment to equilibrium in some basic economic models.
- Construct spreadsheets to plot the time path of dependent variables in economic models with simple lag structures.
- Set up and solve linear first-order difference equations.
- Apply the difference equation solution method to the cobweb, Keynesian and Bertrand models involving a single lag.
- Identify the stability conditions in the above models.

13.1 Dynamic economic analysis

In earlier chapters much of the economic analysis used has been comparative statics. This entails the comparison of different (static) equilibrium situations, with no mention of the mechanism by which price and quantity adjust to their new equilibrium values. The branch of economics that looks at how variables adjust between equilibrium values is known as 'dynamics', and this chapter gives an introduction to some simple dynamic economic models.

The ways in which markets adjust over time vary tremendously. In commodity exchanges, prices are changed by the minute and adjustments to new equilibrium prices are almost instantaneous. In other markets the adjustment process may be a slow trial and error process over several years, in some cases so slow that price and quantity hardly ever reach their proper equilibrium values because supply and demand schedules shift before equilibrium has been reached. There is therefore no one economic model that can explain the dynamic adjustment process in all markets.

The simple dynamic adjustment models explained here will give you an idea of how adjustments can take place between equilibria and how mathematics can be used to calculate the values of variables at different points in time during the adjustment process. They are only very basic models, however, designed to give you an introduction to this branch of economics. The mathematics required to analyse more complex dynamic models goes beyond that covered in this text.

In this chapter, time is considered as a discrete variable and the dynamic adjustment process between equilibria is seen as a step-by-step process. (The distinction between discrete and continuous variables was explained in Section 7.1.) This enables us to calculate different values of the variables that are adjusting to new equilibrium levels:

(i) using a spreadsheet, and
(ii) using the mathematical concept of 'difference equations'.

Models that assume a process of continual adjustment are considered in Chapter 14, using 'differential equations'.

13.2 The cobweb: iterative solutions

In some markets, particularly agricultural markets, supply cannot immediately expand to meet increased demand. Crops have to be planted and grown and livestock takes time to raise. Some manufactured products can also take a while to produce when orders suddenly increase. The cobweb model takes into account this delayed response on the supply side of a market by assuming that quantity supplied now (Q_t^s) depends on the ruling price in the previous time period (P_{t-1}), i.e.

$$Q_t^s = f(P_{t-1})$$

where the subscripts denote the time period. Consumer demand for the same product (Q_t^d), however, is assumed to depend on the current price, i.e.

$$Q_t^d = f(P_t)$$

This is a reasonable picture of many agricultural markets. The quantity offered for sale this year depends on what was planted at the start of the growing season, which in turn depends on last year's price. Consumers look at current prices, though, when deciding what to buy.

The cobweb model also assumes that:

- the market is perfectly competitive
- supply and demand are both linear schedules.

Before we go any further, it must be stressed that this model does *not* explain how price adjusts in all competitive markets, or even in all perfectly competitive agricultural markets. It is a simple model with some highly restrictive assumptions that can only explain how price adjusts in these particular circumstances. Some markets may have a more complex lag structure, e.g. $Q_t^s = f(P_{t-1}, P_{t-2}, P_{t-3})$, or may not have linear demand and supply. You should also not forget that intervention in agricultural markets, such as the EU Common Agricultural Policy, usually means that price is not competitively determined and hence the cobweb assumptions do not apply. Having said all this, the cobweb model can still give a fair idea of how price and quantity adjust in many markets with a delayed supply.

The assumptions of the cobweb model mean that the demand and supply functions can be specified in the format

$$Q_t^d = a + bP_t \quad \text{and} \quad Q_t^s = c + dP_{t-1}$$

where a, b, c and d are parameters specific to individual markets.

Note that, as demand schedules slope down from left to right, the value of b is expected to be negative. As supply schedules usually cut the price axis at a positive value (and therefore the quantity axis at a negative value if the line were theoretically allowed to continue into negative quantities), the value of c will also usually be negative. Remember that these functions have Q as the dependent variable but in supply and demand analysis Q is usually measured along the horizontal axis.

Although desired quantity demanded only equals desired quantity supplied when a market is in equilibrium, it is always true that actual quantity bought equals quantity sold. In the cobweb model it is assumed that in any one time period producers supply a given amount Q_t^s. Thus there is effectively a vertical short-run supply schedule at the amount determined by the previous time period's price. Price then adjusts so that all the produce supplied is bought by consumers. This adjustment means that

$$Q_t^d = Q_t^s$$

Therefore

$$a + bP_t = c + dP_{t-1}$$
$$bP_t = c - a + dP_{t-1}$$
$$P_t = \frac{c - a}{b} + \frac{d}{b}P_{t-1} \tag{1}$$

This is what is known as a '*linear first-order difference equation*'. A difference equation expresses the value of a variable in one time period as a function of its value in earlier periods; in this case

$$P_t = f(P_{t-1})$$

It is clearly a linear relationship as the terms $(c - a)/b$ and d/b will each take a single numerical value in an actual example. It is 'first order' because only a single lag on the previous time period is built into the model and the coefficient of P_{t-1} is a simple constant. In the next section we will see how this difference equation can be used to derive an expression for P_t in terms of t.

Before doing this, let us first get a picture of how the cobweb price adjustment mechanism operates using a numerical example.

Example 13.1

In an agricultural market where the assumptions of the cobweb model apply, the demand and supply schedules are

$$Q_t^d = 400 - 20P_t \quad \text{and} \quad Q_t^s = -50 + 10P_{t-1}$$

A long-run equilibrium has been established for several years but then one year there is an unexpectedly good crop and output rises to 160. Explain how price will behave over the next few years following this one-off 'shock' to the market.

(Note: In this example and in most other examples in this chapter, no specific units of measurement for P or Q are given in order to keep the analysis as simple as possible. In actual applications, of course, price will usually be in £ and quantity in physical units, e.g. thousands of tonnes.)

Solution

In long-run equilibrium, price and quantity will remain unchanged each time period. This means that:

the long-run equilibrium price $P^* = P_t = P_{t-1}$

and the long-run equilibrium quantity $Q^* = Q_t^d = Q_t^s$

Therefore, when the market is in equilibrium

$$Q^* = 400 - 20P^* \quad \text{and} \quad Q^* = -50 + 10P^*$$

Equating to solve for P^* and Q^* gives

$$400 - 20P^* = -50 + 10P^*$$
$$450 = 30P^*$$
$$15 = P^*$$
$$Q^* = 400 - 20P^* = 400 - 300 = 100$$

These values correspond to the point where the supply and demand schedules intersect, as illustrated in Figure 13.1.

If an unexpectedly good crop causes an amount of 160 to be supplied onto the market one year, then this means that the short-run supply schedule effectively becomes the vertical line S_0 in Figure 13.1. To sell this amount the price has to be reduced to P_0, corresponding to the point A where S_0 cuts the demand schedule.

Producers will then plan production for the next time period on the assumption that P_0 is the ruling price. The amount supplied will therefore be Q_1, corresponding to point B. However, in the next time period when this reduced supply quantity Q_1 is put onto the market it will sell

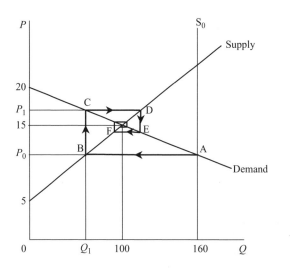

Figure 13.1

for price P_1, corresponding to point C. Further adjustments in quantity and price are shown by points D, E, F, etc. These trace out a cobweb pattern (hence the 'cobweb' name) which converges on the long-run equilibrium where the supply and demand schedules intersect.

In some markets, price will not always return towards its long-run equilibrium level, as we shall see later when some other examples are considered. However, first let us concentrate on finding the actual pattern of price adjustment in this particular example.

Approximate values for the first few prices could be read off the graph in Figure 13.1, but as price converges towards the centre of the cobweb it gets difficult to read values accurately. We shall therefore calculate the first few values of P manually, so that you can become familiar with the mechanics of the cobweb model, and then set up a spreadsheet that can rapidly calculate patterns of price adjustment over a much longer period.

Quantity supplied in each time period is calculated by simply entering the previously ruling price into the market's supply function

$$Q_t^s = -50 + 10P_{t-1}$$

but how is this price calculated? There are two ways:

(a) from first principles, using the given supply and demand schedules, and
(b) using a difference equation, in the format (1) derived earlier.

(a) The demand function

$$Q_t^d = 400 - 20P_t$$

can be rearranged to give the inverse demand function

$$P_t = 20 - 0.05Q_t^d$$

The model assumes that a fixed quantity arrives on the market each time period and then price adjusts until $Q_t^d = Q_t^s$. Thus, P_t can be found by inserting the current quantity supplied, Q_t^s, into the function for P_t. Assuming that the initial disturbance to the system when Q^s rises to 160 occurs in time period 0, the values of P and Q over the next three time periods can be calculated as follows:

$$Q_0^s = 160 \quad \text{(initial given value, inserted into inverse demand function)}$$
$$P_0 = 20 - 0.05Q_0^s = 20 - 0.05(160) = 20 - 8 = 12$$

This price in period 0 then determines quantity supplied in period 1, which is

$$Q_1^s = -50 + 10P_0 = -50 + 10(12) = -50 + 120 = 70$$

This quantity then determines the market-clearing price, which is

$$P_1 = 20 - 0.05Q_1^s = 20 - 0.05(70) = 20 - 3.5 = 16.5$$

The same adjustment process then continues for future time periods, giving

$$Q_2^s = -50 + 10P_1 = -50 + 10(16.5) = -50 + 165 = 115$$
$$P_2 = 20 - 0.05Q_2^s = 20 - 0.05(115) = 20 - 5.75 = 14.25$$

$$Q_3^s = -50 + 10P_2 = -50 + 10(14.25) = -50 + 142.5 = 92.5$$

$$P_3 = 20 - 0.05Q_3^s = 20 - 0.05(92.5) = 20 - 4.625 = 15.375$$

The pattern of price adjustment is therefore 12, 16.5, 14.25, 15.375, etc., corresponding to the cobweb graph in Figure 13.1. Price initially falls below its long-run equilibrium value of 15 and then converges back towards this equilibrium, alternating above and below it but with the magnitude of the difference becoming smaller each period.

(b) The same pattern of price adjustment can be obtained by using the difference equation

$$P_t = \frac{c - a}{b} + \frac{d}{b}P_{t-1} \tag{1}$$

and substituting in the given values of a, b, c and d to get

$$P_t = \frac{(-50) - 400}{-20} + \frac{10}{-20}P_{t-1}$$

$$P_t = 22.5 - 0.5P_{t-1} \tag{2}$$

The original price P_0 still has to be derived by inserting the shock quantity 160 into the demand function, as already explained, which gives

$$P_0 = 20 - 0.05(160) = 12$$

Then subsequent prices can be determined using the difference equation (2), giving

$$P_1 = 22.5 - 0.5P_0 = 22.5 - 0.5(12) = 16.5$$

$$P_2 = 22.5 - 0.5P_1 = 22.5 - 0.5(16.5) = 14.25$$

$$P_3 = 22.5 - 0.5P_2 = 22.5 - 0.5(14.25) = 15.375 \text{ etc.}$$

These prices are the same as those calculated by method (a), as expected.

Table 13.1

	A	B	C	D	E	F	G	H
1	Ex.	COBWEB	MODEL					
2	13.1			Qd=a+bPt		Qs=c+dPt		
3								
4		Parameter	a =	400	c =	-50		
5		values	b =	-20	d =	10		
6		Initial shock	Quantity =	160				
7						Equilibrium	Price =	15
8	Time	Quantity	Price	Change		Equilibrium	Quantity =	100
9	t	Qt	Pt	in Pt				
10	0	160	12.00			Stability =>	STABLE	
11	1	70	16.50	4.50				
12	2	115	14.25	-2.25				
13	3	92.5	15.38	1.13				
14	4	103.75	14.81	-0.56				
15	5	98.125	15.09	0.28				
16	6	100.9375	14.95	-0.14				
17	7	99.53125	15.02	0.07				
18	8	100.23438	14.99	-0.04				
19	9	99.882813	15.01	0.02				
20	10	100.05859	15.00	-0.01				

Table 13.2

CELL	Enter	Explanation
As in *Table 13.1*	*Enter all labels and column headings shown in* Table 13.1	Note: do not enter for the word "STABLE" in cell G10. The stability condition will be deduced by the spreadsheet.
D4	400	These are the parameter values for this example.
D5	-20	
F4	-50	
F5	10	
D6	160	This is initial "shock" quantity in time period 0.
A10 to A20	*Enter numbers from 0 to 10*	These are the time periods.
B10	=D6	Quantity in time period 0 is initial "shock" value.
C10	=(B10-D$4)/D$5	Calculates P_0, the initial market clearing price. Given that $Q^d_t = a + bP_t$ then $P_t = (Q^d_t - a)/b$. Note the $ on cells D4 and D5. Format to 2 dp.
C11 to C20	*Copy formula from C10 down column.*	Will calculate price in each time period (when all quantities in column B are calculated)
B11	=F$4+F$5*C10	Calculates quantity in year 1 based on price in previous time period according to supply function $Q^s_t = c + dP_{t-1}$. Format to 2 dp.
D11	=C11-C10	Calculates change in price between time periods.
B12 to B20	*Copy formula from B11 down column.*	Calculates quantity supplied in each time period.
D12 to D20	*Copy formula from D12 down column.*	Calculates price change since previous time period
H7	=(F4-D4)/(D5-F5)	Calculates equilibrium price using the formula $P^* = (c - a)/(b - d)$
H8	=F4+F5*H7	Calculates equilibrium quantity $Q^* = a + bP^*$
G10	*Enter the formula below*	This uses the Excel "IF" logic function to determine whether $d/(-b)$ is less than 1, greater than 1, or equals 1. This stability criterion is explained later.
=IF(-F5/D5<1,"STABLE",IF(-F5/D5>1,"UNSTABLE","OSCILLATING"))		

A spreadsheet can be set up to calculate price over a large number of time periods. Instructions are given in Table 13.2 for constructing the Excel spreadsheet shown in Table 13.1. This calculates price for each period from first principles, but you can also try to construct your own spreadsheet based on the difference equation approach.

This spreadsheet shows a series of prices and quantities converging on the equilibrium values of 15 for price and 100 for quantity. The first few values can be checked against the manually calculated values and are, as expected, the same. To bring home the point that each price adjustment is smaller than the previous one, the change in price from the previous time period is also calculated. (The price columns are formatted to 2 decimal places so price is calculated to the nearest penny.)

Although the stability of this example is obvious from the way that price converges on its equilibrium value of 15, a stability check is entered which may be useful when this spreadsheet is used for other examples. Assuming that b is always negative and d is positive, the market will be stable if $d/-b < 1$ and unstable (i.e. price will not converge back to its equilibrium) if $d/-b > 1$. (The reasons for this rule are explained later in Section 13.3.)

When you have constructed this spreadsheet yourself, save it so that it can be used for other examples.

To understand why price may not always return to its long-run equilibrium level in markets where the cobweb model applies, consider Example 13.2.

Example 13.2

In a market where the assumptions of the cobweb model apply, the demand and supply functions are

$$Q_t^d = 120 - 4P_t \quad \text{and} \quad Q_t^s = -80 + 16P_{t-1}$$

If in one time period the long-run equilibrium is disturbed by output unexpectedly rising to a level of 90, explain how price will adjust over the next few time periods.

Solution

The long-run equilibrium price can be determined from the formula

$$P^* = \frac{c-a}{b-d} = \frac{(-80)-120}{-4-16} = \frac{-200}{-20} = 10$$

Thus, the long-run equilibrium quantity is

$$Q^* = 120 - 4P^* = 120 - 4(10) = 80$$

You could use the spreadsheet developed for Example 13.1 above to trace out the subsequent pattern of price adjustment but if a few values are calculated manually it can be seen that calculations after period 2 are irrelevant.

Using the standard cobweb model difference equation

$$P_t = \frac{c-a}{b} + \frac{d}{b}P_{t-1} \tag{1}$$

and substituting the known values, we get

$$P_t = \frac{(-80)-120}{-4} + \frac{16}{-4}P_{t-1} = 50 - 4P_{t-1} \tag{2}$$

The initial price P_0 can be found by inserting the shock quantity of 90 into the demand function. Thus

$$Q_0^s = 90 = 120 - 4P_0$$
$$4P_0 = 30$$
$$P_0 = 7.5$$

Putting this value into the difference equation (2) above we get

$$P_1 = 50 - 4P_0 = 50 - 4(7.5) = 20$$
$$P_2 = 50 - 4P_1 = 50 - 4(20) = -30$$

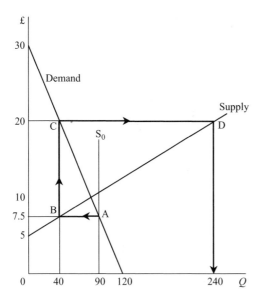

Figure 13.2

There is not much point in going any further with the calculations. Assuming that producers will not pay consumers to take goods off their hands, negative prices cannot exist. What has happened is that price has followed the path *ABCD* traced out in Figure 13.2.

The initial quantity 90 put onto the market causes price to drop to 7.5. Suppliers then reduce supply for the next period to

$$Q_1^s = -80 + 16P_0 = -80 + 16(7.5) = 40$$

This sells for price $P_1 = 20$ and so supply for the following period is increased to

$$Q_2^s = -80 + 16P_1 = -80 + 16(20) = 240$$

Consumers would only consume 120 even if price were zero (where the demand schedule hits the axis) and so, when this quantity of 240 is put onto the market, price will collapse to zero and there will still be unsold produce. Producers will not wish to supply anything for the next time period if they expect a price of zero and so no further production will take place.

This is clearly an unstable market, but why is there a difference between this market and the stable market considered in Example 13.1? It depends on the slopes of the supply and demand schedules. If the absolute value of the slope of the demand schedule is less than the absolute value of the slope of the supply schedule then the market is stable, and vice versa. These slopes are inversely related to parameters b and d, since the vertical axis measures p rather than q. Thus the stability conditions are

Stable: $|d/b| < 1$ Unstable: $|d/b| > 1$

A formal proof of these conditions, based on the difference equation solution method, plus an explanation of what happens when $|d/b| = 1$, is given in Section 13.3.

Although in theoretical models of unstable markets (such as Example 13.2) price 'explodes' and the market collapses, this may not happen in reality if:

- producers learn from experience and do not simply base production plans for the next period on the current price,
- supply and demand schedules are not linear along their entire length,
- government intervention takes place to support production.

Another example of an exploding market is Example 13.3 below, which is solved using the spreadsheet developed for Example 13.1.

Example 13.3

In an agricultural market where the cobweb assumptions hold and

$$Q_t^d = 360 - 8P_t \quad \text{and} \quad Q_t^s = -120 + 12P_{t-1}$$

a long-run equilibrium is disturbed by an unexpectedly good crop of 175 units. Use a spreadsheet to trace out the subsequent path of price adjustment.

Solution

When the given parameters and shock quantity are entered, your spreadsheet should look like Table 13.3. This is clearly unstable as both the automatic stability check and the pattern

Table 13.3

	A	B	C	D	E	F	G	H
1	*Ex.*	COBWEB	MODEL					
2	13.3			Qd=a+bPt		Qs=c+dPt		
3								
4		Parameter	a =	360	c =	-120		
5		Values	b =	-8	d =	12		
6		Initial shock	quantity =	175				
7						Equilibrium	Price =	24
8	Time	Quantity	Price	Change		Equilibrium	Quantity =	168
9	t	Qt	Pt	in Pt				
10	0	175	23.13			Stability =>	UNSTABLE	
11	1	157.5	25.31	2.19				
12	2	183.75	22.03	-3.28				
13	3	144.375	26.95	4.92				
14	4	203.4375	19.57	-7.38				
15	5	114.84375	30.64	11.07				
16	6	247.73438	14.03	-16.61				
17	7	48.398438	38.95	24.92				
18	8	347.40234	1.57	-37.38				
19	9	-101.10352	57.64	56.06				
20	10	571.65527	-26.46	-84.09				

of price adjustments show. According to these figures, the market will continue to operate until the eighth time period following the initial shock. In period 9 nothing will be produced (mathematically the model gives a negative quantity) and the market collapses.

Test Yourself, Exercise 13.1

(In all these questions, assume that the assumptions of the cobweb model apply to each market.)

1. The agricultural market whose demand and supply schedules are

$$Q_t^d = 240 - 20P_t \quad \text{and} \quad Q_t^s = -33\tfrac{1}{3} + 16\tfrac{2}{3}P_{t-1}$$

   is initially in long-run equilibrium. Quantity then falls to 50% of its previous level as a result of an unexpectedly poor harvest. How many time periods will it take for price to return to within 1% of its long-run equilibrium level?

2. In an unstable market, the demand and supply schedules are

$$Q_t^d = 200 - 12.5P_t \quad \text{and} \quad Q_t^s = -60 + 20P_{t-1}$$

   A shock reduction of quantity to 80 throws the system out of equilibrium. How long will it take for the market to collapse completely?

3. By tracing out the pattern of price adjustment after an initial shock that disturbs the previously ruling long-run equilibrium, say whether or not the following markets are stable.

   (a) $Q_t^d = 150 - 1.5P_t \quad \text{and} \quad Q_t^s = -30 + 3P_{t-1}$
   (b) $Q_t^d = 180 - 125P_t \quad \text{and} \quad Q_t^s = -20 + P_{t-1}$

13.3 The cobweb: difference equation solutions

Solving the cobweb difference equation

$$P_t = \frac{c - a}{b} + \frac{d}{b}P_{t-1} \tag{1}$$

means putting it into the format

$$P_t = f(t)$$

so that the value of P_t at any given time t can be immediately calculated without the need to calculate all the preceding values of P_t.

There are two parts to the solution of this cobweb difference equation:

(i) the new long-run equilibrium price, and
(ii) the *complementary function* that tells us how much price diverges from this equilibrium level at different points in time.

A similar format applies to the solution of any linear first-order difference equation. The equilibrium solution (i) is also known as the **particular solution** (PS). In general, the particular solution is a constant value about which adjustments in the variable in question take place over time.

The **complementary function** (CF) tells us how the variable in question, i.e. price in the cobweb model, varies from the equilibrium solution as time changes.

These two elements together give what is called the **general solution** (GS) to a difference equation, which is the full solution. Thus we can write

$$GS = PS + CF$$

Finding the particular solution is straightforward. In the long run the equilibrium price P^* holds in each time period and so

$$P^* = P_t = P_{t-1}$$

Substituting P^* into the difference equation

$$P_t = \frac{c - a}{b} + \frac{d}{b}P_{t-1} \tag{1}$$

we get

$$P^* = \frac{c - a}{b} + \frac{d}{b}P^*$$
$$bP^* = c - a + dP^*$$
$$a - c = (d - b)P^*$$
$$\frac{a - c}{d - b} = P^* \tag{2}$$

This, of course, is the same equilibrium value for price that would be derived in the single-time-period linear supply and demand model

$$Q^{\mathrm{d}} = a + bP \quad \text{and} \quad Q^{\mathrm{s}} = c + dP$$

To find the complementary function, we return to the difference equation (1) but ignore the first term, which is a constant that does not vary over time, i.e. we just consider the equation

$$P_t = \frac{d}{b}P_{t-1} \tag{3}$$

This may seem rather a strange procedure, but it works, as we shall see later when some numerical examples are tackled.

We then assume that P_t depends on t according to the function

$$P_t = Ak^t \tag{4}$$

where A and k are some (as yet) unknown constants. (Note that in this formula, t denotes the power to which k is raised and is not just a time superscript.) This function applies to all values of t, which means that

$$P_{t-1} = Ak^{t-1} \tag{5}$$

Substituting the formulations (4) and (5) for P_t and P_{t-1} back into equation (3) above we get

$$Ak^t = \frac{d}{b}Ak^{t-1}$$

Dividing through by Ak^{t-1} gives

$$k = \frac{d}{b}$$

Putting this result into (4) gives the complementary function as

$$P_t = A\left(\frac{d}{b}\right)^t \tag{6}$$

The value of A cannot be ascertained unless the actual value of P_t is known for a specific value of t. (See the following numerical examples.)

The general solution to the cobweb difference equation therefore becomes

$$P_t = \text{particular solution} + \text{complementary function} = (2) \text{ plus } (6), \text{ giving}$$

$$P_t = \frac{a-c}{d-b} + A\left(\frac{d}{b}\right)^t$$

Stability

From this solution we can see that the stability of the model depends on the value of d/b. If A is a non-zero constant, then there are three possibilities

(i) If $\left|\dfrac{d}{b}\right| < 1$ then $\left(\dfrac{d}{b}\right)^t \to 0$ as $t \to \infty$

This occurs in a stable market. Whatever value the constant A takes the value of the complementary function gets smaller over time. Therefore the divergence of price from its equilibrium also approaches zero. (Note that it is the absolute value of $|d/b|$ that we consider because b will usually be a negative number.)

(ii) If $\left|\dfrac{d}{b}\right| > 1$ then $\left|\left(\dfrac{d}{b}\right)^t\right| \to \infty$ as $t \to \infty$

This occurs in an unstable market. After an initial disturbance, as t increases, price will diverge from its equilibrium level by greater and greater amounts.

(iii) If $\left|\dfrac{d}{b}\right| = 1$ then $\left|\left(\dfrac{d}{b}\right)^t\right| = 1$ as $t \to \infty$

Price will neither return to its equilibrium nor 'explode'. Normally, $b < 0$ and $d > 0$, so $d/b < 0$, which means that $d/b = -1$. Therefore, $(d/b)^t$ will oscillate between $+1$ and -1 depending on whether or not t is an even or odd number. Price will continually fluctuate between two levels (see Example 13.6 below).

We can now use this method of obtaining difference equation solutions to answer some specific numerical cobweb model problems.

Example 13.4

Use the cobweb difference equation solution to answer the question in Example 13.1 above, i.e. what happens in the market where

$$Q_t^d = 400 - 20P_t \quad \text{and} \quad Q_t^s = -50 + 10P_{t-1}$$

if there is a sudden one-off change in Q_t^s to 160?

Solution

Substituting the values for this market $a = 400$, $b = -20$, $c = -50$ and $d = 10$ into the general cobweb difference equation solution

$$P_t = \frac{a - c}{d - b} + A \left(\frac{d}{b} \right)^t \tag{1}$$

gives

$$P_t = \frac{400 - (-50)}{10 - (-20)} + A \left(\frac{10}{-20} \right)^t$$

$$= \frac{450}{30} + A(-0.5)^t$$

$$= 15 + A(-0.5)^t \tag{2}$$

To find the value of A we then substitute in the known value of P_0.

The question tells us that the initial 'shock' output level Q_0 is 160 and so, as price adjusts until all output is sold, P_0 can be calculated by substituting this quantity into the demand schedule. Thus

$$Q_0^d = 160 = 400 - 20P_0$$

$$20P_0 = 240$$

$$P_0 = 12$$

Substituting this value into the general difference equation solution (2) above gives, for time period 0,

$$12 = 15 + A(-0.5)^0$$

$$12 = 15 + A, \text{ since } (-0.5)^0 = 1$$

$$A = -3$$

Thus the complete solution to the difference equation in this example is

$$P_t = 15 - 3(-0.5)^t$$

This is usually called the **definite solution** or the **specific solution** because it relates to a specific initial value.

We can use this solution to calculate the first few values of P_t and compare with those we obtained when answering Example 13.1.

$$P_1 = 15 - 3(-0.5)^1 = 15 + 1.5 = 16.5$$

$$P_2 = 15 - 3(-0.5)^2 = 15 - 3(0.25) = 14.25$$

$$P_3 = 15 - 3(-0.5)^3 = 15 - 3(-0.125) = 15.375$$

As expected, these values are identical to those calculated by the iterative method.

In this particular example, price converges fairly quickly towards its long-run equilibrium level of 15. By time period 9, price will be

$$P_9 = 15 - 3(-0.5)^9$$
$$= 15 - 3(-0.0019531)$$
$$= 15 + 0.0058594 = 15.01 \text{ (to 2 dp)}$$

This is clearly a stable solution. In this difference equation solution

$$P_t = 15 - 3(-0.5)^t$$

and so we can see that, as t gets larger, the value of $(-0.5)^t$ approaches zero. This is because

$$|d/b| = |-0.5| = 0.5 < 1$$

and so the stability condition outlined above is satisfied.

Note that, because $-0.5 < 0$, the direction of the divergence from the equilibrium value alternates between time periods. This is because for any negative quantity $-x$, it will always be true that

$$x < 0, \quad (-x)^2 > 0, \quad (-x)^3 < 0, \quad (-x)^4 > 0, \text{ etc.}$$

Thus for odd-numbered time periods (in this example) price will be above its equilibrium value, and for even-numbered time periods price will be below its equilibrium value.

Although in this example price converges towards its long-run equilibrium value, it would never actually reach it if price and quantity were divisible into infinitesimally small units. Theoretically, this is a bit like the case of the 'hopping frog' back in Chapter 7 when infinite geometric series were examined. The distance from the equilibrium gets smaller and smaller each time period but it never actually reaches zero. For practical purposes, a reasonable cut-off point can be decided upon to define when a full return to equilibrium has been reached. In this numerical example the difference from the equilibrium is less than 0.01 by time period 9, which is for all intents and purposes a full return to equilibrium if P is measured in £.

The above example explained the method of solution of difference equations applied to a simple problem where the answers could be checked against iterative solutions. In other cases, one may need to calculate values for more distant time periods, which are more difficult to calculate manually. The method of solution of difference equations will also be useful for those of you who go on to study intermediate economic theory where some models, particularly in macroeconomics, are based on difference equations in an algebraic format which cannot be solved using a spreadsheet.

We shall now consider another cobweb example which is rather different from Example 13.4 in that

 (i) price does not return towards its equilibrium level and
 (ii) the process of adjustment is more gradual over time.

Example 13.5

In a market where the assumptions of the cobweb model hold

$$Q_t^d = 200 - 8P_t \quad \text{and} \quad Q_t^s = -43 + 8.2P_{t-1}$$

The long-run equilibrium is disturbed when quantity suddenly changes to 90. What happens to price in the following time periods?

Solution
In long-run equilibrium

$$Q^* = Q_t^d = Q_t^s$$

and

$$P^* = P_t = P_{t-1}$$

Substituting these equilibrium values and equating demand and supply we can find the new equilibrium price. Thus

$$200 - 8P^* = Q^* = -43 + 8.2P^*$$
$$243 = 16.2P^*$$
$$15 = P^*$$

This will be an unstable equilibrium as

$$\left| \frac{d}{b} \right| = \left| \frac{8.2}{-8} \right| = 1.025 > 1$$

The difference equation that describes the relationship between price in one period and the next will take the usual cobweb model format

$$P_t = \frac{c - a}{b} + \frac{d}{b}P_{t-1} \tag{1}$$

where $a = 200$, $b = -8$, $c = -43$ and $d = 8.2$, giving

$$P_t = \frac{-43 - 200}{-8} + \frac{8.2}{-8}P_{t-1}$$
$$= 30.375 - 1.025P_{t-1}$$

Using the formula derived above, the solution to this difference equation will therefore be

$$P_t = \frac{a-c}{d-b} + A\left(\frac{d}{b}\right)^t$$

$$= \frac{200-(-43)}{8.2-(-8)} + A\left(\frac{8.2}{-8}\right)^t$$

$$= \frac{243}{16.2} + A(-1.025)^t$$

$$= 15 + A(-1.025)^t \tag{2}$$

The first part of this solution is of course the equilibrium value of price which has already been calculated above. To derive the value of A, we need to find price in period 0. The quantity supplied is 90 in period 0 and so, to find the price that this quantity will sell for, this value is substituted into the demand function. Thus

$$Q_0^d = 90 = 200 - 8P_0$$

$$8P_0 = 110$$

$$P_0 = 13.75$$

Substituting this value into the general solution (2) we get

$$P_0 = 13.75 = 15 + A(-1.025)^0$$

$$13.75 = 15 + A$$

$$-1.25 = A$$

Note that, as in Example 13.1 above, the value of parameter A is the difference between the equilibrium value of price and the value it initially takes when quantity is disturbed from its equilibrium level, i.e.

$$A = P_0 - P^* = 13.75 - 15 = -1.25$$

Putting this value of A into the general solution (2), the specific solution to the difference equation in this example now becomes

$$P_t = 15 - 1.25(-1.025)^t$$

Using this formula to calculate the first few values of P_t gives

$$P_0 = 15 - 1.25(1.025)^0 = 13.75$$

$$P_1 = 15 + 1.25(1.025)^1 = 16.28$$

$$P_2 = 15 - 1.25(1.025)^2 = 13.69$$

$$P_3 = 15 + 1.25(1.025)^3 = 16.35$$

We can see that, although price is gradually moving away from its long-run equilibrium value of 15, it is a very slow process. By period 10, price is still above 13.00, as

$$P_{10} = 15 - 1.25(1.025)^{10} = 13.40$$

and it takes until time period 102 before price becomes negative, as the figures below show:

$$P_{100} = 15 - 1.25(1.025)^{100} = 0.23$$

$$P_{101} = 15 + 1.25(1.025)^{101} = 30.14$$

$$P_{102} = 15 - 1.25(1.025)^{102} = -0.51$$

This example is not a particularly realistic picture of an agricultural market as many changes in supply and demand conditions would take place over a 100-year time period. (Also, quantity becomes negative in time period 85 when the market would collapse – check this yourself using a spreadsheet.) However, it illustrates the usefulness of the difference equation solution in immediately computing values for distant time periods without first needing to compute all the preceding values.

The following example illustrates what happens when a market is neither stable nor unstable.

Example 13.6

The cobweb model assumptions hold in a market where

$$Q_t^d = 160 - 2P_t \quad \text{and} \quad Q_t^s = -20 + 2P_{t-1}$$

If the previously ruling long-run equilibrium is disturbed by an unexpectedly low output of 50 in one time period, what will happen to price in the following time periods?

Solution

Substituting the values $a = 160, b = -2, c = -20$ and $d = 2$ for this market into the cobweb difference equation general solution

$$P_t = \frac{a - c}{d - b} + A\left(\frac{d}{b}\right)^t \tag{1}$$

gives

$$P_t = \frac{160 - (-20)}{2 - (-2)} + A\left(\frac{2}{-2}\right)^t$$

$$= \frac{180}{4} + A(-1)^t$$

$$= 45 + A(-1)^t \tag{2}$$

To determine the value of A, first substitute the given value of 50 for Q_0 into the demand function so that

$$160 - 2P_0 = 50 = Q_0$$

$$110 = 2P_0$$

$$55 = P_0$$

Now substitute this value for P_0 into the general solution (2) above, so that

$$P_0 = 55 = 45 + A(-1)^0$$
$$55 = 45 + A$$
$$10 = A$$

The specific solution to the difference equation for this example is therefore

$$P_t = 45 + 10(-1)^t$$

Using this formula to calculate the first few values of P_t we see that

$$P_0 = 45 + 10(-1)^0 = 45 + 10 = 55$$
$$P_1 = 45 + 10(-1)^1 = 45 - 10 = 35$$
$$P_2 = 45 + 10(-1)^2 = 45 + 10 = 55$$
$$P_3 = 45 + 10(-1)^3 = 45 - 10 = 35$$
$$P_4 = 45 + 10(-1)^4 = 45 + 10 = 55 \text{ etc.}$$

Price therefore continually fluctuates between 35 and 55.

This is the third possibility in the stability conditions examined earlier. In this example

$$\left|\frac{d}{b}\right| = \left|\frac{2}{-2}\right| = |-1| = 1$$

Therefore, as $t \to \infty$, P_t neither converges on its equilibrium level nor explodes until the market collapses. This fluctuation between two price levels from year to year is sometimes observed in certain agricultural markets.

Test Yourself, Exercise 13.2

(Assume that the usual cobweb assumptions apply in these questions.)

1. In a market where

$$Q_t^d = 160 - 20P_t \quad \text{and} \quad Q_t^s = -80 + 40P_{t-1}$$

   quantity unexpectedly drops from its equilibrium value to 75. Derive the difference equation which will calculate price in the time periods following this event.
2. If $Q_t^d = 180 - 0.9P_t$ and $Q_t^s = -24 + 0.8P_{t-1}$ say whether or not the long-run equilibrium price is stable and then use the difference equation method to calculate price in the thirtieth time period after a sudden one-off increase in quantity to 117.

3.   Given the demand and supply schedules

$$Q_t^d = 3450 - 6P_t \quad \text{and} \quad Q_t^s = -729 + 4.5P_{t-1}$$

use difference equations to predict what price will be in the tenth time period after an unexpected drop in quantity to 354, assuming that the market was previously in long-run equilibrium.

13.4   The lagged Keynesian macroeconomic model

In the basic Keynesian model of the determination of national income, if foreign trade and government taxation and expenditure are excluded, the model reduces to the accounting identity,

$$Y = C + I \tag{1}$$

and the consumption function

$$C = a + bY \tag{2}$$

To determine the equilibrium level of national income Y^* we substitute (2) into (1), giving

$$Y^* = a + bY^* + I$$
$$Y^*(1 - b) = a + I$$
$$Y^* = \frac{a + I}{1 - b}$$

This can be evaluated for given values of parameters a and b and exogenously determined investment I.

If there is a disturbance from this equilibrium, e.g. exogenous investment I alters, then the adjustment to a new equilibrium will not be instantaneous. This is the basis of the well-known multiplier effect. An initial injection of expenditure will become income for another sector of the economy. A proportion of this will be passed on as a further round of expenditure, and so on until the 'ripple effect' dies away.

Because consumer expenditure may not adjust instantaneously to new levels of income, a lagged effect may be introduced. If it is assumed that consumers' expenditure in one time period depends on the income that they received in the previous time period, then the consumption function becomes

$$C_t = a + bY_{t-1} \tag{3}$$

where the subscripts denote the time period.

National income, however, will still be determined by the sum of all expenditure within the current time period. Therefore the accounting identity (1), when time subscripts are introduced, can be written as

$$Y_t = C_t + I_t \tag{4}$$

From (3) and (4) we can derive a difference equation that explains how Y_t depends on Y_{t-1}. Substituting (3) into (4) we get

$$Y_t = (a + bY_{t-1}) + I_t$$
$$Y_t = bY_{t-1} + a + I_t \qquad (5)$$

This difference equation (5) can be solved using the method explained in Section 13.3 above. However, let us first illustrate how this lagged effect works using a numerical example.

Example 13.7

In a basic Keynesian macroeconomic model it is assumed that initially

$$Y_t = C_t + I_t$$

where $I_t = 134$ is exogenously determined, and

$$C_t = 40 + 0.6Y_{t-1}$$

The level of investment I_t then falls to 110 and remains at this level each time period. Trace out the pattern of adjustment to the new equilibrium value of Y, assuming that the model was initially in equilibrium.

Solution

Although this pattern of adjustment can best be viewed using a spreadsheet, let us first work out the first few steps of the process manually and relate them to the familiar 45°-line income-expenditure graph (illustrated in Figure 13.3) often used to show how Y is determined in introductory economics texts.

If the system is initially in equilibrium then income in one time period is equal to expenditure in the previous time period, and income is the same each time period. Thus

$$Y_t = Y_{t-1} = Y^*$$

where Y^* is the equilibrium level of Y. Therefore, when the original value of I_t of 134 is inserted into the accounting identity the model becomes

$$Y^* = C_t + 134 \qquad (1)$$
$$C_t = 40 + 0.6Y^* \qquad (2)$$

By substitution of (2) into (1)

$$Y^* = (40 + 0.6Y^*) + 134$$
$$Y^*(1 - 0.6) = 40 + 134$$
$$0.4Y^* = 174$$
$$Y^* = 435$$

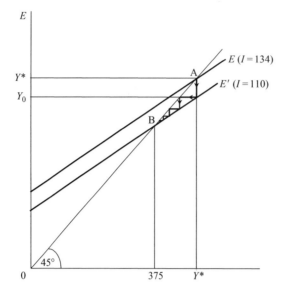

Figure 13.3

This is the initial equilibrium value of Y before the change in I.

   Assume time period 0 is the one in which the drop in I to 110 occurs. Consumption in time period 0 will be based on income earned the previous time period, i.e. when Y was still at the old equilibrium level of 435. Thus

$$C_0 = 40 + 0.6(435) = 40 + 261 = 301$$

Therefore

$$Y_0 = C_0 + I_0 = 301 + 110 = 411$$

   In the next time period, the lagged consumption function means that C_1 will be based on Y_0. Thus

$$Y_1 = C_1 + 110$$
$$= (40 + 0.6Y_0) + 110$$
$$= 40 + 0.6(411) + 110$$
$$= 40 + 246.6 + 110 = 396.6$$

The value of Y for other time periods can be calculated in a similar fashion:

$$Y_2 = C_2 + I_2$$
$$= (40 + 0.6Y_1) + 110$$
$$= 40 + 0.6(396.6) + 110$$
$$= 387.96$$

$$Y_3 = C_3 + I_3$$
$$= (40 + 0.6Y_2) + 110$$
$$= 40 + 0.6(387.96) + 110$$
$$= 382.776$$

and so on. It can be seen that in each time period Y decreases by smaller and smaller amounts as it readjusts towards the new equilibrium value. This new equilibrium value can easily be calculated using the same method as that used above to work out the initial equilibrium.

When $I = 110$ and $Y_t = Y_{t-1} = Y^*$ then the model becomes

$$Y^* = C_t + I = C_t + 110$$
$$C_t = 40 + 0.6Y^*$$

By substitution

$$Y^* = (40 + 0.6Y) + 110$$
$$(1 - 0.6)Y^* = 150$$
$$Y^* = \frac{150}{0.4} = 375$$

This path of adjustment is illustrated in Figure 13.3 by the zigzag line with arrows which joins the old equilibrium at A with the new equilibrium at B. (Note that this diagram is not to scale and just shows the direction and relative magnitude of the steps in the adjustment process.)

Unlike the cobweb model described earlier, the adjustment in this Keynesian model is always in the same direction, instead of alternating on either side of the final equilibrium. Successive values of Y just approach the equilibrium by smaller and smaller increments because the ratio in the complementary function to the difference equation (explained below) is not negative as it was in the cobweb model. If the initial equilibrium had been below the new equilibrium then, of course, Y would have approached its new equilibrium from below instead of from above.

Further steps in the adjustment of Y in this model are shown in the Excel spreadsheet in Table 13.4, which is constructed as explained in Table 13.5. This clearly shows Y closing in on its new equilibrium as time increases.

Difference equation solution

Let us now return to the problem of how to solve the difference equation

$$Y_t = bY_{t-1} + a + I_t \tag{1}$$

The general solution can then be applied to numerical problems, such as Example 13.7 above. By 'solving' this difference equation we mean putting it in the format

$$Y_t = f(t)$$

so that the value of Y_t can be determined for any given value of t.

Table 13.4

	A	B	C	D	E
1	Ex.	LAGGED	KEYNESIAN	MODEL	
2	13.7		where	Yt = Ct + It	
3				Ct = a + bYt-1	
4	Parameters				
5	a =	40		Old I value =	134
6	b =	0.6		New I value =	110
7				Old Equil Y =	435
8	Time			New Equil Y =	375
9	t	C	Y		
10	0	301.00	411.00		
11	1	286.60	396.60		
12	2	277.96	387.96		
13	3	272.78	382.78		
14	4	269.67	379.67		
15	5	267.80	377.80		
16	6	266.68	376.68		
17	7	266.01	376.01		
18	8	265.60	375.60		
19	9	265.36	375.36		
20	10	265.22	375.22		
21	11	265.13	375.13		
22	12	265.08	375.08		
23	13	265.05	375.05		
24	14	265.03	375.03		
25	15	265.02	375.02		
26	16	265.01	375.01		
27	17	265.01	375.01		
28	18	265.00	375.00		

The basic method is the same as that explained earlier, i.e. the solution is split into two components: the equilibrium or particular solution and the complementary function. We first need to find the particular solution, which will be the new equilibrium value of Y^*. When this is equilibrium achieved

$$Y_t = Y_{t-1} = Y^*$$

In equilibrium, the single lag Keynesian model

$$C_t = a + bY_{t-1} \tag{2}$$

and

$$Y_t = C_t + I_t \tag{3}$$

can therefore be written as

$$C_t = a + bY^*$$
$$Y^* = C_t + I_t$$

Table 13.5

CELL	Enter	Explanation
As in Table 13.4	*Enter all labels and column headings*	
B5	40	These are given parameter values for
B6	0.6	consumption function in this example.
E5	134	Original given investment level.
E6	110	New investment level.
D6	160	This is initial "shock" quantity in time period 0.
A10 to A28	*Enter numbers from 0 to 18*	These are the time periods used.
E7	=(B5+E5)/(1-B6)	Calculates initial equilibrium value of *Y* using formula $Y = (a + I)/(1 - b)$.
E8	=(B5+E6)/(1-B6)	Same formula calculates new equilibrium value of *Y*, using new value of *I* in cell E6.
B10	=B5+B6*E7	Calculates consumption in time period 0 using formula $C = a + bY_{t-1}$ where Y_{t-1} is the old equilibrium value in cell E7.
C10	=B10+E$6	Calculates *Y* in time period 0 as sum of current consumption value in cell B10 and new investment value. Note the $ on cell E6 to anchor when copied.
B11	=B$5+B$6*C10	Calculates consumption in time period 1 based on Y_0 value in cell C10. Note the $ on cells B5 and B6 to anchor when copied.
B12 to B28	Copy formula from B11 down column.	Calculates consumption in each time period.
C11 to C28	Copy formula from C10 down column.	Calculates national income Y_t in each time period.

By substitution

$$Y^* = a + bY^* + I_t$$

$$(1 - b)Y^* = a + I_t$$

$$Y^* = \frac{a + I_t}{1 - b} \qquad (4)$$

If the given values of a, b and I_t are put into (4) then the equilibrium value of Y is determined. This is the first part of the difference equation solution.

Returning to the difference equation (1) which we are trying to solve

$$Y_t = bY_{t-1} + a + I_t \qquad (1)$$

If the two constant terms a and I_t are removed then this becomes

$$Y_t = bY_{t-1} \qquad (5)$$

To find the complementary function we use the standard method and assume that this solution is in the format

$$Y_t = Ak^t \qquad (6)$$

where A and k are unknown parameters. This means that

$$Y_{t-1} = Ak^{t-1} \tag{7}$$

Substituting (6) and (7) into (5) gives

$$Ak^t = bAk^{t-1}$$
$$k = b$$

Thus the complementary function is

$$Y_t = Ab^t \tag{8}$$

The general solution to the difference equation is the sum of the particular solution (4) and the complementary function (8). Hence

$$Y_t = \frac{a + I_t}{1 - b} + Ab^t \tag{9}$$

If t is increased, then the value of b^t in the general solution (9) will diminish as long as $|b| < 1$. This condition will be met since b is the marginal propensity to consume which has been estimated to lie between 0 and 1 in empirical studies. Therefore Y_t will always head towards its new equilibrium value.

The value of the constant A can be determined if an initial value Y_0 is known. Substituting into (9), this gives

$$Y_0 = \frac{a + I_t}{1 - b} + Ab^0$$

Remembering that $b^0 = 1$, this means that

$$A = Y_0 - \frac{a + I_t}{1 - b} \tag{10}$$

Thus A is the value of the difference between the initial level of income Y_0, immediately after the shock, and its final equilibrium value Y^*.

Putting this result into (9) above, the general solution to our difference equation becomes

$$Y_t = \frac{a + I_t}{1 - b} + \left(Y_0 - \frac{a + I_t}{1 - b} \right) b^t \tag{11}$$

This may seem to be a rather cumbersome formula but it is straightforward to use. If you remember that

$$\frac{a + I_t}{1 - b} = Y^*$$

is the equilibrium value of Y_t and rewrite (11) as

$$Y_t = Y^* + (Y_0 - Y^*)b^t \tag{12}$$

you will find it easier to work with.

We can now check that this solution to the lagged Keynesian model difference equation works with the numerical Example 13.7 considered above. This model assumed

$$Y_t = C_t + I_t$$

where I_t was initially 134 and

$$C_t = 40 + 0.6Y_{t-1}$$

which corresponded to an initial equilibrium of Y_t of 435.

When I_t was exogenously decreased to 110, the adjustment path towards the new equilibrium value of Y of 375 was worked out by an iterative method. Now let us see what values our difference equation will give.

We have to be careful in determining the initial value Y_0, immediately *after* the increase in investment has taken place. This depends on I_0, which will be the *new* level of investment of 110, and C_0. The level of consumption in period 0 depends on the previously existing equilibrium level of Y_t which was 435 in time period 'minus one'. Therefore

$$C_0 = a + bY_{t-1} = 40 + 0.6(435) = 301$$

$$Y_0 = C_0 + I_0 = 301 + 110 = 411 \tag{13}$$

This is the same initial value Y_0 as that calculated in Example 13.7. The new equilibrium value of income is

$$Y^* = \frac{a + I_t}{1 - b} = \frac{40 + 110}{1 - 0.6} = \frac{150}{0.4} = 375 \tag{14}$$

Substituting (13) and (14) into the formula for the general solution to the difference equation derived above

$$Y_t = Y^* + (Y_0 - Y^*)b^t \tag{15}$$

the general solution for this numerical example becomes

$$Y_t = 375 + (411 - 375)0.6^t$$
$$= 375 + 36(0.6)^t$$

The first few values of Y are thus

$$Y_1 = 375 + 36(0.6) = 375 + 21.6 = 396.6$$

$$Y_2 = 375 + 36(0.6)^2 = 375 + 12.96 = 387.96$$

$$Y_3 = 375 + 36(0.6)^3 = 375 + 7.776 = 382.776$$

These are exactly the same as the answers computed by the iterative method in Example 13.7 and also the same as those produced by the spreadsheet in Table 13.4, which is what one would expect.

This difference equation solution can now be used to calculate Y_t in any given time period. For example, in time period 9 it will be

$$Y_9 = 375 + 36(0.6)^9 = 375 + 0.3628 = 375.3628$$

As t increases in value, eventually the value of $(0.6)^t$ becomes so small as to make the second term negligible. In the above example we can say that for all intents and purposes Y_t has effectively reached its equilibrium value of 375 by the ninth time period, although theoretically Y_t would never actually reach 375 if infinitesimally small increments were allowed.

By now, many of you may be thinking that this difference equation method of computing the different values of Y in the adjustment process in a Keynesian macroeconomic model is extremely long-winded and it would be much quicker to compute the values by the iterative method, particularly if a spreadsheet can be used.

In many cases you may be right. However, you must remember that this chapter is only intended to give you an insight into the methods that can be used to trace out the time path of adjustment in dynamic economic models. The mathematical methods of solution explained here can be adapted to tackle more complex problems that cannot be illustrated on a spreadsheet. Also, economists need to set up mathematical formulations for functional relationships in order to estimate the parameters of these functions. Those of you who study econometrics after the first year of your course will discover that the algebraic solutions to difference equations can help in the setting up of models for testing certain dynamic economic relationships.

Now that the general solution to the lagged Keynesian macroeconomic model has been derived, it can be applied to other numerical examples and may even allow you to compute answers more quickly than by switching on your computer and setting up a spreadsheet.

Example 13.8

There is initially an equilibrium in the basic Keynesian model

$$Y_t = C_t + I_t$$
$$C_t = 650 + 0.5Y_{t-1}$$

with I_t remaining at 300. Then I_t suddenly increases to 420 and remains there. What will be the actual level of Y six time periods after this change?

Solution

The initial equilibrium in period 'minus 1' before the change is

$$Y^*_{-1} = \frac{a + I_t}{1 - b} = \frac{650 + 300}{1 - 0.5} = \frac{950}{0.5} = 1,900$$

Therefore the value of C in time period 0 when the increase in I takes place is

$$C_0 = 650 + 0.5(1,900) = 650 + 950 = 1,600$$

and so the value of Y_t immediately after this shock is

$$Y_0 = C_0 + I_0 = 1,600 + 420 = 2,020$$

The new equilibrium level of Y is

$$Y^* = \frac{a + I_t}{1 - b} = \frac{650 + 420}{1 - 0.5} = \frac{1,070}{0.5} = 2,140$$

Substituting these values into the general solution for the lagged Keynesian macroeconomic model difference equation, we get the general solution for this example, which is

$$Y_t = Y^* + (Y_0 - Y^*)b^t$$
$$= 2,140 + (2,020 - 2,140)0.5^t$$
$$= 2,140 - 120(0.5)^t$$

Therefore, six time periods after the increase in investment

$$Y_6 = 2,140 - 120(0.5)^6$$
$$= 2,140 - 1.875 = 2,138.125$$

Example 13.9

How many time periods will it take Y_t to reach 2,130 in the preceding example?

Solution

We know that $Y_t = 2,130$ and we wish to find t. Thus substituting this value and the initial value Y_0 and the new equilibrium Y^* calculated in Example 13.8, into the Keynesian model general solution formula

$$Y_t = Y^* + (Y_0 - Y^*)b^t$$

we get

$$2,130 = 2,140 + (2,020 - 2,140)0.5^t$$
$$-10 = -120(0.5)^t$$
$$0.08333 = (0.5)^t$$

To get t, put this into the log form, which gives

$$\log 0.08333 = t \log 0.5$$
$$\frac{\log 0.083333}{\log 0.5} = t$$
$$3.585 = t$$

Therefore Y will have exceeded 2,130 by the end of the fourth time period.

Only the most basic lagged Keynesian model has been considered so far in this section. Other possible formulations have been suggested for the ways in which past income levels can determine current expenditure. For example

$$C_t = a + bY_{t-2}$$

or

$$C_t = a + b_1 Y_{t-1} + b_2 Y_{t-2}$$

The latter example is known as a 'distributed lag' model. The solutions of these more complex models require more advanced mathematical methods than are explained in this basic mathematics text. You should, however, be able to adapt the spreadsheet set up in Table 13.4 to trace out the adjustment path of income in a distributed lag model with given parameters.

Example 13.10

Use a spreadsheet to estimate Y_t for the twelve time periods after I_t is increased to 140, assuming that Y_t is determined by the distributed lag Keynesian model

$$Y_t = C_t + I_t$$
$$C_t = 320 + 0.5 Y_{t-1} + 0.3 Y_{t-2}$$

and that the system had previously been in equilibrium with I at 90.

Solution

This is a spreadsheet exercise that you can do yourself by making the necessary adjustments to the formulae that were used to set up the spreadsheet in Table 13.4 when tackling Example 13.7. Be careful in setting up the initial values, however, as C will depend on the old equilibrium level of Y up to period 1. Some of the initial values are calculated manually below for you to check against.

The initial equilibrium level Y^* would have satisfied the equations

$$Y^* = C_t + 90 \tag{1}$$
$$C_t = 320 + 0.5 Y^* + 0.3 Y^* = 320 + 0.8 Y^*$$

By substitution into (1)

$$Y^* = 320 + 0.8 Y^* + 90$$
$$0.2 Y^* = 410$$
$$Y^* = 2,050 = Y_{t-1} = Y_{t-2}$$

Thus

$$C_0 = 320 + 0.5(2,050) + 0.3(2,050) = 1,960$$
$$Y_0 = C_0 + I_0 = 1,960 + 140 = 2,100$$
$$C_1 = 320 + 0.5(2,100) + 0.3(2,050) = 1,985$$
$$Y_1 = C_1 + I_1 = 1,985 + 140 = 2,125 \text{ etc.}$$

Your spreadsheet should show Y_t converging on 2,300.

Test Yourself, Exercise 13.3

1. A Keynesian macroeconomic model with a single-time-period lag on the consumption function, as described below, is initially in equilibrium with the level of I_t given at 500.

    $$Y_t = C_t + I_t$$
    $$C_t = 750 + 0.5Y_{t-1}$$

    I_t is then increased to 650. Use difference equation analysis to find the value of Y_t in the fourth time period after this disturbance to the system. Will it then be within 1% of its new equilibrium level?

2. There is initially an equilibrium in the macroeconomic model

    $$Y_t = C_t + I_t$$
    $$C_t = 2,500 + 0.9\, Y_{t-1}$$

    with the level of I_t set at 1,100. Investment is then increased to 1,500 where it remains for future time periods. Calculate what the level of Y_t will be in the fortieth time period after this investment increase.

3. In a basic Keynesian model with a government sector

    $$Y_t = C_t + I_t + G_t$$

    where $I_t = 269$, $G_t = 310$ (exogenously determined Government expenditure), and

    $$C_t = 80 + 0.8Y_{t-1}^D$$

    where Y_t^D is disposable (after-tax) income. Assume that all income is taxed at a rate of 25%. Government expenditure is then increased to 450 and kept at this level. What tax revenue can the government expect to raise five time periods after this initial rise in expenditure?

4. Use a spreadsheet to trace out the pattern of adjustment of Y_t towards its new equilibrium value in the model

    $$Y_t = C_t + I_t$$
    $$C_t = 310 + 0.7Y_{t-1}$$

    if I_t is exogenously increased from 240 to 350 and then kept at this new level. Assume the system was initially in equilibrium. What is the value of C in the fourteenth time period after this increase in I_t?

13.5 Duopoly price adjustment

An oligopoly is a market with a small number of sellers. It is difficult for economists to predict price and output in oligopoly because firms' reactions to their rivals' actions can vary depending on the strategy they adopt. Firms may naively assume that rivals will not react to whatever pricing policy they themselves operate, they may try to outguess their rivals, or they may collude. You will learn more about these different models in your economics course. Here we will just examine how price may adjust over time in one of the simpler 'naive' models applied to a duopoly, which is an oligopoly with only two sellers.

Two models, the Cournot model and the Bertrand model, assume that firms do not think ahead. The Cournot model assumes that firms think that their rivals will not change their output in response to their own output decisions and the Bertrand model assumes that firms think that rivals will not change their price.

Without going into the details of the model, the predictions of the Bertrand model can be summarized in terms of the 'reactions functions' shown in Figure 13.4 for two duopolists X and Y. These show the price that will maximize one firm's profits given the value of the other firm's price read off the other axis. For example, X's reaction function R^X slopes up from right to left. If Y's price is higher, then X can get away with a higher price, but if Y lowers its price then X also has to reduce its price otherwise it will lose sales.

This model assumes that both firms have identical cost structures and so the reaction functions are symmetrical, intersecting where they both cross the 45°-line representing equal prices. The prediction is that prices will eventually settle at levels P_*^X and P_*^Y, which are equal. The path of adjustment from an initial price P_0^X is shown in Figure 13.4. In time period 1, Y reacts to P_0^X by setting price P_1^Y; then X sets price P_2^X in time period 2, and so on until P_*^X and P_*^Y are reached. Note that we are assuming that the firms take it in turns to adjust price and so each firm only sets a new price every other time period. More recent applications of this basic model based on Game Theory assume that forms go straight to the intersection point, which is called the 'Nash equilibrium'.

Let us now use our knowledge of difference equations to derive a function that will tell us what the price of one of the firms will be in any given time period with the aid of a numerical example.

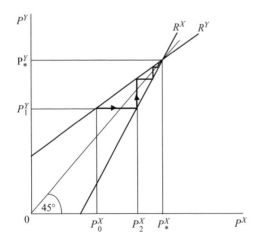

Figure 13.4

Example 13.11

Two duopolists, firms X and Y, have the reaction functions

$$P_t^X = 45 + 0.8P_{t-1}^Y \tag{1}$$

$$P_t^Y = 45 + 0.8P_{t-1}^X \tag{2}$$

If the assumptions of the Bertrand model hold, derive a difference equation for P_t^X and calculate what P_t^X will be in time period 10 if firm X starts off in time period 0 by setting a price of 300.

Solution

Substituting Y's reaction function (2) into X's reaction function (1), we get

$$P_t^X = 45 + 0.8(45 + 0.8P_{t-2}^X)$$

$$P_t^X = 45 + 36 + 0.64P_{t-2}^X$$

$$P_t^X = 81 + 0.64P_{t-2}^X \tag{3}$$

Note that the value P_{t-2}^X appears because we have substituted in the reaction function of Y for period $t-1$ to correspond to the value of P_{t-1}^Y in X's reaction function, i.e. X is reacting to what Y did in the previous time period which, in turn, depends on what X did in the period before that one and so we have substituted in the equation

$$P_{t-1}^Y = 45 + 0.8P_{t-2}^X$$

The particular solution to the difference equation (3) will be where price no longer changes and

$$P_{t-2}^X = P_t^X = P_*^X$$

which is the equilibrium value of P_t^X. Substituting P_*^X into (3)

$$P_*^X = 81 + 0.64P_*^X$$

$$0.36P_*^X = 81$$

$$P_*^X = 225$$

To find the complementary function we use the standard method of ignoring the constant term and assuming that

$$P_t^X = Ak^t$$

where A and k are constant parameters. Substituting into the difference equation

$$P_t^X = 81 + 0.64P_{t-2}^* \tag{4}$$

where the constant 81 is ignored, gives

$$Ak^t = 0.64Ak^{t-2}$$

Cancelling the common term Ak^{t-2}, we get

$$k^2 = 0.64$$
$$k = 0.8$$

The complementary function is therefore

$$P_t^X = A(0.8)^t$$

Adding the complementary function and particular solution, the general solution to difference equation (1) becomes

$$P_t^X = A(0.8)^t + 225$$

To find the value of A we need to know the specific value of P_t^X at some point in time. The question specifies that initially (i.e. in time period 0) P_t^X is 300 and so

$$P_0^X = 300 = A(0.8)^0 + 225$$
$$300 = A + 225$$
$$A = 75$$

Thus, as in the previous difference equation applications, A is the difference between the final equilibrium and the initial value of the variable in question. The general solution to the difference equation (1) is therefore

$$P_t^X = 75(0.8)^t + 225$$

This can be used to calculate the value of P_t^X every *alternate* time period. (Remember that in the intervening time periods X keeps price constant while Y adjusts price.) The price adjustment path will therefore be

$$P_0^X = 75(0.8)^0 + 225 = 75 + 225 = 300$$
$$P_2^X = 75(0.8)^2 + 225 = 48 + 225 = 273$$
$$P_4^X = 75(0.8)^4 + 225 = 30.72 + 225 = 255.72 \text{ etc.}$$

The question asks what price will be in time period 10 and so this can be calculated as

$$P_{10}^X = 75(0.8)^{10} + 225 = 8.05 + 225 = 233.05$$

Example 13.12

Two firms X and Y in an oligopolistic market take a shortsighted view of their situation and set price on the basis of their rivals' price in the previous time period according to the reaction functions

$$P_t^X = 300 + 0.75 P_{t-1}^Y$$

$$P_t^Y = 300 + 0.75 P_{t-1}^X$$

Assume that each adjusts its price every other time period. The market is initially in equilibrium with $P_t^X = P_t^Y = 1{,}200$. Firm X then decides to try to improve its profits by raising price to 1,650. Taking into account the reactions to rivals' price changes described in the above functions, calculate what X's price will be in the eighth time period after its breakaway price rise.

Solution

$$
\begin{aligned}
P_t^X &= 300 + 0.75 P_{t-1}^Y \\
&= 300 + 0.75(300 + 0.75 P_{t-2}^X) \\
&= 300 + 225 + 0.5625 P_{t-2}^X \\
&= 525 + 0.5625 P_{t-2}^X \quad\quad\quad\quad\quad\quad\quad\quad\quad\quad (1)
\end{aligned}
$$

This difference equation is in the same format as that found in Example 13.11 above. Its solution will therefore also be in the same format, i.e.

$$P_t^X = Ak^t + P_*^X$$

The equilibrium value P_*^X is given in the question as 1,200. This can easily be checked in our difference equation (1) because in equilibrium

$$P_t^X = P_{t-2}^X = P_*^X$$

and so

$$P_*^X = 525 + 0.5625 P_*^X$$

$$0.4375 P_*^X = 525$$

$$P_*^X = 1{,}200$$

To find the complementary function, let $P_t^X = Ak^t$ and substitute into the difference equation (1) after dropping the constant 525. Thus

$$Ak^t = 0.5625 Ak^{t-2}$$

Cancelling Ak^{t-2} this gives

$$k^2 = 0.5625$$

$$k = 0.75$$

Adding the complementary function and particular solution, the general solution to the difference equation becomes

$$P_t^X = A(0.75)^t + 1,200$$

The value of A can be found when the initial $P_0^X = 1,650$ is substituted. Thus

$$P_0^X = 1,650 = A(0.75)^0 + 1,200$$
$$1,650 = A + 1,200$$
$$450 = A$$

The definite solution to this difference equation is therefore

$$P_t^X = 450(0.75)^t + 1,200$$

and so in the eighth period after the initial price rise

$$P_8^X = 450(0.75)^8 + 1,200$$
$$= 45.05 + 1,200 = 1,245.05$$

Test Yourself, Exercise 13.4

1. Two duopolists X and Y react to each others' prices according to the functions

   $$P_t^X = 240 + 0.9P_{t-1}^Y$$
   $$P_t^Y = 240 + 0.9P_{t-1}^X$$

   If firm X sets an initial price of 2,900, what will its price be twenty time periods later? Assume that each firm adjusts price every alternate time period.

2. In an oligopolistic market, the two firms X and Y have the following price reaction functions:

   $$P_t^X = 800 + 0.6P_{t-1}^Y$$
   $$P_t^Y = 800 + 0.6P_{t-1}^X$$

   The usual assumptions of the Bertrand model apply and price is initially in equilibrium at a level of 2,000 for both firms. Firm X then decides to cut price to 1,500 to try to steal Y's market share. We know from the analysis of this model that X's price reduction will be short-lived and price will creep back towards its equilibrium level, but how short-lived? Calculate whether or not P^X will be back within 1% of its equilibrium value within six time periods. (Be careful how you calculate the value of A in the difference equation as this time the initial value is below the equilibrium value of P^X.)

3. In a duopoly where the assumptions of the Bertrand model hold, the two firms' reaction functions are

$$P_t^X = 95.54 + 0.83 P_{t-1}^Y$$
$$P_t^Y = 95.54 + 0.83 P_{t-1}^X$$

If firm X unexpectedly changes price to 499, derive the solution to the difference equation that determines P_t^X and use it to predict P_t^X in the twelfth time period after the initial change.

14 Exponential functions, continuous growth and differential equations

Learning objectives

After completing this chapter students should be able to:

- Use the exponential function and natural logarithms to derive the final sum, initial sum and growth rate when continuous growth takes place.
- Compare and contrast continuous and discrete growth rates.
- Set up and solve linear first-order differential equations.
- Use differential equation solutions to predict values in basic market and macro-economic models.
- Comment on the stability of economic models where growth is continuous.

14.1 Continuous growth and the exponential function

In Chapter 7, growth was treated as a process taking place at discrete time intervals. In this chapter we shall analyse growth as a continuous process, but it is first necessary to understand the concepts of exponential functions and natural logarithms. The term 'exponential function' is usually used to describe the specific natural exponential function explained below. However, it can also be used to describe any function in the format

$$y = A^x \quad \text{where } A \text{ is a constant and } A > 1$$

This is known as an exponential function to base A. When x increases in value this function obviously increases in value very rapidly if A is a number substantially greater than 1. On the other hand, the value of A^x approaches zero if x takes on larger and larger negative values. For all values of A it can be deduced from the general rules for exponents (explained in Chapter 2) that $A^0 = 1$ and $A^1 = A$.

Example 14.1

Find the values of $y = A^x$ when A is 2 and x takes the following values:
(a) 0.5, (b) 1, (c) 3, (d) 10, (e) 0 , (f) −0.5, (g) −1, and (h) −3

Solution

(a) $A^{0.5} = 1.41$ (b) $A^1 = 2$ (c) $A^3 = 8$

(d) $A^{10} = 1024$ (e) $A^0 = 1$ (f) $A^{-0.5} = 0.71$

(g) $A^{-1} = 0.50$ (h) $A^{-3} = 0.13$

The natural exponential function

In mathematics there is a special number which when used as a base for an exponential function yields several useful results. This number is

2.7182818   (to 7 dp)

and is usually represented by the letter 'e'. You should be able to get this number on your calculator by entering 1 and then using the $[e^x]$ function key.

To find e^x for any value of x on a calculator the usual procedure is to enter the number (x) and then press the $[e^x]$ function key. To check that you can do this, try using your calculator to obtain the following exponential values:

$$e^{0.5} = 1.6487213$$

$$e^4 = 54.59815$$

$$e^{-2.624} = 0.0725122$$

If you do not get these values, ask your tutor for assistance. If your calculator does not have an $[e^x]$ function key then it is probably worth buying a new calculator, or you can use the EXP function in Excel. There also exist tables of exponential values which were used by students before calculators with exponential function keys became available.

In economics, exponential functions to the base e are particularly useful for analysing growth rates. This number, e, is also used as a base for natural logarithms, explained later in Section 14.4. Although it has already been pointed out that, strictly speaking, the specific function $y = e^x$ should be known as the 'natural exponential function', from now on we shall adopt the usual convention and refer to it simply as the 'exponential function'.

To understand how this rather awkward value for e is derived, we return to the method used for calculating the value of an investment developed in Chapter 7. You will recall that the final value (F) of an initial investment (A) deposited for t discrete time periods at an interest rate of i can be calculated from the formula

$$F = A(1 + i)^t$$

If the interest rate is 100% then $i = 1$ and the formula becomes

$$F = A(1 + 1)^t = A(2)^t$$

Assume the initial sum invested $A = 1$. If interest is paid at the end of each year, then after 1 year the final sum will be

$$F_1 = (1 + 1)^1 = 2$$

In Chapter 7 it was also explained how interest paid monthly at the annual rate divided by 12 will give a larger final return than this nominal annual rate because the interest credited each month will be reinvested. When the nominal annual rate of interest is 100% ($i = 1$) and the initial sum invested is assumed to be 1, the final sum after 12 months invested at a monthly interest rate of $\frac{1}{12}(100\%)$ will be

$$F_{12} = \left(1 + \frac{1}{12}\right)^{12} = 2.6130353$$

If interest was to be credited daily at the rate of $\frac{1}{365}(100\%)$ then the final sum would be

$$F_{365} = \left(1 + \frac{1}{365}\right)^{365} = 2.7145677$$

If interest was to be credited by the hour at a rate of $\frac{1}{8760}(100\%)$ (given that there are 8,760 hours in a 365-day year) then the final sum would be

$$F_{8,760} = \left(1 + \frac{1}{8,760}\right)^{8,760} = 2.7181267$$

From the above calculations we can see that the more frequently that interest is credited the closer the value of the final sum accumulated gets to 2.7182818, the value of e. When interest at a nominal annual rate of 100% is credited at infinitesimally small time intervals then growth is continuous and e is equal to the final sum credited. Thus

$$e = \left(1 + \frac{1}{n}\right)^{n} = 2.7182818 \quad \text{where } n \to \infty$$

This result means that a sum A invested for one year at a nominal annual interest rate of 100% credited continuously will accumulate to the final sum of

$$F = eA = 2.7182818A$$

This translates into the annual equivalent rate of

$$\text{AER} = 2.7182818 - 1 = 1.7182818 = 171.83\% \text{ (to 2 dp)}$$

Although bank interest may not actually be paid instantaneously so that a sum invested grows continually every second, the crediting of interest on a daily basis, which is quite common, gives an equivalent annual rate that is practically the same as the continuous rate. (One has to go to the 4th decimal place to find a difference between the two.) Continuous growth also occurs in other variables relevant to economics, e.g. population, the amount of natural materials mined. Other variables may continuously decline in value over time, e.g. the stock of a non-renewable natural resource.

14.2   Accumulated final values after continuous growth

To derive a formula that will give the final sum accumulated after a period of continuous growth, we first assume that growth occurs at several discrete time intervals throughout a year. We also assume that A is the initial sum, r is the nominal annual rate of growth, n is the number of times per year that increments are accumulated and y is the final value. Using the final sum formula developed in Chapter 7, this means that after t years of growth the final sum will be

$$y = A \left(1 + \frac{r}{n}\right)^{nt}$$

To reduce this to a simpler formulation, multiply top and bottom of the exponent by r so that

$$y = A \left(1 + \frac{r}{n}\right)^{\left(\frac{n}{r}\right)rt} \tag{1}$$

If we let $m = \frac{n}{r}$ then $\frac{1}{m} = \frac{r}{n}$ and so (1) can be written as

$$y = A \left(1 + \frac{1}{m}\right)^{mrt} = A \left[\left(1 + \frac{1}{m}\right)^{m}\right]^{rt} \tag{2}$$

Growth becomes continuous as the number of times per year that increments in growth are accumulated increases towards infinity. When $n \to \infty$ then $\frac{n}{r} = m \to \infty$.

Therefore, using the result derived in Section 14.1 above,

$$\left(1 + \frac{1}{m}\right)^{m} \to e \quad \text{as } m \to \infty$$

Substituting this result back into (2) above gives

$$y = Ae^{rt}$$

This formula can be used to find the final value of any variable growing continuously at a known annual rate from a given original value.

Example 14.2

Population in a developing country is growing continuously at an annual rate of 3%. If the population is now 4.5 million, what will it be in 15 years' time?

Solution

The final value of the population (in millions) is found by using the formula $y = Ae^{rt}$ and substituting the given numbers: initial value $A = 4.5$; rate of growth $r = 3\% = 0.03$; number of time periods $t = 15$, giving

$$y = 4.5e^{0.03(15)} = 4.5e^{0.45} = 4.5 \times 1.5683122 = 7.0574048 \text{ million}$$

Thus the predicted final population is 7,057,405.

Example 14.3

An economy is forecast to grow continuously at an annual rate of 2.5%. If its GNP is currently €56 billion, what will the forecast for GNP be at the end of the third quarter the year after next?

Solution

In this example: $t = 1.75$ years, $r = 2.5\% = 0.025$, $A = 56$ (€ billion). Therefore, the final value of GNP will be

$$y = Ae^{rt} = 56e^{0.025(1.75)} = 56e^{0.04375} = 58.504384$$

Thus the forecast for GNP is €58,504,384,000.

So far we have only considered positive growth, but the exponential function can also be used to analyse continuous decay if the rate of decline is treated as a negative rate of growth.

Example 14.4

A river flow through a hydroelectric dam is 18 million gallons a day and shrinking continuously at an annual rate of 4%. What will the flow be in 6 years' time?

Solution

The 4% rate of decline becomes the negative growth rate $r = -4\% = -0.04$. We also know the initial values $A = 18$ and $t = 6$. Thus the final value is

$$y = Ae^{rt} = 18e^{-0.04(6)} = 18e^{-0.24} = 14.16$$

Therefore, the river flow will shrink to 14.16 million gallons per day.

Continuous and discrete growth rates compared

In Section 14.1 it was explained how interest at a rate of 100% credited continuously through-out a year gives an annual equivalent rate of $r = e - 1 = 1.7182818 = 171.83\%$, a difference of 71.83%. However, in practice interest is usually credited at much lower annual rates. This means that the difference between the nominal and annual equivalent rates when interest is credited continuously will be much smaller. This is illustrated in Table 14.1 for the case when the nominal annual rate of interest is 6%.

These figures show that the annual equivalent rate when interest is credited continuously is the same as that when interest is credited on a daily basis, if rounded to two decimal places, although there will be a slight difference if this rounding does not take place.

Table 14.1

Interest credited	Frequency rate per annum (n)	Nominal rate $\left(\frac{i}{n}\right)$	Annual equivalent rate $\left(1 + \frac{i}{n}\right)^n - 1$
Annually	1	6%	6%
6 monthly	2	3%	$(1.03)^2 - 1 = 0.0609 = 6.09\%$
3 monthly	4	1.5%	$(1.015)^4 - 1 = 0.06136 = 6.14\%$
Monthly	12	0.5%	$(1.005)^{12} - 1 = 0.06167 = 6.17\%$
Daily	365	0.0164%	$(1.00016)^{365} - 1 = 0.061831 = 6.18\%$
Continuously	$\rightarrow \infty$	$\rightarrow 0$	$e^{0.06} - 1 = 0.0618365 = 6.18\%$

Test Yourself, Exercise 14.1

1. A country's population is currently 32 million and is growing continuously at an annual rate of 3.5%. What will the population be in 20 years' time if this rate of growth persists?
2. A company launched a successful new product last year. The current weekly sales level is 56,000 units. If sales are expected to grow continually at an annual rate of 12.5%, what will be the expected level of sales 36 weeks from now? (Assume that 1 year is exactly 52 weeks.)
3. Current stocks of mineral M are 250 million tonnes. If these stocks are continually being used up at an annual rate of 9%, what amount of M will remain after 30 years?
4. A renewable natural resource R will allow an estimated maximum consumption rate of 200 million units per annum. Current annual usage is 65 million units. If the annual level of usage grows continually at an annual rate of 7.5% will there be sufficient R to satisfy annual demand after (a) 5 years, (b) 10 years, (c) 15 years, (d) 20 years?
5. Stocks of resource R are shrinking continually at an annual rate of 8.5%. How much will remain in 30 years' time if current stocks are 725,000 units?
6. If €25,000 is deposited in an account where interest is credited on a daily basis that can be approximated to the continuous accumulation of interest at a nominal annual rate of 4.5%, what will the final sum be after five years?

14.3   Continuous growth rates and initial amounts

Derivation of continuous rates of growth

The growth rate r can simply be read off from the exponent of a continuous growth function in the format $y = Ae^{rt}$.

To prove that this is the growth rate we can use calculus to derive the rate of change of this exponential growth function.

If variable y changes over time according to the function $y = Ae^{rt}$ then rate of change of y with respect to t will be the derivative dy/dt. However, it is not a straightforward exercise to differentiate this function. For the time being let us accept the result (explained below in

Section 14.4) that

$$\text{if } y = e^t \quad \text{then } \frac{dy}{dt} = e^t$$

i.e. the derivative of an exponential function is the function itself.

Thus, using the chain rule,

$$\text{when } y = Ae^{rt} \quad \text{then } \frac{dy}{dt} = rAe^{rt}$$

This derivative approximates to the *absolute* amount by which y increases when there is a one unit increment in time t, but when analysing growth rates we are usually interested in the *proportional* increase in y with respect to its original value. The *rate* of growth is therefore

$$\frac{\frac{dy}{dt}}{y} = \frac{rAe^{rt}}{Ae^{rt}} = r$$

Even though r is the instantaneous rate of growth at any given moment in time, it must be expressed with reference to a time interval, usually a year in economic applications, e.g. 4.5% per annum. It is rather like saying that the slope of a curve is, say, 1.78 at point X. A slope of 1.78 means that height increases by 1.78 units for every 1 unit increase along the horizontal axis, but at a single point on a curve there is no actual movement along either axis.

Example 14.5

Owing to continuous improvements in technology and efficiency in production, an empirical study found a factory's output of product Q at any moment in time to be determined by the function

$$Q = 40e^{0.03t}$$

where t is the number of years from the base year in the empirical study and Q is the output per year in tonnes. What is the annual growth rate of production?

Solution

When the accumulated amount from continuous growth is expressed by a function in the format $y = Ae^{rt}$ then the growth rate r can simply be read off from the function. Thus when

$$Q = 40e^{0.03t}$$

the rate of growth is

$$r = 0.03 = 3\%$$

Initial amounts

What if you wished to find the initial amount A that would grow to a given final sum y after t time periods at continuous growth rate r? Given the continuous growth final sum formula

$$y = Ae^{rt}$$

then, by dividing both sides by e^{rt}, we can derive the **initial sum formula**

$$A = ye^{-rt}$$

Example 14.6

A parent wants to ensure that their young child will have a fund of £35,000 to finance his/her study at university, which is expected to commence in 12 years' time. They wish to do this by investing a lump sum now. How much will they need to invest if this investment can be expected to grow continuously at an annual rate of 5%?

Solution

Given values are: final amount $y = 35,000$, continuous growth rate $r = 5\% = 0.05$, and time period $t = 12$. Thus the initial sum, using the formula derived above, will be

$$A = ye^{-rt} = 35,000\,e^{-0.05(12)} = 35,000\,e^{-0.6}$$
$$= 35,000 \times 0.5488116 = £19,208.41$$

Example 14.7

A manager of a wildlife sanctuary wants to ensure that in ten years' time the number of animals of a particular species in the sanctuary will total 900. How many animals will she need to start with now if this particular animal population grows continuously at an annual rate of 8.5%?

Solution

Given the final amount of $y = 900$, continuous growth rate $r = 8.5\% = 0.085$, and time period $t = 10$, then using the initial sum formula

$$A = ye^{-rt} = 900\,e^{-0.085(10)} = 900\,e^{-0.85} = 900 \times 0.4274149 = 384.67$$

Therefore, she will need to start with 385 animals, as you cannot have a fraction of an animal!

Test Yourself, Exercise 14.2

1. A statistician estimates that a country's population N is growing continuously and can be determined by the function

    $$N = 3,620,000e^{0.02t}$$

    where t is the number of years after 2000. What is the population growth rate? Will population reach 10 million by the year 2050?

2. Assuming that oil stocks will continue to be depleted at the same continuous rate (in proportion to the amount remaining), the amount of oil remaining in an oilfield (B), measured in barrels of oil, has been estimated as

    $$B = 2,430,000,000e^{-0.09t}$$

    where t is the number of years after 2000. What proportion of the oil stock is extracted each year? How much oil will remain by 2020?

3. An individual wants to ensure that in 15 years' time, when they plan to retire, they will have a pension fund of £240,000. They wish to achieve this by investing a lump sum now, rather than making regular annual contributions. If their investment is expected to grow continuously at an annual rate of 4.5%, how much will they need to invest now?

4. The owner of an artificial lake, which has been created with the main aim of making a commercial return from recreational fishing, has to decide how many fish to stock the lake with. Allowing for the natural rate of growth of the fish population and the depletion caused by fishing, the number of fish in the lake is expected to shrink continuously by 3.2% a year. How many fish should the owner stock the lake with if they wish to ensure that the fish population will still be 500 in 5 years' time, given that it will not be viable to add more fish after the initial stock is introduced?

14.4   Natural logarithms

In Chapter 2, we saw how logarithms to base 10 were defined and utilized in mathematical problems. You will recall that the logarithm of a number to base X is the power to which X must be raised in order to equal that number. Logarithms to the base 'e' have several useful properties and applications in mathematics. These are known as 'natural logarithms', and the usual notation is 'ln' (as opposed to 'log' for logarithms to base 10).

As with values of the exponential function, natural logarithms can be found on a mathematical calculator. Using the [LN] function key on your calculator, check that you can derive the following values:

$$\ln 1 = 0$$
$$\ln 2.6 = 0.9555114$$
$$\ln 0.45 = -0.7985$$

The rules for using natural logarithms are the same as for logarithms to any other base. For example, to multiply two numbers, their logarithms are added. But how do you then transform the sum of the logarithms back to a number, i.e. what is the 'antilog' of a natural logarithm?

To answer this question, consider the exponential function

$$y = e^x \tag{1}$$

By definition, the natural logarithm of y will be x because that is the power to which e is taken to equal x. Thus we can write

$$\ln y = x \tag{2}$$

If we only know the value of the natural logarithm $\ln y$ and wish to find y then, by substituting (2) into (1), it must be true that

$$y = e^x = e^{\ln y}$$

Therefore y can be found from the natural logarithm $\ln y$ by finding the exponential of $\ln y$. For example, if

$$\ln y = 3.214$$

then

$$y = e^{\ln y} = e^{3.214} = 24.8784$$

We can check that this is correct by finding the natural logarithm of our answer. Thus

$$\ln y = \ln 24.8784 = 3.214$$

Although you would not normally need to actually use natural logarithms for basic numeric problems, the example below illustrates how natural logarithms can be used for multiplication.

Example 14.8

Multiply 5,623.76 by 441.873 using natural logarithms.

Solution

Taking natural logarithms and performing multiplication by adding them:

$$\ln 5,623.760 = 8.6347558+$$
$$\underline{\ln 441.873 = 6.0910225}$$
$$14.725778 \quad \text{(to 6 dp)}$$

To transform this logarithm back to its corresponding number we find

$$e^{14.725778} = 2,484,987.7$$

This answer can be verified by carrying out a straightforward multiplication on your calculator.

Determination of continuous growth rates using natural logarithms

To understand how natural logarithms can help determine rates of continuous growth, consider the following example.

Example 14.9

The consumption of natural mineral resource M has risen from 38 million tonnes (per annum) to 68.4 million tonnes over the last 12 years. If it is assumed that growth in consumption has been continuous, what is the annual rate of growth?

Solution

If growth is continuous then the final consumption level of M will be determined by the exponential function:

$$M = M_0 e^{rt} \qquad (1)$$

This time the known values are: the final value $M = 68.4$, the initial consumption value $M_0 = 38$, and $t = 12$, with the rate of growth r being the unknown value that we are trying to determine.

Substituting these known values into (1) gives

$$68.4 = 38 e^{12r}$$

$$1.8 = e^{12r} \qquad (2)$$

In (2), the power to which e must be raised to equal 1.8 is $12r$. Therefore,

$$\ln 1.8 = 12r$$

$$r = \frac{\ln 1.8}{12} = \frac{0.5877867}{12} = 0.0489822$$

and so consumption has risen at an annual rate of 4.9%.

A general formula for finding a continuous rate of growth when y, A and t are all known can be derived from the final sum formula. Given

$$y = A e^{rt}$$

then

$$\frac{y}{A} = e^{rt}$$

taking natural logs $\ln\left(\frac{y}{A}\right) = rt$

giving the **rate of growth formula**

$$\frac{1}{t} \ln\left(\frac{y}{A}\right) = r$$

Example 14.10

Over the last 15 years a country's population has risen continuously at the same annual growth rate from 8.2 million to 11.9 million. What is this rate of growth?

Solution

Using the formula for finding a continuous growth rate and entering the known values gives the rate of growth as

$$r = \frac{1}{t} \ln \left(\frac{y}{A} \right) = \frac{1}{15} \ln \left(\frac{11.9}{8.2} \right)$$

$$= \frac{1}{15} \ln(1.45122) = \frac{1}{15} (0.3724) = 0.02483 = 2.48\%$$

Natural logarithms can also be used to work out rates of decay, which are negative rates of growth.

Example 14.11

The annual catch of fish from a specific sea area is declining continually at a constant rate. Ten years ago the total catch was 940 tonnes and this year the total catch is 784 tonnes. What is the rate of decline?

Solution

If the decline is continuous then the catch C at any point in time will be determined by the function

$$C = C_0 e^{rt}$$

where C_0 is the catch in the initial time period.
    Substituting the known values into this function gives

$$784 = 940 e^{10r}$$

$$0.8340426 = e^{10r}$$

$$\ln 0.8340426 = 10r$$

$$-0.1814708 = 10r$$

$$-0.0181471 = r$$

Therefore the rate of decline is 1.8%.

Rates of growth and decay can also be determined over time periods of less than a year by employing the same method.

Example 14.12

Consumption of mineral M is known to be increasing continually at a constant rate per annum. The daily rate of consumption was 46.4 tonnes on 1 March and had risen to 47.2 tonnes by 1 June. What is the annual growth rate for consumption of this mineral?

Solution

From 1 March to 1 June is 3 months, or a quarter of a year. Thus using the standard final sum formula for continuous growth

$$47.2 = 46.4e^{r(0.25)}$$

$$1.017241 = e^{0.25r}$$

$$\ln 1.017241 = 0.25r$$

$$0.0170944 = 0.25r$$

$$0.0683778 = r$$

Therefore the annual growth rate is 6.84%.

Comparison of discrete and continuous growth

A direct comparison of the continuous growth rate r and the discrete growth rate i that would accumulate the same final sum F over 1 year for a given initial sum A can be found using natural logarithms, as follows:

Continuous growth final sum $F = Ae^r$

Discrete growth final sum $F = A(1 + i)$

Therefore

$$Ae^r = A(1 + i)$$
$$e^r = (1 + i)$$

Taking logs gives the required function for r in terms of i

$$r = \ln(1 + i) \tag{1}$$

To get i as a function of r, the exponential of each side of (1) is taken, giving

$$e^r = e^{\ln(1+i)}$$
$$e^r = 1 + i$$
$$e^r - 1 = i$$

Example 14.13

(i) Find the continuous growth rate that would correspond over a discrete growth annual rate of:
(a) 0% (b) 10% (c) 50% (d) 100%
(ii) Find the discrete annual growth rates that would correspond to the continuous growth rates (a), (b), (c) and (d) in (i) above.

Give all answers to 2 significant decimal places.

Solution

(i) Using the formula $r = \ln(1 + i)$ the answers are:

(a) $i = 0\% = 0$ $\qquad r = \ln(1 + 0) = \ln 1 = 0\%$

(b) $i = 10\% = 0.1$ $\quad r = \ln(1 + 0.1) = \ln 1.1 = 0.09531 = 9.53\%$

(c) $i = 50\% = 0.5$ $\quad r = \ln(1 + 0.5) = \ln 1.5 = 0.405465 = 40.55\%$

(d) $i = 100\% = 1$ $\quad r = \ln(1 + 1) = \ln 2 = 0.6931472 = 69.31\%$

(ii) Using the formula $i = e^r - 1$ the answers are

(a) $r = 0\% = 0$ $\qquad i = e^0 - 1 = 1 - 1 = 0\%$

(b) $r = 10\% = 0.1$ $\quad i = e^{0.1} - 1 = 1.10517 - 1 = 0.10517 = 10.52\%$

(c) $r = 50\% = 0.5$ $\quad i = e^{0.5} - 1 = 1.64872 - 1 = 0.64872 = 64.87\%$

(d) $r = 100\% = 1$ $\quad i = e^1 - 1 = 2.7182818 - 1 = 1.7182818 = 171.83\%$

Test Yourself, Exercise 14.3

1. In an advanced industrial economy, population is observed to have grown at a steady rate from 50 to 55 million over the last 20 years. What is the annual rate of growth?

2. If the average quantity of petrol used per week by a typical private motorist has increased from 32.1 litres to 48.4 litres over the last 20 years, what has been the average annual growth rate in petrol consumption assuming that this increase in petrol consumption has been continuous? If, over the same time period, petrol consumption for a typical private car has fallen from 8.75 litres per 100 km to 6.56 litres per 100 km, what has been the average annual growth rate in the distance covered each week by a typical motorist?

3. World reserves of mineral M are observed to have declined from 830 million tonnes to 675 million tonnes over the last 25 years. Assuming this decline to have been continuous, calculate the annual rate of decline and then predict what reserves will be left in 10 years' time.

4. An economy's GNP grows from €5,682 million to €5,727 million during the first quarter of a new government's term of office. If this growth rate persisted through

its entire term of office of 4 years, what would GNP be at the time of the next election?

5.   If the number of a protected species of animal in a reserve increased continually from 600 in 1992 to 1,450 in 2002, what was the annual growth rate?

6.   Over a 12-month period what continuous growth rate is equivalent to a discrete growth rate of 6%?

7.   What discrete annual growth rate is equivalent to a continuous growth rate of 6% persisting over 12 months?

8.   What interest rate would you prefer to be used to add interest to your savings: 8% applied on a continuous basis or 9% applied once a year?

14.5   Differentiation of logarithmic functions

We have already used the rule that if $y = e^t$ then $dy/dt = e^t$. This result can be derived if we accept as given the rule for differentiation of the natural logarithm function. This rule says that if

$$f(y) = \ln y$$

then $\dfrac{df}{dy} = \dfrac{1}{y}$

This rule can be proved mathematically but the proof is rather complex. It is not necessary for you to understand it to follow the economic applications in this basic mathematics text and so you are just asked to accept it as given. Note that this rule also implies that

$$\int \frac{1}{x} = \ln x$$

This was the exceptional case in integration not dealt with in Chapter 12.

Returning to the exponential function, we write this with the two sides of the equality side swapped around. Thus

$$e^t = y$$

As the natural logarithm of y is the exponent of e, by definition, then

$$t = \ln y$$

Therefore, using the rule for differentiating natural logarithmic functions stated above

$$\frac{dt}{dy} = \frac{1}{y} \tag{1}$$

The inverse function rule in calculus states that

$$\frac{dy}{dt} = \frac{1}{\left(\dfrac{dt}{dy}\right)} \tag{2}$$

Substituting (1) into (2) gives

$$\frac{dy}{dt} = \frac{1}{\left(\dfrac{1}{y}\right)} = y$$

Thus we have shown that when $y = e^t$ then

$$\frac{dy}{dt} = y$$

which is the result we wished to prove.

14.6 Continuous time and differential equations

We have already seen how continuous growth rates can be determined and how continuous growth affects the final sum accumulated, but to analyse certain economic models where continuous dynamic adjustment occurs we also need to understand what differential equations are and how they can be solved.

Differential equations contain the derivative of an unknown function. For example

$$\frac{dy}{dt} = 6y + 27$$

Solving a differential equation in this format entails finding the function y in terms of t. This will enable us to find the value of y for any given value of t.

There are many forms that differential equations can take, but we will confine the analysis here to the case of linear first-order differential equations. First-order means that only first-order derivatives are included. Thus a first-order differential equation may contain terms in dy/dt but not higher order derivatives such as d^2y/dt^2. Linear means that a differential equation does not contain a product such as $y\,(dy/dt)$. More advanced mathematical economics texts will cover the analysis of higher-order and non-linear differential equations.

As well as containing the first-order derivative, a first-order differential equation will usually also contain the unknown function (y) itself. Thus a first-order differential equation may contain:

- a constant (although this may be zero)
- the unknown function y
- the first-order derivative dy/dt

At first sight you might think integration would be the way to find the unknown function. However, as a differential equation will include terms in y rather than t the solution is not quite so straightforward. For example, if we had started with a basic derivative such as

$$\frac{dy}{dt} = t \qquad\qquad \text{then we could use integration to find}$$

$$y = \int t \cdot dt = 0.5t^2 + C \quad \text{where } C \text{ is an unknown constant.}$$

But if we start with a differential equation such as

$$\frac{\mathrm{d}y}{\mathrm{d}t} = 6y + 27$$

this method cannot be used to find y.

The next two sections explain how to find solutions to linear first-order differential equations. Firstly, the *homogeneous* case is considered, where there is no constant term and the differential equation to be solved takes the format

$$\frac{\mathrm{d}y}{\mathrm{d}t} = by \quad \text{where } b \text{ is a constant parameter}$$

Secondly, the *non-homogeneous* case is considered, where there is a non-zero constant term c and the differential equation to be solved takes the format

$$\frac{\mathrm{d}y}{\mathrm{d}t} = by + c$$

The information in these forms of differential equations corresponds to some not uncommon situations in economics. We may know the rate at which an economic variable is increasing and its value at a specific time but may not know the direct relationship between its value and the time period.

14.7   Solution of homogeneous differential equations

The exponential function can help us to derive the solution to a differential equation. In Section 14.4 we learned that the exponential function has the property that

$$\text{if } y = \mathrm{e}^t \quad \text{then} \quad \frac{\mathrm{d}y}{\mathrm{d}t} = \mathrm{e}^t$$

Thus, using the chain rule for differentiation, for any constant b,

$$\text{if } y = \mathrm{e}^{bt} \quad \text{then} \quad \frac{\mathrm{d}y}{\mathrm{d}t} = b\mathrm{e}^{bt}$$

Therefore, if the differential equation to be solved has no constant term and has the format

$$\frac{\mathrm{d}y}{\mathrm{d}t} = by$$

then a possible solution is

$$y = \mathrm{e}^{bt}$$

because this would give

$$\frac{\mathrm{d}y}{\mathrm{d}t} = b\mathrm{e}^{bt} = by$$

For example, if the differential equation to be solved is

$$\frac{dy}{dt} = 5y$$

then one possible solution is

$$y = e^{5t}$$

as this gives

$$\frac{dy}{dt} = 5e^{5t} = 5y$$

However, there are other possible solutions. For example,

$$\text{if } y = 3e^{5t} \quad \text{then} \quad \frac{dy}{dt} = 5(3e^{5t}) = 5y$$

$$\text{if } y = 7e^{5t} \quad \text{then} \quad \frac{dy}{dt} = 5(7e^{5t}) = 5y$$

In fact, we can multiply the original solution of e^{5t} by any constant parameter and still get the same solution after differentiation.

Therefore, for any differential equation in the format

$$\frac{dy}{dt} = by$$

the **general solution** can be specified as

$$y = Ae^{bt} \quad \text{where } A \text{ is an arbitrary constant.}$$

This must be so since

$$\frac{dy}{dt} = bAe^{bt} = by$$

The actual value of A can be found if the value for y is known for a specific value of t. This will enable us to find the **definite solution**. This is easiest to evaluate when the value of y is known for $t = 0$ as any number taken to the power zero is the number itself.

For example, the general solution to the differential equation

$$\frac{dy}{dt} = 5y$$

will be

$$y_t = Ae^{5t} \tag{1}$$

where y has been given the subscript t to denote the time period that it corresponds to. If it is known that when $t = 0$ then $y_0 = 12$ then by substituting these values into (1) we get

$$y_0 = 12 = Ae^0$$

As we know that $e^0 = 1$ then

$$12 = A$$

Substituting this value into the general solution (1) we get the definite solution

$$y_t = 12e^{5t}$$

This definite solution can now be used to predict y_t for any value of t. For example, when $t = 3$, then

$$y_3 = 12e^{5(3)} = 12e^{15} = 12(3,269,017.4) = 39,228,208$$

Example 14.14

Solve the differential equation $dy/dt = 1.5y$ if the value of y is 34 when $t = 0$ and then use the solution to predict the value of y when $t = 7$.

Solution

Using the method explained above, the general solution to this differential equation will be

$$y_t = Ae^{1.5t}$$

When $t = 0$ then

$$y_0 = 34 = Ae^0$$

Therefore

$$34 = A$$

The definite solution is thus

$$y_t = 34e^{1.5t}$$

When $t = 7$ then using this definite solution we can predict

$$y_7 = 34e^{1.5(7)} = 34e^{10.5} = 34(36,315.5) = 1,234,727$$

Table 14.2

t	$y_t = 8e^{0.2t}$	Change in y per time period $dy/dt = 0.2y = 1.6e^{0.2t}$	$(dy/dt)/y_t = r$
0	8.00	1.6	0.2
1	9.77	1.954	0.2
2	11.93	2.387	0.2
3	14.58	2.915	0.2
4	17.80	3.561	0.2
5	21.75	4.349	0.2

Differential equation solutions and growth rates

You may have noticed that the solutions to these differential equations have the same format as the functions encountered in Section 14.2 which gave final values after continuous growth for a given time period. This is because what we have done this time is to derive the relationship between y and t, starting from the knowledge that $dy/dt = ry$, i.e. that the rate of increase of y (over time) depends on the growth rate r and the specific value of y. This can be a difficult point to grasp, because there are actually two rates involved and it is easy to confuse them.

(i) dy/dt is the rate of increase of y with respect to time t (but over a specified time period it will be a quantity of y rather than a ratio)

(ii) r is the rate of increase of y with respect to its own current value

When y increases in magnitude over time, larger and larger increases in the value of y each time period will be necessary to maintain the same proportional rate of growth r. In other words, the value of dy/dt must get bigger as t increases. Table 14.2 illustrates how this happens for the function $y_t = 8e^{0.2t}$, assuming an initial value of 8. To keep the ratio of the increase in y to its current value constant at the 20% rate of growth implicit in this function, the value of dy/dt has to keep increasing. You can check that this must be so by differentiating. Since $dy/dt = 0.2y_t$ it is obvious that dy/dt must increase if y_t does.

Test Yourself, Exercise 14.4

1. For each of the differential equations below (i) derive the definite solution, and (ii) use this solution to predict the value of y when $t = 10$.

(a) $\dfrac{dy}{dt} = 0.2y$ with initial value $y_0 = 200$

(b) $\dfrac{dy}{dt} = 1.2y$ with initial value $y_0 = 45$

(c) $\dfrac{dy}{dt} = -0.4y$ with initial value $y_0 = 14$

(d) $\dfrac{dy}{dt} = 1.32y$ with initial value $y_0 = 40$

(e) $\dfrac{dy}{dt} = -0.025y$ with initial value $y_0 = 128$

2.   The function $y_t = 3e^{0.1t}$ gives the value of y_t at any given time t. When $t = 8$

(a)   what is the rate of growth of y with respect to itself?
(b)   what is the rate of growth of y with respect to time?

14.8   Solution of non-homogeneous differential equations

When the constant is not zero and a differential equation takes the format

$$\frac{dy}{dt} = by + c$$

the solution is derived in two parts:

(i)   The complementary function, and
(ii)   The particular solution.

The **complementary function** (CF) is the same as the solution derived above for the case with no constant, i.e. $y_t = Ae^{bt}$.

The **particular solution** (PS) is any one particular solution to the complete differential equation. It is also sometimes called the particular integral. For most economic applications you can normally use the final equilibrium value of the unknown function as the particular solution.

Thus the full solution, which is called the *general solution* (GS), is the sum of these two components, i.e.

$$GS = CF + PS$$

This will be in the format

$$y_t = Ae^{bt} + PS$$

The value of the arbitrary constant A can be calculated if a value for y is known for a given value of t. A specific value for A will turn the general solution into a **definite solution** (DS).

In an economic model this definite solution can usually be interpreted as

$$y = \{\textit{Function that shows divergence from equilibrium}\} + \{\textit{Equilibrium value}\}$$

The example below explains how this method of solution works.

Example 14.15

Solve the differential equation $dy/dt = 6y + 27$ if the value of y is 18 when $t = 0$.

Solution

To derive the complementary function from the differential equation in the question, we first consider the 'reduced equation' (RE) without the constant term. Thus in this example the elimination of the constant gives the reduced equation

$$\frac{dy}{dt} = 6y \qquad\qquad\qquad\text{(RE)}$$

Using the result derived in the previous section that for any differential equation in the format

$$\frac{dy}{dt} = by \quad \text{then } y_t = Ae^{bt}$$

the solution to the (RE) above in this example will therefore be the complementary function

$$y_t = Ae^{6t} \tag{CF}$$

To derive the particular solution we consider the situation where the function y reaches its equilibrium value and will not change any more if t increases and so

$$\frac{dy}{dt} = 0$$

The value of y for which this result holds will be a constant, which we can denote by the letter K. This will be the particular solution to the differential equation. In this example, given that

$$\frac{dy}{dt} = 6y + 27$$

if y is constant at value K then

$$\frac{dy}{dt} = 6K + 27 = 0$$
$$K = -4.5 \tag{PS}$$

Putting this PS together with the CF already derived, the general solution will be

$$y_t = Ae^{6t} - 4.5 \tag{GS}$$

As the initial value of y is 18 when $t = 0$ then (remembering that $e^0 = 1$)

$$y_0 = 18 = Ae^0 - 4.5$$
$$18 = A - 4.5$$
$$22.5 = A$$

Putting this value for A into the GS gives the definite solution

$$y_t = 22.5e^{6t} - 4.5 \tag{DS}$$

If you enter a few values for t you will see that the value of y in this function rapidly becomes extremely large. For example, when $t = 3$ then

$$y_3 = 22.5e^{6(3)} - 4.5 = 22.5e^{18} - 4.5 = 22.5(65,659,969) - 4.5 = 1,477,349,303$$

Before we investigate the usefulness of this method for the analysis of dynamic economic models, we will work through another example just to make sure that you understand how to arrive at a solution.

Example 14.16

Given the differential equation $dy/dt = -1.5y + 12$ derive a function for y in terms of t given the initial value $y_0 = 10$.

Solution

The reduced equation without the constant is

$$\frac{dy}{dt} = -1.5y \tag{RE}$$

This means that the complementary function will be

$$y_t = Ae^{-1.5t} \tag{CF}$$

If y is assumed to equal a constant K then

$$\frac{dy}{dt} = -1.5K + 12 = 0$$

giving the particular solution

$$K = 8 \tag{PS}$$

Putting (CF) and (PS) together, the general solution is therefore

$$y_t = Ae^{-1.5t} + 8 \tag{GS}$$

Given the initial value for y_0 we can find A as

$$y_0 = 10 = Ae^0 + 8$$
$$2 = A$$

Putting this value for A into the GS gives the definite solution

$$y_t = 2e^{-1.5t} + 8 \tag{DS}$$

Convergence and stability

If the solution to Example 14.16 above is used to calculate a few values of y, it can be seen that these values converge on the equilibrium value of 8 as t gets larger, as shown in Table 14.3.

Why does this set of values differ from the pattern in Example 14.15 where the values of y_t increased exponentially? The answer is that in any differential equation with a solution in the format

$$y_t = Ae^{bt} + PS$$

Table 14.3

t	$y_t = 2e^{-1.5t} + 8$
0	10
1	8.44626
2	8.099574
3	8.022218
4	8.004958
5	8.001106

Table 14.4

t	$y = e^t$	$y = e^{-t}$
0	1	1
1	2.718	0.367879
2	7.389	0.135335
3	20.086	0.049787
4	54.598	0.018316
5	148.413	0.006738
6	403.429	0.002479

it is the value of the exponent b that determines convergence or divergence.

Convergence towards the particular solution occurs if $b < 0$

Divergence away from the particular solution occurs if $b > 0$

The reason for this becomes obvious when we compare what happens to the basic exponential functions $y = e^t$ and $y = e^{-t}$ when t increases. As Table 14.4 illustrates, the function e^t expands at an increasing rate whilst the function e^{-t} rapidly diminishes. If any positive value of b multiplies t then the function will be a multiple of the expanding values in the e^t column above. On the other hand, a negative value for b will mean that the function will be a multiple of the diminishing values in the e^{-t} column. As the CF normally shows the divergence of an economic variable from equilibrium, if the CF diminishes towards zero then the function as a whole approaches its equilibrium value.

Checking differential equation solutions with Excel

If you wish to check that you have derived the correct solution to a differential equation you can use a spreadsheet to calculate a set of values. (Table 14.3 did this for an earlier example.) Just enter a series of values for t in one column and then enter the formula for the solution in the first cell in the next column, using the Excel EXP formula, and then copy it down the column. For example, if the first value for $t = 0$ is in cell A5 then the formula to enter for the first value of the function $y = 2e^{-1.5t} + 8$ from Example 14.16 above will be:

$$= 2{}^*\text{EXP}(-1.5{}^*\text{A5}) + 8$$

When $t = 0$ the formula should give the given initial value of y_0 and if the exponent of e is negative then the values of y should converge on the particular solution. If they do not then you may have made some mistake in your derivation of the solution and it is worth checking through your calculations again.

Test Yourself, Exercise 14.5

For each of the differential equations below:
(a) derive the definite solution,
(b) use this solution to predict the value of y when t is 5, and
(c) say whether values of y converge or diverge as t increases.

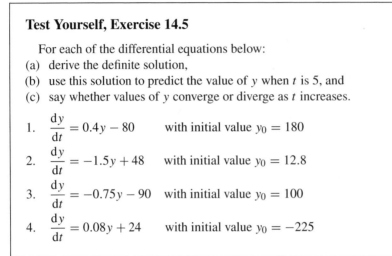

1. $\dfrac{dy}{dt} = 0.4y - 80$ with initial value $y_0 = 180$

2. $\dfrac{dy}{dt} = -1.5y + 48$ with initial value $y_0 = 12.8$

3. $\dfrac{dy}{dt} = -0.75y - 90$ with initial value $y_0 = 100$

4. $\dfrac{dy}{dt} = 0.08y + 24$ with initial value $y_0 = -225$

14.9   Continuous adjustment of market price

Assume that in a perfectly competitive market the speed with which price P adjusts towards its equilibrium value depends on how much excess demand there is. This is quite a reasonable proposition. If consumers wish to purchase a lot more produce than suppliers are willing to sell at the current price, then there will be great pressure for price to rise, but if there is only a slight shortfall then price adjustment may be sluggish. If excess demand is negative this means that quantity supplied exceeds quantity demanded, in which case price would tend to fall.

To derive the differential equation that describes this process, assume that the demand and supply functions are

$$Q_d = a + bP \quad \text{and} \quad Q_s = c + dP$$

with the parameters $a, d > 0$ and $b, c < 0$.

If r represents the rate of adjustment of P in proportion to excess demand then we can write

$$\frac{dP}{dt} = r(Q_d - Q_s)$$

Substituting the demand and supply functions for Q_d and Q_s gives

$$\frac{dP}{dt} = r[(a + bP) - (c + dP)]$$
$$= r(a - c + bP - dP)$$
$$= r(b - d)P + r(a - c)$$

As r, a, b, c and d are all constant parameters this is effectively a first-order linear differential equation with one term in P plus a constant term. This format is similar to the ones in the previous examples, except that it is P that changes over time rather than y, and so the same method of solution can be employed, as the following examples illustrate.

Example 14.17

A perfectly competitive market has the demand and supply functions

$$Q_d = 170 - 8P \quad \text{and} \quad Q_s = -10 + 4P$$

When the market is out of equilibrium the rate of adjustment of price is a function of excess demand such that

$$\frac{dP}{dt} = 0.5(Q_d - Q_s)$$

In the initial time period price P_0 is 10, which is not its equilibrium value. Derive a function for P in terms of t, and comment on the stability of this market.

Solution

Substituting the functions for Q_d and Q_s into the rate of price change function gives

$$\frac{dP}{dt} = 0.5[(170 - 8P) - (-10 + 4P)] = 0.5(-8 - 4)P + 0.5(170 + 10)$$

which simplifies to

$$\frac{dP}{dt} = -6P + 90$$

To solve this linear first-order differential equation we first consider the reduced equation without the constant term

$$\frac{dP}{dt} = -6P \tag{RE}$$

The complementary function that is the solution to this RE will be

$$P_t = Ae^{-6t} \tag{CF}$$

The particular solution is found by assuming P is equal to a constant K so that

$$\frac{dP}{dt} = -6K + 90 = 0$$

$$K = 15 \tag{PS}$$

This is the market equilibrium price. (Check this yourself using the supply and demand functions and basic linear algebra.)

Putting (CF) and (PS) together gives the general solution

$$P_t = Ae^{-6t} + 15 \tag{GS}$$

The value of A can be determined by putting the initial value of 10 for P_0 into the GS. Thus

$$P_0 = 10 = Ae^0 + 15$$
$$-5 = A$$

Using this value for A in (GS) gives the definite solution to this differential equation, which is

$$P_t = -5e^{-6t} + 15 \tag{DS}$$

The coefficient of t in this exponential function is the negative number -6. This means that the first term in (DS), i.e. the complementary function, will get closer to zero as t gets larger and so P_t will converge on its equilibrium value of 15. This market is therefore stable.

We can check this by using the above solution to calculate P_t. For example, when

$$t = 2 \quad \text{then } P_2 = -5e^{-6(2)} + 15 = -5e^{-12} + 15 = 14.99997$$

This is extremely close to the equilibrium price of 15 and so we can say that price returns to its equilibrium value within the first few time periods in this particular market.

In other markets the rate of adjustment may not be so rapid, as the following example demonstrates.

Example 14.18

If the demand and supply functions in a competitive market are

$$Q_d = 50 - 0.2P \qquad Q_s = -10 + 0.3P$$

and the rate of adjustment of price when the market is out of equilibrium is

$$\frac{dP}{dt} = 0.4(Q_d - Q_s)$$

derive and solve the relevant difference equation to get a function for P in terms of t given that price is 100 in time period 0. Comment on the stability of this market.

Solution

Substituting the demand and supply functions into the rate of change function gives

$$\frac{dP}{dt} = 0.4[(50 - 0.2P) - (-10 + 0.3P)] = 0.4(-0.2 - 0.3)P + 0.4(50 + 10)$$

$$\frac{dP}{dt} = -0.2P + 24$$

The reduced equation without the constant term is

$$\frac{dP}{dt} = -0.2P \tag{RE}$$

The complementary function will therefore be

$$P_t = Ae^{-0.2t} \qquad \text{(CF)}$$

To find the particular solution we assume P is equal to a constant K at the equilibrium price and so

$$\frac{\mathrm{d}P}{\mathrm{d}t} = -0.2K + 24 = 0$$

$$K = 120 \qquad \text{(PS)}$$

The CF and PS together give the general solution

$$P_t = Ae^{-0.2t} + 120 \qquad \text{(GS)}$$

As price is 100 in time period 0 then

$$P_0 = 100 = Ae^0 + 120$$

$$-20 = A$$

and so the definite solution to this differential equation is

$$P_t = -20\,e^{-0.2t} + 120 \qquad \text{(DS)}$$

We can tell that this market is stable as the coefficient of t in the exponential function is the negative number -0.2. However, the sample values calculated below show that the convergence of P_t on its equilibrium value of 120 is relatively slow.

t	$P_t = -20\,e^{-0.2t} + 120$
0	100
5	112.642
15	119.004

If a spreadsheet is used to calculate P_t for values of t from 0 to 21 and these are plotted on a graph using the Excel Chart Wizard function, it will look like Figure 14.1. You should note that this differs from the pattern in the cobweb model considered in Chapter 13, where price alternated above and below its final equilibrium value by smaller and smaller amounts if the market was stable. This time, price gradually approaches its equilibrium value *from one direction only*. A similar time path will occur in other similar market models with

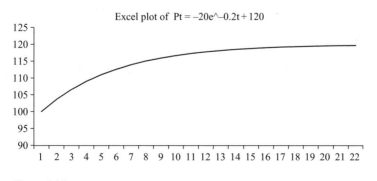

Figure 14.1

continuous price adjustment, although if the initial value is above the equilibrium then price will, obviously, approach this equilibrium from above rather than from below.

Test Yourself, Exercise 14.6

1. If the demand and supply functions in a competitive market are

$$Q_d = 35 - 0.5P \quad \text{and} \quad Q_s = -4 + 0.8P$$

and the rate of adjustment of price when the market is out of equilibrium is

$$\frac{dP}{dt} = 0.25(Q_d - Q_s)$$

derive and solve the relevant differential equation to get a function for P in terms of t given that price is 37 in time period 0. Comment on the stability of this market.

2. The demand and supply functions in a competitive market are

$$Q_d = 280 - 4P \quad \text{and} \quad Q_s = -35 + 8P$$

and price is currently 19. When not at equilibrium the rate of adjustment of price is

$$\frac{dP}{dt} = 0.08(Q_d - Q_s)$$

Derive and solve the relevant differential equation to get a function for P in terms of t and use the solution to explain how close price will be to its equilibrium value after seven time periods.

3. A raw commodity is traded in a market where it has been reliably estimated that

$$Q_d = 95 - 1.8P \quad \text{and} \quad Q_s = -12.4 + 2.1P$$

Its price adjusts in proportion to excess demand at the rate

$$\frac{dP}{dt} = 0.28(Q_d - Q_s)$$

where t is measured in months. The current spot price is $29.35 a tonne. In 4 months' time your company will need to buy a large amount of this commodity. If someone offers you a forward contract and guarantees to supply the amount you need at a price of $24.75 would it be worth signing this contract?

4. In a competitive market where $Q_d = 560 - 6P$ and $Q_s = -46 + 28.7P$ the initial price P_0 is £50. Derive a function for the time path of P and use it to predict price in time period 5 given that price adjusts in proportion to excess demand at the rate

$$\frac{dP}{dt} = 0.01(Q_d - Q_s)$$

How many time periods would you have to wait for the price to drop by £20?

5. A price of \$65 per tonne is currently being quoted for a mineral traded in a competitive commodity market where $Q_d = 243 - 3.5P$ and $Q_s = -7.8 + 2.2P$. This price adjusts in proportion to excess demand at the rate

$$\frac{dP}{dt} = 0.16(Q_d - Q_s)$$

What is your forecast for price when t is 8?

14.10 Continuous adjustment in a Keynesian macroeconomic model

In a basic Keynesian macroeconomic model, with no foreign trade and no government sector, total expenditure (E) will be the sum of consumer expenditure (C) and exogenously determined investment (I). This model can therefore be specified as

where consumption $\quad E = C + I$

$$C = a + bY$$

In equilibrium $\quad E = Y$

and so $\quad Y = C + I$

However, this macroeconomic system may not always be in equilibrium. For example, if there is an exogenous increase in I, it may take a while before all the knock-on multiplier effects work through. Let us assume that the speed with which Y adjusts is directly proportional to the difference between total expenditure E and current income Y at ratio r. This relationship can be written as

$$\frac{dY}{dt} = r(E - Y) = r(C + I - Y)$$
$$= r(a + bY + I - Y)$$
$$= r(b - 1)Y + r(a + I)$$

As r, a, b and I are all constant this is effectively a differential equation with one term in Y with the constant coefficient $r(b - 1)$ plus another constant term $r(a + I)$. The standard method of solution for first-order linear differential equations can therefore be employed, as shown in the following examples.

Example 14.19

In a basic Keynesian macroeconomic model

$$C = 360 + 0.8Y$$

$$I = 120$$

When the system is out of equilibrium the rate of adjustment of Y is

$$\frac{dY}{dt} = 0.25(E - Y) = 0.25(C + I - Y)$$

If national income is initially 2,000, derive a function for Y in terms of t and comment on the stability of this system.

Solution

Substituting the consumption function C and the given level of investment I into the rate of adjustment function gives the differential equation to be solved

$$\frac{dY}{dt} = 0.25(360 + 0.8Y + 120 - Y)$$

$$= 0.25(480 - 0.2Y)$$

$$= 120 - 0.05Y$$

The relevant reduced equation without the constant term is

$$\frac{dY}{dt} = -0.05Y \tag{RE}$$

The corresponding complementary function will therefore be

$$Y_t = Ae^{-0.05t} \tag{CF}$$

Assuming Y equals a constant value K in equilibrium to determine the particular solution

$$\frac{dY}{dt} = 120 - 0.05K = 0$$

$$K = 2,400 \tag{PS}$$

The CF and PS together give the general solution

$$Y_t = Ae^{-0.05t} + 2,400 \tag{GS}$$

As Y is 2,000 in the initial time period 0 then substituting this known value into the GS gives

$$Y_0 = 2,000 = Ae^0 + 2,400$$

$$-400 = A$$

The definite solution given this initial value is therefore

$$Y_t = -400e^{-0.05t} + 2,400 \tag{DS}$$

This market is stable because the coefficient of t in the exponential function is negative. The convergence on the equilibrium value of 2,400 is very slow though, as the values below illustrate.

t	$Y_t = -400e^{-0.05t} + 2,400$
10	2157.388
20	2252.848
50	2367.166

You will also note that as t increases the values of Y_t approach equilibrium from one direction only.

Now that you are familiar with the different steps of the solution method, we will work through another similar example, but with the initial value above the final equilibrium value.

Example 14.20

In a macroeconomic model

$$C = 200 + 0.75Y$$

$$E = C + I \quad \text{and} \quad I = 80$$

$$\frac{dY}{dt} = 0.8(E - Y)$$

If $Y_0 = 1,200$, derive a function for Y in terms of t and comment on the stability of this model.

Solution

Substituting C and I for E in the rate of adjustment function gives

$$\frac{dY}{dt} = 0.8(C + I - Y) = 0.8(200 + 0.75Y + 80 - Y)$$

$$= 0.8(280 - 0.25Y)$$

$$= 224 - 0.2Y$$

To solve this differential equation we first set up the reduced equation

$$\frac{dY}{dt} = -0.2Y \tag{RE}$$

The complementary function is therefore

$$Y_t = Ae^{-0.2t} \tag{CF}$$

Assuming Y equals a constant value K in equilibrium, the particular solution will be

$$\frac{dY}{dt} = 224 - 0.2K = 0$$

$$K = 1,120 \tag{PS}$$

The CF and PS together give the general solution

$$Y_t = Ae^{-0.2t} + 1,120 \tag{GS}$$

As Y is 1,200 in the initial time period 0 then

$$Y_0 = 1,200 = Ae^0 + 1,120$$
$$80 = A$$

The definite solution given this initial value is therefore

$$Y_t = 80e^{-0.2t} + 1,120 \tag{DS}$$

This market is stable because the coefficient of t in the exponential function is negative. The initial value was higher than the final equilibrium of 1,120 and so values of Y approach equilibrium from above, as the values below illustrate.

t	$Y_t = 80e^{-0.2t} + 1,120$
5	1149.43
10	1130.83
20	1121.47

Test Yourself, Exercise 14.7

Given that $E = C + I$, derive a function for Y in terms of t for each of the following macroeconomic models and then use it to predict Y when t is 10.

1. $C = 50 + 0.6Y$ and $\dfrac{dY}{dt} = 0.5(E - Y)$ given $I = 22$ and $Y_0 = 205$.

2. $C = 360 + 0.7Y$ and $\dfrac{dY}{dt} = 0.65(E - Y)$ given $I = 115$ and $Y_0 = 1520$.

3. $C = 275 + 0.82Y$ and $\dfrac{dY}{dt} = 0.2(E - Y)$ given $I = 90$ and $Y_0 = 2040$.

4. $C = 48 + 0.53Y$ and $\dfrac{dY}{dt} = 0.48(E - Y)$ given $I = 18.5$ and $Y_0 = 132$.

5. $C = 90 + 0.61Y$ and $\dfrac{dY}{dt} = 0.45(E - Y)$ given $I = 45$ and $Y_0 = 328$.

15 Matrix algebra

Learning objectives

After completing this chapter students should be able to:

- Formulate multi-variable economic models in matrix format.
- Add and subtract matrices.
- Multiply matrices by a scalar value and by another matrix.
- Calculate determinants and cofactors.
- Derive the inverse of a matrix.
- Use the matrix inverse to solve a system of simultaneous equations both manually and using a spreadsheet.
- Derive the Hessian matrix of second-order derivatives and use it to check the second-order conditions in an unconstrained optimization problem.
- Derive the bordered Hessian matrix and use it to check the second-order conditions in a constrained optimization problem.

15.1 Introduction to matrices and vectors

Suppose that you are responsible for hiring cars for your company's staff to use. The weekly hire rates for the five different sizes of car that are available are: Compact: £139, Intermediate: £160, Large: £205, 7-Seater People Carrier: £340 and Luxury limousine: £430. For next week you know that your car hire requirements will be: 4 Compact, 3 Intermediate, 12 Large, 2 People Carrier and 1 Luxury limousine. How would you work out the total car hire bill?

If you worked out total expenditure as

$$4 \times £139 + 3 \times £160 + 12 \times £205 + 2 \times £340 + 1 \times £430 = £4,606$$

then you would be correct. You would have also already done a matrix multiplication problem, although you may not have realized it! Before we look at the formal theory of matrices, let us continue with this example for a while longer.

If you know that your car hire requirements will change from week to week, it can help make calculations clearer if the number of cars required in each category are set out in tabular form, as in Table 15.1.

Table 15.1

Cars required	Week 1	Week 2	Week 3
Compact	4	7	2
Intermediate	3	5	5
Large	12	9	5
People carrier	2	1	3
Luxury limousine	1	1	2

The total car hire bill for each week can then be calculated by multiplying the number of cars to be hired in each category by the corresponding price.

A **matrix** is defined as an array of numbers (or algebraic symbols) set out in rows and columns. Therefore, the car hire requirements for the 3-week period in this example can be set out as the matrix

$$
A = \begin{bmatrix} 4 & 7 & 2 \\ 3 & 5 & 5 \\ 12 & 9 & 5 \\ 2 & 1 & 3 \\ 1 & 1 & 2 \end{bmatrix}
$$

where each row corresponds to a size of car and each column corresponds to a week. The usual notation system is to denote matrices by a capital letter in bold type, as for matrix **A** above, and to enclose the elements of a matrix in a set of squared brackets, i.e. [ ].

Matrices may also be specified with algebraic terms instead of numbers. Each entry is usually known as an 'element'. The elements in each matrix must form a complete rectangle, without any blank spaces. For example, if there are 5 rows and 3 columns there must be 3 elements in each row and 5 elements in each column. An element may be zero though.

The size of a matrix is called its 'order'. The **order** is specified as:

(number of rows) × (number of columns)

For example, the matrix **A** above has 5 rows and 3 columns and so its order is 5 × 3.

Matrices with only one column or row are known as **vectors**. These are usually represented by lower case letters, in bold. For example, the set of car rental prices we started this chapter with can be specified (in £) as the 1 × 5 **row vector**

$$
p = \begin{bmatrix} 139 & 160 & 205 & 340 & 430 \end{bmatrix}
$$

and the car hire requirements in week 1 can be specified as the 5 × 1 **column vector**

$$
q = \begin{bmatrix} 4 \\ 3 \\ 12 \\ 2 \\ 1 \end{bmatrix}
$$

Matrix addition and subtraction

Matrices that have the same order can be added together, or subtracted. The addition, or subtraction, is performed on each of the corresponding elements.

Example 15.1

A retailer sells two products, Q and R, in two shops A and B. The number of items sold for the last 4 weeks in each shop are shown in the two matrices **A** and **B** below, where the columns represent weeks and the rows correspond to products Q and R, respectively.

$$\mathbf{A} = \begin{bmatrix} 5 & 4 & 12 & 7 \\ 10 & 12 & 9 & 14 \end{bmatrix} \quad \text{and} \quad \mathbf{B} = \begin{bmatrix} 8 & 9 & 3 & 4 \\ 8 & 18 & 21 & 5 \end{bmatrix}$$

Derive a matrix for total sales for this retailer for these two products over the last 4 weeks.

Solution

Total sales for each week will simply be the sum of the corresponding elements in matrices **A** and **B**. For example, in week 1 the total sales of product Q will be 5 plus 8. Total combined sales for Q and R can therefore be represented by the matrix

$$\mathbf{T} = \mathbf{A} + \mathbf{B} = \begin{bmatrix} 5 & 4 & 12 & 7 \\ 10 & 12 & 9 & 14 \end{bmatrix} + \begin{bmatrix} 8 & 9 & 3 & 4 \\ 8 & 18 & 21 & 5 \end{bmatrix}$$

$$= \begin{bmatrix} 5+8 & 4+9 & 12+3 & 7+4 \\ 10+8 & 12+18 & 9+21 & 14+5 \end{bmatrix} = \begin{bmatrix} 13 & 13 & 15 & 11 \\ 18 & 30 & 30 & 19 \end{bmatrix}$$

An element of a matrix can be a negative number, as in the solution to the example below.

Example 15.2

If $\mathbf{A} = \begin{bmatrix} 12 & 30 \\ 8 & 15 \end{bmatrix}$ and $\mathbf{B} = \begin{bmatrix} 7 & 35 \\ 4 & 8 \end{bmatrix}$ what is $\mathbf{A} - \mathbf{B}$?

Solution

$$\mathbf{A} - \mathbf{B} = \begin{bmatrix} 12 & 30 \\ 8 & 15 \end{bmatrix} - \begin{bmatrix} 7 & 35 \\ 4 & 8 \end{bmatrix} = \begin{bmatrix} 12-7 & 30-35 \\ 8-4 & 15-8 \end{bmatrix} = \begin{bmatrix} 5 & -5 \\ 4 & 7 \end{bmatrix}$$

Scalar multiplication

There are two forms of multiplication that can be performed on matrices. A matrix can be multiplied by a specific value, such as a number (scalar multiplication) or by another matrix (matrix multiplication). Scalar multiplication simply involves the multiplication of each element in a matrix by the scalar value, as in Example 15.3. Matrix multiplication is rather more complex and is explained later, in Section 15.2.

Example 15.3

The number of units of a product sold by a retailer for the last 2 weeks are shown in matrix **A** below, where the columns represent weeks and the rows correspond to the two different shop units that sold them.

$$A = \begin{bmatrix} 12 & 30 \\ 8 & 15 \end{bmatrix}$$

If each item sells for £4, derive a matrix for total sales revenue for this retailer for these two shop units over this two-week period.

Solution

Total revenue is calculated by multiplying each element in matrix of sales quantities **A** by the scalar value 4, the price that each unit is sold at. Thus total revenue can be represented (in £) by the matrix

$$R = 4A = \begin{bmatrix} 4 \times 12 & 4 \times 30 \\ 4 \times 8 & 4 \times 15 \end{bmatrix} = \begin{bmatrix} 48 & 120 \\ 32 & 60 \end{bmatrix}$$

The scalar value that a matrix is multiplied by may be an algebraic term rather than a specific number value. For example, if the product price in Example 15.3 above was specified as p instead of £4 then the total revenue matrix would become

$$R = \begin{bmatrix} 12p & 30p \\ 8p & 15p \end{bmatrix}$$

Scalar division works in the same way as scalar multiplication, but with each element divided by the relevant scalar value.

Example 15.4

If the set of car rental prices in the vector $\mathbf{p} = \begin{bmatrix} 139 & 160 & 205 & 340 & 430 \end{bmatrix}$ includes VAT (Value Added Tax) at 17.5% and your company can claim this tax back, what is the vector **v** of prices without this tax?

Solution

First of all we need to find the scalar value used to scale down the original vector element values. As the tax rate is 17.5% then the quoted prices will be 117.5% times the basic price. Therefore a quoted price divided by 1.175 will be the basic price and so the vector of prices (in £) without the tax will be

$$v = \left(\frac{1}{1.175} \right) p = \left(\frac{1}{1.175} \right) \begin{bmatrix} 139 & 160 & 205 & 340 & 430 \end{bmatrix}$$

$$= \begin{bmatrix} \left(\frac{1}{1.175} \right) 139 & \left(\frac{1}{1.175} \right) 160 & \left(\frac{1}{1.175} \right) 205 & \left(\frac{1}{1.175} \right) 340 & \left(\frac{1}{1.175} \right) 430 \end{bmatrix}$$

$$= \begin{bmatrix} 118.30 & 136.17 & 174.47 & 289.36 & 365.96 \end{bmatrix}$$

Test Yourself, Exercise 15.1

1. A firm uses 3 different inputs K, L and R to make two final products X and Y. Each unit of X produced requires 2 units of K, 8 units of L and 23 units of R. Each unit of Y produced requires 3 units of K, 5 units of L and 26 units of R. Set up these input requirements in matrix format.

2. 'A vector is a special form of matrix but a matrix is not a special form of vector'. Is this statement true?

3. For the pairs of matrices below say whether it is possible to add them together and then, where it is possible, derive the matrix $\mathbf{C} = \mathbf{A} + \mathbf{B}$.

(a) $\mathbf{A} = \begin{bmatrix} 2 & 35 \\ 18 & 15 \end{bmatrix}$ and $\mathbf{B} = \begin{bmatrix} 4 & 35 \\ 9 & 8 \end{bmatrix}$

(b) $\mathbf{A} = \begin{bmatrix} 5 & 3 \\ 8 & 1 \end{bmatrix}$ and $\mathbf{B} = \begin{bmatrix} 7 & 0 & 2 \\ 8 & 8 & 1 \end{bmatrix}$

(c) $\mathbf{A} = \begin{bmatrix} 10 \\ 3 \\ 12 \\ 6 \\ 1 \end{bmatrix}$ and $\mathbf{B} = \begin{bmatrix} 4 \\ 0 \\ 2 \\ -9 \\ 1 \end{bmatrix}$

4. A company sells 4 products and the sales revenue (in £m.) from each product sold through the company's three retail outlets in a year are given in the matrix

$$\mathbf{R} = \begin{bmatrix} 7 & 3 & 1 & 4 \\ 6 & 3 & 8 & 2.5 \\ 4 & 1.2 & 2 & 0 \end{bmatrix}$$

If profit earned is always 20% of sales revenue, use scalar multiplication to derive a matrix showing profit on each product for each retail outlet.

15.2 Basic principles of matrix multiplication

If one matrix is multiplied by another matrix, the basic rule is to multiply elements along the rows of the first matrix by the corresponding elements down the columns of the second matrix. The easiest way to understand how this operation works is to first work through some examples that only involve matrices with one row or column, i.e. vectors.

Returning to our car hire example, consider the two vectors

$$\mathbf{p} = \begin{bmatrix} 139 & 160 & 205 & 340 & 430 \end{bmatrix} \quad \text{and} \quad \mathbf{q} = \begin{bmatrix} 4 \\ 3 \\ 12 \\ 2 \\ 1 \end{bmatrix}$$

The row vector $\mathbf{p}$ contains the prices of hire cars in each category and the column vector $\mathbf{q}$ contains the quantities of cars in each category that your company wishes to hire for the week.

At the start of this chapter we worked out the total car hire bill as

$$139 \times 4 + 160 \times 3 + 205 \times 12 + 340 \times 2 + 430 \times 1 = £4,606$$

In terms of these two vectors, what we have done is multiply the first element in the row vector **p** by the first element in the column vector **q**. Then, going across the row, the second element of **p** is multiplied by the second element down the column of **q**. The same procedure is followed for the other elements until we get to the end of the row and the bottom of the column.

Now consider the situation where the car hire prices are still shown by the vector

$$\mathbf{p} = \begin{bmatrix} 139 & 160 & 205 & 340 & 430 \end{bmatrix}$$

but there are now three weeks of different car hire requirements, shown by the columns of matrix

$$\mathbf{A} = \begin{bmatrix} 4 & 7 & 2 \\ 3 & 5 & 5 \\ 12 & 9 & 5 \\ 2 & 1 & 3 \\ 1 & 1 & 2 \end{bmatrix}$$

To calculate the total car hire bill for each of the three weeks, we need to find the vector

$$\mathbf{t} = \mathbf{pA}$$

This should have the order 1×3, as there will be one element (i.e. the bill) for each of the three weeks. The first element of **t** is the bill for the first week, which we have already found in the example above. The car hire bill for the second week is worked out using the same method, but this time the elements across the row vector **p** multiply the elements down the second column of matrix **A**, giving

$$139 \times 7 + 160 \times 5 + 205 \times 9 + 340 \times 1 + 430 \times 1 = £4,388$$

The third element is calculated in the same manner, but working down the third column of **A**. The result of this matrix multiplication exercise is therefore

$$\mathbf{t} = \mathbf{pA} = \begin{bmatrix} 139 & 160 & 205 & 340 & 430 \end{bmatrix} \begin{bmatrix} 4 & 7 & 2 \\ 3 & 5 & 5 \\ 12 & 9 & 5 \\ 2 & 1 & 3 \\ 1 & 1 & 2 \end{bmatrix}$$

$$= \begin{bmatrix} 4606 & 4388 & 3983 \end{bmatrix}$$

The above examples have shown how the basic principle of matrix multiplication involves the elements across a row vector multiplying the elements down the columns of the matrix being multiplied, and then summing all the products obtained. If the first matrix has more than one row (i.e. it is not a vector) then the same procedure is followed across each row. This means that the number of rows in the final product matrix will correspond to the number of rows in the first matrix.

Example 15.5

Multiply the two matrices $\mathbf{A} = \begin{bmatrix} 2 & 3 \\ 8 & 1 \end{bmatrix}$ and $\mathbf{B} = \begin{bmatrix} 7 & 5 & 2 \\ 4 & 8 & 1 \end{bmatrix}$

Solution

Using the method explained above, the product matrix will be

$$\mathbf{AB} = \begin{bmatrix} 2 & 3 \\ 8 & 1 \end{bmatrix} \begin{bmatrix} 7 & 5 & 2 \\ 4 & 8 & 1 \end{bmatrix}$$

$$= \begin{bmatrix} 2 \times 7 + 3 \times 4 & 2 \times 5 + 3 \times 8 & 2 \times 2 + 3 \times 1 \\ 8 \times 7 + 1 \times 4 & 8 \times 5 + 1 \times 8 & 8 \times 2 + 1 \times 1 \end{bmatrix} = \begin{bmatrix} 26 & 34 & 7 \\ 60 & 48 & 17 \end{bmatrix}$$

You now may be wondering what happens if the number of elements along the rows of the first matrix (or vector) does not equal the number of elements in the columns of the matrix that it is multiplying. The answer to this question is that it is *not* possible to multiply two matrices if the number of columns in the first matrix does not equal the number of rows in the second matrix. Therefore, if a matrix $\mathbf{A}$ has order $m \times n$ and another matrix $\mathbf{B}$ has order $r \times s$, then the multiplication $\mathbf{AB}$ can only be performed if $n = r$, in which case the resulting matrix $\mathbf{C} = \mathbf{AB}$ will have order $(m \times s)$.

This principle is illustrated in Example 15.5 above. Matrix $\mathbf{A}$ has order 2×2 and matrix $\mathbf{B}$ has order 2×3 and so the product matrix $\mathbf{AB}$ has order 2×3. Some other examples of how the order of different matrices affects the order of the product matrix when they are multiplied are given in Table 15.2.

Table 15.2

A	B	Order of product matrix AB
5×3	3×2	5×2
1×8	8×1	1×1
3×5	2×4	Matrix multiplication not possible
3×4	4×3	3×3
4×3	4×3	Matrix multiplication not possible

Test Yourself, Exercise 15.2

1. Given the vector $\mathbf{v} = \begin{bmatrix} 2 & 5 \end{bmatrix}$ and matrix $\mathbf{A} = \begin{bmatrix} 6 & 2 \\ 3 & 7 \end{bmatrix}$ find the product matrix $\mathbf{vA}$.
2. For the pairs of matrices below say if it is possible to derive the product matrix $\mathbf{C} = \mathbf{AB}$ and, when this is possible, calculate the elements of this product matrix.

(a) $\mathbf{A} = \begin{bmatrix} 2 & 10 \\ 7 & 15 \end{bmatrix}$ and $\mathbf{B} = \begin{bmatrix} 4 & 2 \\ 9 & 8 \end{bmatrix}$

(b) $\mathbf{A} = \begin{bmatrix} 5 & 3 \\ 8 & 1 \end{bmatrix}$ and $\mathbf{B} = \begin{bmatrix} 7 & 0 & 2 \\ 12 & 8 & 1 \end{bmatrix}$

(c) $\mathbf{A} = \begin{bmatrix} 9 \\ 3 \\ 12 \\ 6 \\ 1 \end{bmatrix}$ and $\mathbf{B} = \begin{bmatrix} 4 \\ 0 \\ 2 \\ -9 \\ 1 \end{bmatrix}$

3. A company's input requirements over the next four weeks for the three inputs X, Y and Z are given (in numbers of units of each input) by the matrix

$$\mathbf{R} = \begin{bmatrix} 2 & 0.5 & 1 & 7 \\ 6 & 3 & 8 & 2.5 \\ 4 & 5 & 2 & 0 \end{bmatrix}$$

The company can buy these inputs from two suppliers, whose prices for the three inputs X, Y and Z are in given (in £) by the matrix

$$\mathbf{P} = \begin{bmatrix} 4 & 6 & 2 \\ 5 & 8 & 1 \end{bmatrix}$$

where the two rows represent the suppliers and the three columns represent the input prices. Use matrix multiplication to derive a matrix that will give the total input bill for the next four weeks for both suppliers.

15.3   Matrix multiplication – the general case

Now that the basic principles have been explained with some straightforward examples, we can set out a general formula for matrix multiplication that can be applied to more complex matrix multiplication exercises. The general $m \times n$ matrix with any number of rows m and columns n can be written as

$$\mathbf{A} = \begin{bmatrix} a_{11} & a_{12} & \cdots & a_{1n} \\ a_{21} & a_{22} & \cdots & a_{2n} \\ \vdots & \vdots & \vdots & \vdots \\ a_{m1} & a_{m2} & \cdots & a_{mn} \end{bmatrix}$$

For each element a_{ij} the subscript i denotes the row number and the subscript j denotes the column number. For example

a_{11} = element in row 1, column 1

a_{12} = element in row 1, column 2

a_{1n} = element in row 1, column n

a_{mn} = element in row m, column n

If this general $m \times n$ matrix **A** multiplies the general $n \times r$ matrix **B** then the product will be the $m \times r$ matrix **C**. Thus we can write

$$\mathbf{AB} = \begin{bmatrix} a_{11} & a_{12} & \cdots & a_{1n} \\ a_{21} & a_{22} & \cdots & a_{2n} \\ \vdots & \vdots & \vdots & \vdots \\ a_{m1} & a_{m2} & \cdots & a_{mn} \end{bmatrix} \begin{bmatrix} b_{11} & b_{12} & \cdots & b_{1r} \\ b_{21} & b_{22} & \cdots & b_{2r} \\ \vdots & \vdots & \vdots & \vdots \\ b_{n1} & b_{n2} & \cdots & b_{nr} \end{bmatrix}$$

$$= \begin{bmatrix} c_{11} & c_{12} & \cdots & \cdots & c_{1r} \\ c_{21} & c_{22} & \cdots & \cdots & c_{2r} \\ \vdots & \vdots & \vdots & \vdots & \vdots \\ c_{m1} & c_{m2} & \cdots & \cdots & c_{mr} \end{bmatrix} = \mathbf{C}$$

where

$$c_{11} = a_{11}b_{11} + a_{12}b_{21} + \cdots + a_{1n}b_{n1}$$
$$c_{12} = a_{11}b_{12} + a_{12}b_{22} + \cdots + a_{1n}b_{n2}$$

$$\vdots \qquad \vdots \qquad \vdots \qquad\qquad \vdots$$

$$c_{mr} = a_{m1}b_{1r} + a_{m2}b_{2r} + \cdots + a_{mn}b_{nr}$$

Example 15.6

Find the product matrix $\mathbf{C} = \mathbf{AB}$ when

$$\mathbf{A} = \begin{bmatrix} 4 & 2 & 12 \\ 6 & 0 & 20 \\ 1 & 8 & 5 \end{bmatrix} \quad \text{and} \quad \mathbf{B} = \begin{bmatrix} 10 & 0.5 & 1 & 7 \\ 6 & 3 & 8 & 2.5 \\ 4 & 4 & 2 & 0 \end{bmatrix}$$

Solution

Using the general matrix multiplication formula, the elements of the first two rows of the product matrix **C** can be calculated as:

$$c_{11} = 4 \times 10 + 2 \times 6 + 12 \times 4 = 40 + 12 + 48 = 100$$
$$c_{12} = 4 \times 0.5 + 2 \times 3 + 12 \times 4 = 2 + 6 + 48 = 56$$
$$c_{13} = 4 \times 1 + 2 \times 8 + 12 \times 2 = 4 + 16 + 24 = 44$$
$$c_{14} = 4 \times 7 + 2 \times 2.5 + 12 \times 0 = 28 + 5 + 0 = 33$$
$$c_{21} = 6 \times 10 + 0 \times 6 + 20 \times 4 = 60 + 0 + 80 = 140$$
$$c_{22} = 6 \times 0.5 + 0 \times 3 + 20 \times 4 = 3 + 0 + 80 = 83$$
$$c_{23} = 6 \times 1 + 0 \times 8 + 20 \times 2 = 6 + 0 + 40 = 46$$
$$c_{24} = 6 \times 7 + 0 \times 2.5 + 20 \times 0 = 42 + 0 + 0 = 42$$

Now try and calculate the elements of the final row yourself. You should get the values

$$c_{31} = 78, c_{32} = 44.5, c_{33} = 75, c_{34} = 27$$

The complete product matrix will therefore be

$$\mathbf{C} = \mathbf{AB} = \begin{bmatrix} 100 & 56 & 44 & 33 \\ 140 & 83 & 46 & 42 \\ 78 & 44.5 & 75 & 27 \end{bmatrix}$$

Although the calculations for matrix multiplication of small matrices can be done manually fairly quickly, it is now becoming obvious that for large matrices the calculations will be very tedious and time-consuming. Several economics applications involving matrix multiplication do not actually require you to calculate all the elements of the product matrix. For occasions when you do need to calculate all these elements, an Excel spreadsheet can be used.

Using Excel for matrix multiplication

The best way to explain how to use the Excel MMULT formula to multiply two matrices **A** and **B** is to work through an example.

Example 15.7

Given the two matrices

$$\mathbf{A} = \begin{bmatrix} 8 & 4 & 3 \\ 4 & 5 & 6 \end{bmatrix} \quad \text{and} \quad \mathbf{B} = \begin{bmatrix} 0.8 & 0.3 & 0.1 \\ 0.5 & 0.2 & 0.4 \\ 0.3 & 0.2 & 0.1 \end{bmatrix}$$

find the product matrix **AB** using an Excel spreadsheet.

Solution

(a) Enter the values of matrices **A** and **B** on a spreadsheet. For example, put the elements of **A** in cells (A2; C3) and the elements of **B** in cells (E2; G4). You can also enter labels for the matrix names in the rows of cells above.
(b) Highlight the cells where you want the calculated **AB** matrix to go. Since the order of **A** is 2 × 3 and the order of **B** is 3 × 3 the product matrix **AB** must have order 2 × 3. You therefore need to highlight a block of cells with 2 rows and 3 columns, such as (A6; C7).
(c) With this cell range still highlighted, enter the formula = MMULT(A2;C3,E2;G4) or use whatever cells range applies for your matrices to be multiplied, or use mouse to mark out matrices to be multiplied with dotted lines.
(d) Hold down the CNTRL and SHIFT keys together and press ENTER (if you do not do this then the formula will not treat all the highlighted cells as part of an array, i.e. a matrix).

Your spreadsheet and the computed product matrix **AB** should now be as shown in Table 15.3. In the simple example above you can check the answers manually. However, once you are satisfied that you can use the Excel MMULT formula properly then you can use it for more complex examples where manual computation would be too time-consuming.

Table 15.3

	A	B	C	D	E	F	G	H
1	**Matrix**	**A**			**Matrix**	**B**		
2	8	4	3		0.8	0.3	0.1	
3	4	5	6		0.5	0.2	0.4	
4					0.3	0.2	0.1	
5	**Matrix**	**AB**						
6	9	3.8	2.7					
7	8	3.4	3					

Table 15.4

	A	B	C	D	E	F	G	H	I	J	K	L	M	N	O
1	**Matrix A**							**Matrix B**							
2	120	160	195	220	285	350		8	9	10	11	12	14	3	2
3	125	165	200	225	290	355		12	13	14	15	16	9	12	5
4	130	170	205	230	150	360		4	5	6	5	6	9	3	7
5	135	175	210	235	200	380		8	9	10	11	12	3	3	4
6	140	180	215	240	110	500		5	6	7	0	3	0	2	3
7								2	3	4	1	4	1	2	0
8								**Product**							
9								**Matrix**	**AB**						
10								7545	8875	10205	7465	10065	5885	4795	4140
11								7740	9100	10460	7680	10330	6065	4920	4245
12								7210	8455	9700	7895	10160	6245	4755	3915
13								7660	8995	10330	8125	10620	6440	5000	4155
14								7610	8995	10380	8455	11060	6735	5165	3975

Example 15.8

In the spreadsheet in Table 15.4, the MMULT formula has been used to multiply the 5×6 matrix **A** by the 6×8 matrix **B** to get the 5×8 product matrix **AB**. Try entering the matrices **A** and **B** yourself and see if you can use the Excel MMULT formula to get the same product matrix **AB**.

Vectors of coefficients

In economic models it is common to specify one dependent variable as a function of a vector of explanatory variables, especially when employing econometric analysis to estimate coefficients of these explanatory variables. A typical vector format for a function is $\mathbf{q} = \boldsymbol{\beta}\mathbf{x}$ where $\boldsymbol{\beta}$ is the vector of coefficients for the exogenous explanatory variables in vector $\mathbf{x}$.

For example, assume that the demand for oil in time t is the linear function

$$q^t = \beta_0 + \beta_1 x_1^t + \beta_2 x_2^t + \beta_3 x_3^t + \beta_4 x_4^t + \beta_5 x_5^t$$

where the superscript t denotes the time period (rather than an exponent) for all variables and

x_1 = price of oil

x_2 = average income

x_3 = price of substitute fuel

x_4 = price of complement (e.g. *cars*)

x_5 = population

This linear demand function for oil in time period t may be specified in vector format as

$$q^t = \beta x^t = \begin{bmatrix} \beta_0 & \beta_1 & \beta_2 & \beta_3 & \beta_4 & \beta_5 \end{bmatrix} \begin{bmatrix} 1 \\ x_1^t \\ x_2^t \\ x_3^t \\ x_4^t \\ x_5^t \end{bmatrix}$$

Note that, although the are five independent explanatory variables in this economic model, the vector of coefficients β has the order 1×6 because there is also a constant term, β_0. The vector of values of the explanatory variables also has 6 elements and thus takes the order 6×1. However, because it multiplies the constant, the first element in the column remains as 1 even though the values of other elements (i.e. the explanatory variables) may change for different time periods. The actual values of the coefficients β_0, β_1, β_2, etc. will be estimated by a method such as Ordinary Least Squares, which you should come across in your statistics or econometrics modules.

As vector β has the order 1×6 and the vector of values of the explanatory variables x has the order 6×1 then the product matrix βx will have the order 1×1. This means that it will contain the single element q^t which is the predicted output.

Example 15.9

Assume that the demand for oil (in millions of barrels) can be explained by the model $q = \beta x$ and the vector of coefficients of the explanatory variables has been reliably estimated as

$$\beta = \begin{bmatrix} \beta_0 & \beta_1 & \beta_2 & \beta_3 & \beta_4 & \beta_5 \end{bmatrix} = \begin{bmatrix} 4.2 & -0.1 & 0.4 & 0.2 & -0.1 & 0.2 \end{bmatrix}$$

Calculate the demand for oil when the vector of explanatory variables is

$$x = \begin{bmatrix} 1 \\ x_1^t \\ x_2^t \\ x_3^t \\ x_4^t \\ x_5^t \end{bmatrix} = \begin{bmatrix} \text{Constant} \\ \text{Price} \\ \text{Income} \\ \text{Price of substitute} \\ \text{Price of complement} \\ \text{Population (in m.)} \end{bmatrix} = \begin{bmatrix} 1 \\ 30 \\ 18.5 \\ 52 \\ 12.8 \\ 61 \end{bmatrix}$$

Solution

The demand for oil is calculated as

$$\mathbf{q} = \boldsymbol{\beta}\mathbf{x} = \begin{bmatrix} 4.2 & -0.1 & 0.4 & 0.2 & -0.1 & 0.2 \end{bmatrix} \begin{bmatrix} 1 \\ 30 \\ 18.5 \\ 52 \\ 12.8 \\ 61 \end{bmatrix} = [29.92]$$

Thus the answer is 29.92 million barrels.

You can check the calculations for arriving at this answer manually or using Excel.

Test Yourself, Exercise 15.3

1. For each of the pairs of matrices **A** and **B** below use an Excel spreadsheet to find the product matrix **AB**.

    (a) $\mathbf{A} = \begin{bmatrix} 4 & 1 & 3 \\ 9 & 8 & 2 \end{bmatrix}$ and $\mathbf{B} = \begin{bmatrix} 2 & 10 & 2 \\ 5 & 5 & 8 \\ 1.5 & 0 & 1 \end{bmatrix}$

    (b) $\mathbf{A} = \begin{bmatrix} 7 & 10 & 3 \\ 9 & 5 & 2 \\ 4 & 0 & 5 \end{bmatrix}$ and $\mathbf{B} = \begin{bmatrix} 11 & 2.5 & 1 & 4 \\ 5 & 5 & 8 & 0 \\ 3 & 0 & 1 & 4 \end{bmatrix}$

    (c) $\mathbf{A} = \begin{bmatrix} 45 & 34 & 4 & 8 \\ 6 & 7 & 22 & 10 \\ 70 & 3 & 90 & 5 \\ 2 & 2 & 0 & 23 \\ -6 & 5 & 3 & 9 \end{bmatrix}$ and $\mathbf{B} = \begin{bmatrix} 2 & 5 & 3 & 4 & 32 & 65 \\ 9 & 5 & 0 & 0 & 9 & 2 \\ 8 & 46 & 1 & 7 & 85 & 31 \\ 4 & 0 & 20 & 24 & 3 & 8 \end{bmatrix}$

2. The demand for good G depends on a vector of four explanatory variables **x**. There is a linear relationship, including a constant term, between these explanatory variables and g, the amount of good G demanded such that $\mathbf{g} = \boldsymbol{\beta}\mathbf{x}$ where $\boldsymbol{\beta}$ is the vector of coefficients

    $$\boldsymbol{\beta} = \begin{bmatrix} \beta_0 & \beta_1 & \beta_2 & \beta_3 & \beta_4 \end{bmatrix} = \begin{bmatrix} 36 & -0.4 & 0.02 & 1.2 & 0.3 \end{bmatrix}$$

    Calculate the demand for good G when the vector of values of the explanatory variables is

    $$\mathbf{x} = \begin{bmatrix} 1 \\ 14 \\ 8 \\ 82.5 \\ 3.2 \end{bmatrix} \quad \text{where the element } x_1 \text{ refers to the constant}$$

15.4   The matrix inverse and the solution of simultaneous equations

The concept of 'matrix division' is approached in matrix algebra by deriving the inverse of a matrix. One reason for wanting to find a matrix inverse is because it can be used to help solve a set of simultaneous equations specified in matrix format. For example, consider the set of four simultaneous equations:

$$3x_1 + 8x_2 + x_3 + 2x_4 = 96$$

$$20x_1 - 2x_2 + 4x_3 + 0.5x_4 = 69$$

$$11x_1 + 3x_2 + 3x_3 - 5x_4 = 75$$

$$x_1 + 12x_2 + x_3 + 8x_4 = 134$$

These equations can be represented in matrix format by putting:

- the coefficients of the four unknown variables x_1, x_2, x_3 and x_4 into a 4×4 matrix **A**
- the four unknown variables themselves into a 4×1 vector **x**
- the constant terms from the right-hand side of the equations into the 4×1 vector **b**.

These can be written as

$$\mathbf{Ax} = \begin{bmatrix} 3 & 8 & 1 & 2 \\ 20 & -2 & 4 & 0.5 \\ 11 & 3 & 3 & -5 \\ 1 & 12 & 1 & 8 \end{bmatrix} \begin{bmatrix} x_1 \\ x_2 \\ x_3 \\ x_4 \end{bmatrix} = \begin{bmatrix} 96 \\ 69 \\ 75 \\ 134 \end{bmatrix} = \mathbf{b}$$

If this is not immediately obvious to you, try working through the matrix multiplication process to get the product matrix **Ax**. Working across the rows of **A**, each element multiplies the elements down the vector of unknown variables x_1, x_2, x_3 and x_4. If you write out the calculations in full for the four elements of the product matrix **Ax** and equate to the corresponding element in vector **b**, then you should get the same set of simultaneous equations. For example, multiplying the elements across the first row of **A** by the elements down the column vector **x** gives the first element of **Ax** as

$$3x_1 + 8x_2 + x_3 + 2x_4$$

so setting this equal to the first element of the product vector **b**, which is 96, gives us the first of our set of simultaneous equations.

You could of course, solve this set of simultaneous equations by the standard row operations method but there are certain advantages from using the matrix method, as you will find out later on.

The same matrix format as that derived above can be used for the general case. Assume that there are n unknown variables $x_1, x_2, \ldots, x_n$ and n constant values $b_1, b_2, b_3, \ldots, b_n$ such that

$$a_{11}x_1 + a_{12}x_2 + \cdots + a_{1n}x_n = b_1$$

$$a_{21}x_1 + a_{22}x_2 + \cdots + a_{2n}x_n = b_2$$

$$\vdots \qquad \vdots \qquad \qquad \vdots \qquad \vdots$$

$$a_{n1}x_1 + a_{n2}x_2 + \cdots + a_{nn}x_n = b_n$$

This system of n simultaneous equations with n unknowns can be written in matrix format as $\mathbf{Ax} = \mathbf{b}$, where $\mathbf{A}$ is the $n \times n$ matrix of coefficients

$$\mathbf{A} = \begin{bmatrix} a_{11} & a_{12} & \cdots & a_{1n} \\ a_{21} & a_{22} & \cdots & a_{2n} \\ \vdots & \vdots & \vdots & \vdots \\ a_{n1} & a_{n2} & \cdots & a_{nn} \end{bmatrix}$$

and $\mathbf{x}$ is the vector of unknown variables $\quad \mathbf{x} = \begin{bmatrix} x_1 \\ x_2 \\ \vdots \\ x_n \end{bmatrix}$

and $\mathbf{b}$ is the vector of constant parameters $\mathbf{b} = \begin{bmatrix} b_1 \\ b_2 \\ \vdots \\ b_n \end{bmatrix}$

How does this specification of the set of simultaneous equations in the matrix format $\mathbf{Ax} = \mathbf{b}$ help us to solve for the unknown variables in $\mathbf{x}$? If x and A were single terms, instead of vectors and matrices, and $Ax = b$ then basic algebra would suggest that x could be found by simply re-specifying the equation as $x = A^{-1}b$. The same logic is used when $\mathbf{x}$, $\mathbf{A}$ and $\mathbf{b}$ are matrices and we try to find $\mathbf{x} = \mathbf{A}^{-1}\mathbf{b}$.

The derivation of the matrix inverse $\mathbf{A}^{-1}$ is, however, a rather involved procedure and it is explained over the next few sections in this chapter. There is no denying that some students will find it hard work ploughing through the analysis. It is worth it, though, because you will learn:

- How to solve large sets of simultaneous equations in a few seconds by using matrix inversion on a spreadsheet.
- How to use a set of tools that will be invaluable in the analysis of economic models with more than two variables, particularly when checking the second-order conditions of optimization problems.

Conditions for the existence of the matrix inverse

In Chapter 5, it was explained that in a system of linear simultaneous equations the basic rule for a unique solution to exist is that the number of unknowns must equal the number of equations, and linear dependence between equations must not be present. As long as these conditions hold then matrix analysis can be used to solve for any number of unknown variables. Since the number of unknown variables must equal the number of equations the matrix of coefficients $\mathbf{A}$ must be *square*, i.e. the number of rows equals the number of columns. Also, if we know the values for $\mathbf{A}$ and $\mathbf{b}$ and wish to find $\mathbf{x}$ using the formula $\mathbf{x} = \mathbf{A}^{-1}\mathbf{b}$ then we first have to establish whether the inverse matrix $\mathbf{A}^{-1}$ can actually be determined, because in some circumstances it may not exist.

Before we can define what we mean by the inverse of a matrix we need to introduce the concept of the **identity matrix**. This is any square matrix with each element along the diagonal (from top left to bottom right) being equal to 1 and with all other elements being zero. For example, the 3×3 identity matrix is

$$\mathbf{I} = \begin{bmatrix} 1 & 0 & 0 \\ 0 & 1 & 0 \\ 0 & 0 & 1 \end{bmatrix}$$

This identity matrix is the matrix equivalent to the number '1' in standard mathematics. Any matrix multiplied by the identity matrix will give the original matrix. For example

$$\begin{bmatrix} 7 & 2 & 3 \\ 4 & 8 & 1 \\ 5 & 12 & 4 \end{bmatrix} \begin{bmatrix} 1 & 0 & 0 \\ 0 & 1 & 0 \\ 0 & 0 & 1 \end{bmatrix} = \begin{bmatrix} 7 & 2 & 3 \\ 4 & 8 & 1 \\ 5 & 12 & 4 \end{bmatrix}$$

Therefore, a matrix $\mathbf{A}$ can be inverted if there exists an inverse $\mathbf{A}^{-1}$ such that $\mathbf{A}^{-1}\mathbf{A} = \mathbf{I}$, the identity matrix.

Using this definition we can now see that if

$$\mathbf{A}\mathbf{x} = \mathbf{b}$$

multiplying both sides by $\mathbf{A}^{-1}$ gives

$$\mathbf{A}^{-1}\mathbf{A}\mathbf{x} = \mathbf{A}^{-1}\mathbf{b}$$

Since $\mathbf{A}^{-1}\mathbf{A} = \mathbf{I}$ this means that

$$\mathbf{I}\mathbf{x} = \mathbf{A}^{-1}\mathbf{b}$$

As any matrix or vector multiplied by the identity matrix gives the same matrix or vector then

$$\mathbf{x} = \mathbf{A}^{-1}\mathbf{b}$$

There are several instances *when the inverse of a matrix may not exist*:

Firstly, the *zero*, or *null matrix*, which has all its elements equal to zero. There are zero matrices corresponding to each possible order. For example, the 2×2 zero matrix will be

$$\mathbf{0} = \begin{bmatrix} 0 & 0 \\ 0 & 0 \end{bmatrix}$$

Just as it is not possible to determine the inverse of zero in basic arithmetic, the inverse of the zero matrix $\mathbf{0}$ cannot be calculated. However, if we were trying to solve a set of simultaneous equations, we would be unlikely to start of with a matrix of coefficients that were all zero as this would not tell us very much!

Secondly, *linear dependence* of two or more rows (or columns) of a matrix will prevent its inverse being calculated. Linear dependency means that all the terms in one row (or column) are the same scalar multiple of the corresponding elements in another row (or column). The reason for this will become obvious when we have worked through the method for finding the inverse, but we can illustrate the problem with a simple example.

Consider the two simultaneous equations

$$8x + 10y = 120 \tag{1}$$

$$4x + 5y = 60 \tag{2}$$

All the values of (2) are 0.5 of the values in (1). Clearly this pair of simultaneous equations cannot be solved by row operations to find the unknowns x and y. If (2) was multiplied by 2 and subtracted from (1) then we would end up with zero on both sides of the equation, which does not tell us anything. This linear dependency would also lead us down the same dead end if we tried to solve using the matrix inverse.

To actually find the inverse of a matrix, we first need to consider some special concepts associated with *square matrices*, namely:

- The Determinant
- Minors
- Cofactors
- The Adjoint Matrix

These are explained in the following sections.

Test Yourself, Exercise 15.4

1. Identify which of the following sets of simultaneous equations may be suitable for solving by matrix algebra and then put them in appropriate matrix format:

   (a) $5x + 4y + 9z = 95$ (b) $6x + 4y + 8z = 56$
        $2x + y + 4z = 32$ $3x + 2y + 4z = 28$
        $2x + 5y + 4z = 61$ $x - 8y + 2z = 34$

   (c) $5x + 4y + 2z = 95$ (d) $12x + 2y + 3z = 124$
        $9x + 4y = 32$ $6x + 7y + z = 42$
        $2x + 4y + 4z = 61$

2. Which of the following are identity matrices?

   (a) $\begin{bmatrix} 1 & 1 \\ 1 & 1 \end{bmatrix}$ (b) $[1]$ (c) $\begin{bmatrix} 1 & 0 \\ 0 & 1 \end{bmatrix}$ (d) $\begin{bmatrix} 0 & 1 \\ 1 & 0 \end{bmatrix}$ (e) $\begin{bmatrix} 1 & 0 \\ 0 & 1 \\ 1 & 0 \end{bmatrix}$

3. Are there obvious reasons why it may not be possible to derive an inverse for any of the matrices below?

   (a) $\begin{bmatrix} 8 & 6 \\ 3 & 1 \end{bmatrix}$ (b) $\begin{bmatrix} 8 & 1 \\ 4 & 5 \\ 7 & 3 \end{bmatrix}$ (c) $\begin{bmatrix} 4 & 2 \\ 2 & 1 \end{bmatrix}$ (d) $\begin{bmatrix} 9 & 9 \\ 1 & 0 \end{bmatrix}$ (e) $\begin{bmatrix} 5 & 11 & 0 \\ -2 & 4 & 0.2 \\ 0 & -5 & 1 \end{bmatrix}$

15.5 Determinants

For a 2nd order matrix (i.e. order 2×2) the determinant is a number calculated by multiplying the elements in opposite corners and subtracting. The usual notation for a determinant is a set of vertical parallel lines either side of the array of elements, instead of the squared brackets used for a matrix. The determinant of the general 2×2 matrix $\mathbf{A}$, written as $|\mathbf{A}|$, will therefore be:

$$|\mathbf{A}| = \begin{vmatrix} a_{11} & a_{12} \\ a_{21} & a_{22} \end{vmatrix} = a_{11}a_{22} - a_{21}a_{12}$$

Example 15.10

Find the determinant of the matrix $\mathbf{A} = \begin{bmatrix} 5 & 7 \\ 4 & 9 \end{bmatrix}$

Solution

Using the formula defined above, the determinant of matrix $\mathbf{A}$ will be

$$|\mathbf{A}| = \begin{vmatrix} 5 & 7 \\ 4 & 9 \end{vmatrix} = 5 \times 9 - 7 \times 4 = 45 - 28 = 17$$

If any sets of rows or columns of a matrix are *linearly dependent* then the determinant will be zero and we have what is known as a *singular* matrix. For example, if the second row is twice the value of the corresponding element in the first row and

$$\mathbf{A} = \begin{bmatrix} 5 & 8 \\ 10 & 16 \end{bmatrix}$$

then the determinant

$$|\mathbf{A}| = \begin{vmatrix} 5 & 8 \\ 10 & 16 \end{vmatrix} = 5 \times 16 - 8 \times 10 = 80 - 80 = 0$$

The formula for the matrix inverse (which we will derive later) involves division by the determinant. Therefore, a condition for the inverse of a matrix to exist is that the matrix must be *non-singular*, i.e. the determinant must not be zero. This condition applies to determinants of any order.

The determinant of a 3rd order matrix

For the general 3rd order matrix

$$\mathbf{A} = \begin{bmatrix} a_{11} & a_{12} & a_{13} \\ a_{21} & a_{22} & a_{23} \\ a_{31} & a_{32} & a_{33} \end{bmatrix}$$

the determinant $|\mathbf{A}|$ can be calculated as

$$|\mathbf{A}| = a_{11} \begin{vmatrix} a_{22} & a_{23} \\ a_{32} & a_{33} \end{vmatrix} - a_{12} \begin{vmatrix} a_{21} & a_{23} \\ a_{31} & a_{33} \end{vmatrix} + a_{13} \begin{vmatrix} a_{21} & a_{22} \\ a_{31} & a_{32} \end{vmatrix}$$

This entails multiplying each of the elements in the first row by the determinant of the matrix remaining when the corresponding row and column are deleted. For example, the element a_{11} is multiplied by the determinant of the matrix remaining when row 1 and column 1 are deleted from the original 3×3 matrix. If we start from a_{11} then, as we use this method for each element across the row, the sign of each term will be positive and negative alternately. Thus the second term has a negative sign.

Example 15.11

Derive the determinant of matrix $\mathbf{A} = \begin{bmatrix} 4 & 6 & 1 \\ 2 & 5 & 2 \\ 9 & 0 & 4 \end{bmatrix}$

Solution

Expanding across the first row using the above formula, the determinant will be

$$|\mathbf{A}| = 4\begin{vmatrix} 5 & 2 \\ 0 & 4 \end{vmatrix} - 6\begin{vmatrix} 2 & 2 \\ 9 & 4 \end{vmatrix} + \begin{vmatrix} 2 & 5 \\ 9 & 0 \end{vmatrix}$$

$$= 4(20 - 0) - 6(8 - 18) + (0 - 45) = 80 + 60 - 45 = 95$$

Although the determinants of the 3rd order matrices above were found by expanding along the first row, they could also have been found by expanding along any other row or column. The same principle of multiplying each element along the expansion row (or down the expansion column) by the determinant of the matrix remaining when the corresponding row and column are deleted from the original matrix **A** is employed. This can help make the calculations easier if it is possible to expand along a row or column with one or more elements equal to zero, as in the example below.

However, there are rules regarding *the sign of each term*, which must be followed. These are explained for the general case in the next section. For a 3rd order determinant it is sufficient to remember that the first term will be positive if you expand along the 1st or 3rd row or column and the first term will be negative if you expand along the 2nd row or column. The signs of the subsequent terms in the expansion will then alternate.

For example, another way of finding the determinant of the matrix in Example 15.11 above is to expand along the 3rd row, which includes a zero and will therefore require less calculation.

Example 15.11 (reworked)

Derive the determinant of matrix $\mathbf{A} = \begin{bmatrix} 4 & 6 & 1 \\ 2 & 5 & 2 \\ 9 & 0 & 4 \end{bmatrix}$ by expanding along the 3rd row.

Solution

Expanding across the 3rd row, the first term will have a positive sign and so

$$|\mathbf{A}| = 9\begin{vmatrix} 6 & 1 \\ 5 & 2 \end{vmatrix} - 0\begin{vmatrix} 4 & 1 \\ 2 & 2 \end{vmatrix} + 4\begin{vmatrix} 4 & 6 \\ 2 & 5 \end{vmatrix}$$

$$= 9(12 - 5) - 0 + 4(20 - 12) = 63 + 32 = 95$$

Test Yourself, Exercise 15.5

1.   Evaluate the following determinants:

$$|\mathbf{A}| = \begin{vmatrix} 8 & 2 \\ 3 & 1 \end{vmatrix} \qquad |\mathbf{B}| = \begin{vmatrix} 30 & 12 \\ 10 & 4 \end{vmatrix} \qquad |\mathbf{C}| = \begin{vmatrix} 5 & 8 \\ -7 & 0 \end{vmatrix}$$

$$|\mathbf{D}| = \begin{vmatrix} 2 & 5 & 9 \\ 4 & 8 & 3 \\ 1 & 7 & 4 \end{vmatrix} \qquad |\mathbf{E}| = \begin{vmatrix} 4 & 3 & 10 \\ 7 & 0 & 3 \\ 12 & 2 & 5 \end{vmatrix}$$

15.6   Minors, cofactors and the Laplace expansion

The Laplace expansion is a method that can be used to evaluate determinants of any order. Before explaining this method, we need to define a few more concepts (some of which we have actually already started using).

Minors

The minor $|\mathbf{M}_{ij}|$ of matrix $\mathbf{A}$ is the determinant of the matrix left when row i and column j have been deleted.

For example, if the first row and first column are deleted from matrix

$$\mathbf{A} = \begin{bmatrix} a_{11} & a_{12} & a_{13} \\ a_{21} & a_{22} & a_{23} \\ a_{31} & a_{32} & a_{33} \end{bmatrix}$$

the determinant of the remaining matrix will be the minor

$$|\mathbf{M}_{11}| = \begin{vmatrix} a_{22} & a_{23} \\ a_{32} & a_{33} \end{vmatrix}$$

Example 15.12

Find the minor $|\mathbf{M}_{31}|$ of the matrix $\qquad \mathbf{A} = \begin{bmatrix} 8 & 2 & 3 \\ 1 & 9 & 4 \\ 4 & 3 & 6 \end{bmatrix}$

Solution

The minor $|\mathbf{M}_{31}|$ is the determinant of the matrix remaining when the 3rd row and 1st column have been eliminated from matrix $\mathbf{A}$. Therefore

$$|\mathbf{M}_{31}| = \begin{vmatrix} 2 & 3 \\ 9 & 4 \end{vmatrix} = 8 - 27 = -19$$

Using this definition of a minor, the formula for the determinant of a 3rd order matrix expanded across the first row could specified as

$$|\mathbf{A}| = a_{11}|\mathbf{M}_{11}| - a_{12}|\mathbf{M}_{12}| + a_{13}|\mathbf{M}_{13}|$$

Cofactors

A cofactor is the same as a minor, except that its sign is determined by the row and column that it corresponds to. The sign of cofactor $|\mathbf{C}_{ij}|$ is equal to $(-1)^{i+j}$. Thus if the row number and column number sum to an odd number, the sign will be negative. For example, to derive the cofactor $|\mathbf{C}_{12}|$ for the general 3rd order matrix $\mathbf{A}$ we eliminate the 1st row and the 2nd column and then, since $i + j = 3$, we multiply the determinant of the elements that remain by $(-1)^3$. Therefore

$$|\mathbf{C}_{12}| = (-1)^3 \begin{vmatrix} a_{21} & a_{23} \\ a_{31} & a_{33} \end{vmatrix} = (-1) \begin{vmatrix} a_{21} & a_{23} \\ a_{31} & a_{33} \end{vmatrix}$$

Example 15.13

Find the cofactor $|\mathbf{C}_{22}|$ of the matrix $\mathbf{A} = \begin{bmatrix} 8 & 2 & 3 \\ 1 & 9 & 4 \\ 4 & 3 & 6 \end{bmatrix}$

Solution

The cofactor $|\mathbf{C}_{22}|$ is the determinant of the matrix remaining when the 2nd row and 2nd column have been eliminated. It will have the sign $(-1)^4$ since $i + j = 4$. The solution is therefore

$$|\mathbf{C}_{22}| = (-1)^4 \begin{vmatrix} 8 & 3 \\ 4 & 6 \end{vmatrix} = (+1)(48 - 12) = 36$$

The determinant of a 3rd order matrix in terms of its cofactors, expanded across the first row, can now be specified as

$$|\mathbf{A}| = a_{11}|\mathbf{C}_{11}| + a_{12}|\mathbf{C}_{12}| + a_{13}|\mathbf{C}_{13}| \tag{1}$$

Although this looks very similar to the formula for $|\mathbf{A}|$ in terms of its minors, set out above, you should note that the sign of the second term is positive. This is because the cofactor itself will have a negative sign.

The Laplace expansion

For matrices of any order n, using the Laplace expansion, the determinant can specified as

$$|\mathbf{A}| = \sum_{i,j=1}^{i,j=n} a_{ij} |\mathbf{C}_{ij}|$$

where the summation from 1 to n can take place across the rows (i) or down the columns (j). If you check the formula (1) above for determinant of a 3rd order matrix in terms of its cofactors, you will see that this employs the Laplace expansion.

If the original matrix is 4th order or greater, then the first set of cofactors derived by using the Laplace expansion will themselves be 3rd order or greater. Therefore, the Laplace expansion has to be used again to break these cofactors down. This process needs to continue until the determinant is specified in terms of 2nd order cofactors which can then be evaluated.

With larger determinants this method can involve quite a lot of calculations and so it is usually quicker to use a spreadsheet for numerical examples. But first let us work through an example by doing the calculations manually to make sure that you understand how this method works.

Example 15.14

Use the Laplace expansion to find the determinant of matrix $\mathbf{A} = \begin{bmatrix} 8 & 10 & 2 & 3 \\ 0 & 5 & 7 & 10 \\ 2 & 2 & 1 & 4 \\ 3 & 4 & 4 & 0 \end{bmatrix}$

Solution

Expanding down the first column (because there is a zero which means one less set of calculations), the first round of the Laplace expansion gives

$$|\mathbf{A}| = 8 \begin{vmatrix} 5 & 7 & 10 \\ 2 & 1 & 4 \\ 4 & 4 & 0 \end{vmatrix} - 0 \begin{vmatrix} 10 & 2 & 3 \\ 2 & 1 & 4 \\ 4 & 4 & 0 \end{vmatrix} + 2 \begin{vmatrix} 10 & 2 & 3 \\ 5 & 7 & 10 \\ 4 & 4 & 0 \end{vmatrix} - 3 \begin{vmatrix} 10 & 2 & 3 \\ 5 & 7 & 10 \\ 2 & 1 & 4 \end{vmatrix}$$

A second round of the Laplace expansion is then used to break these 3rd order cofactors down into 2nd order cofactors that can be evaluated. The second term is zero and disappears and so this gives

$$|\mathbf{A}| = 8 \left(5 \begin{vmatrix} 1 & 4 \\ 4 & 0 \end{vmatrix} - 2 \begin{vmatrix} 7 & 10 \\ 4 & 0 \end{vmatrix} + 4 \begin{vmatrix} 7 & 10 \\ 1 & 4 \end{vmatrix} \right) + 2 \left(10 \begin{vmatrix} 7 & 10 \\ 4 & 0 \end{vmatrix} - 5 \begin{vmatrix} 2 & 3 \\ 4 & 0 \end{vmatrix} + 4 \begin{vmatrix} 2 & 3 \\ 7 & 10 \end{vmatrix} \right)$$

$$- 3 \left(10 \begin{vmatrix} 7 & 10 \\ 1 & 4 \end{vmatrix} - 5 \begin{vmatrix} 2 & 3 \\ 1 & 4 \end{vmatrix} + 2 \begin{vmatrix} 2 & 3 \\ 7 & 10 \end{vmatrix} \right)$$

$$= 8[5(-16) - 2(-40) + 4(18)] + 2[10(-40) - 5(-12) + 4(-1)]$$

$$\quad - 3[10(18) - 5(5) + 2(-1)]$$

$$= 8[-80 + 80 + 72] + 2[-400 + 60 - 4] - 3[180 - 25 - 2]$$

$$= 8(72) + 2(-344) - 3(153)$$

$$= 576 - 688 - 459$$

$$= -571$$

Using Excel to evaluate determinants

It is very straightforward to use the Excel function MDETERM to evaluate determinants. Just type in the matrix that you want the determinant for and then, in the cell where you want the value of the determinant to appear, enter

=MDETERM (*cell range for matrix*)

The range can either be marked out by holding the left mouse key down after you have typed the first bracket in the formula (and will be enclosed by dotted lines) or you can just type in the cell range. For example, if you had entered the 4 × 4 matrix from Example 15.14 above in cells B2 to E5 and you wanted the determinant to appear in cell G2 you would type = MDETERM (B2:E5) in cell G2.

Test Yourself, Exercise 15.6

1. For the matrix $\mathbf{A} = \begin{bmatrix} 5 & 0 & 4 \\ 8 & 3 & 6 \\ 2 & 7 & 1 \end{bmatrix}$ evaluate the following minors and cofactors:

   (a) $|\mathbf{M}_{11}|$ (b) $|\mathbf{M}_{33}|$ (c) $|\mathbf{M}_{12}|$ (d) $|\mathbf{C}_{21}|$ (e) $|\mathbf{C}_{13}|$ (f) $|\mathbf{C}_{12}|$

2. Manually calculate the values of the determinants of following matrices and then check your answers using Excel:

$$\mathbf{A} = \begin{bmatrix} 2 & 6 & 2 & 3 \\ 10 & 5 & 7 & 25 \\ 0 & 2 & 1 & 5 \\ 4 & -3 & 4 & 9 \end{bmatrix} \quad \mathbf{B} = \begin{bmatrix} 8 & 6 & 2 & 1 \\ 3 & 8 & 7 & -4 \\ 0 & -2 & 1 & 5 \\ 4 & 3 & 3 & 2 \end{bmatrix} \quad \mathbf{C} = \begin{bmatrix} 1 & 5 & 2 & 1 & 0 \\ 6 & 1 & 0 & -4 & 3 \\ 0 & 4 & 7 & 2 & 1 \\ 9 & 2 & 3 & 2 & 2 \\ 0 & 4 & 8 & 0 & 6 \end{bmatrix}$$

15.7 The transpose matrix, the cofactor matrix, the adjoint and the matrix inverse formula

There are still a few more concepts that are needed before we can determine the inverse of a matrix.

The transpose of a matrix

To get the transpose of a matrix, usually written as $\mathbf{A}^\mathrm{T}$, the rows and columns are swapped around, i.e. row 1 becomes column 1 and column 1 becomes row 1, etc. If a matrix is not square then the numbers of rows and columns will alter when it is transposed.

For example, if $\mathbf{A} = \begin{bmatrix} 5 & 20 \\ 16 & 9 \\ 12 & 6 \end{bmatrix}$ then $\mathbf{A}^\mathrm{T} = \begin{bmatrix} 5 & 16 & 12 \\ 20 & 9 & 6 \end{bmatrix}$

The cofactor matrix

If we replace every element in a matrix by its corresponding cofactor then we get the *cofactor matrix*, usually denoted by **C**.

$$\text{For example if } \mathbf{A} = \begin{bmatrix} 2 & 4 & 3 \\ 3 & 5 & 0 \\ 4 & 2 & 5 \end{bmatrix} \quad \text{then } \mathbf{C} = \begin{bmatrix} 25 & -15 & -12 \\ -14 & -2 & 12 \\ -15 & 9 & -2 \end{bmatrix}$$

To make sure you understand how these numbers were calculated, let us work through some of them. The cofactor $|C_{ij}|$ of matrix **A** is the determinant of the matrix remaining when row i and column j have been eliminated, with the sign $(-1)^{i+j}$. Thus, some selected elements of the cofactor matrix are

$$c_{11} = |C_{11}| = (-1)^{(1+1)} \begin{vmatrix} a_{22} & a_{23} \\ a_{32} & a_{33} \end{vmatrix} = (-1)^2 \begin{vmatrix} 5 & 0 \\ 2 & 5 \end{vmatrix} = (25 - 0) = 25$$

$$c_{21} = |C_{21}| = (-1)^{(2+1)} \begin{vmatrix} a_{12} & a_{13} \\ a_{32} & a_{33} \end{vmatrix} = (-1)^3 \begin{vmatrix} 4 & 3 \\ 2 & 5 \end{vmatrix} = (-1)(20 - 6) = -14$$

Check for yourself the calculation of some of the other elements of **C**.

The adjoint matrix

The adjoint matrix, usually denoted by **AdjA**, is the transpose of the cofactor matrix,

$$\text{Thus if } \mathbf{A} = \begin{bmatrix} a_{11} & a_{12} & a_{13} \\ a_{21} & a_{22} & a_{23} \\ a_{31} & a_{32} & a_{33} \end{bmatrix} \quad \text{then } \mathbf{AdjA} = \begin{bmatrix} |C_{11}| & |C_{21}| & |C_{31}| \\ |C_{12}| & |C_{22}| & |C_{32}| \\ |C_{13}| & |C_{23}| & |C_{33}| \end{bmatrix}$$

Using the cofactor example above, we have already shown that for

$$\text{matrix } \mathbf{A} = \begin{bmatrix} 2 & 4 & 3 \\ 3 & 5 & 0 \\ 4 & 2 & 5 \end{bmatrix} \text{ the cofactor matrix is } \mathbf{C} = \begin{bmatrix} 25 & -15 & -12 \\ -14 & -2 & 12 \\ -15 & 9 & -2 \end{bmatrix}$$

$$\text{Therefore the adjoint matrix will be } \mathbf{AdjA} = \mathbf{C}^{\mathrm{T}} = \begin{bmatrix} 25 & -14 & -15 \\ -15 & -2 & 9 \\ -12 & 12 & -2 \end{bmatrix}$$

The inverse matrix

The formula for $\mathbf{A}^{-1}$, the inverse of matrix **A**, can now be stated as

$$\mathbf{A}^{-1} = \frac{\mathbf{AdjA}}{|\mathbf{A}|}$$

as long as the determinant $|\mathbf{A}|$ is non-singular, i.e. it must not be zero.

Example 15.15

$$\text{Find the inverse matrix } \mathbf{A}^{-1} \text{ for matrix } \mathbf{A} = \begin{bmatrix} 2 & 4 & 3 \\ 3 & 5 & 0 \\ 4 & 2 & 5 \end{bmatrix}$$

Solution

We have already determined the adjoint for this particular matrix in the example above. Its determinant $|\mathbf{A}|$ can be evaluated by expanding down the 3rd column as

$$|\mathbf{A}| = 3\begin{vmatrix} 3 & 5 \\ 4 & 2 \end{vmatrix} - 0 + 5\begin{vmatrix} 2 & 4 \\ 3 & 5 \end{vmatrix}$$

$$= 3(6 - 20) + 5(10 - 12)$$

$$= 3(-14) + 5(-2)$$

$$= -42 - 10 = -52$$

Therefore, given that we already know that $\mathbf{AdjA} = \begin{bmatrix} 25 & -14 & -15 \\ -15 & -2 & 9 \\ -12 & 12 & -2 \end{bmatrix}$

the inverse matrix will be

$$\mathbf{A}^{-1} = \frac{\mathbf{AdjA}}{|\mathbf{A}|} = \frac{\begin{bmatrix} 25 & -14 & -15 \\ -15 & -2 & 9 \\ -12 & 12 & -2 \end{bmatrix}}{-52} = \begin{bmatrix} -0.48 & 0.27 & 0.29 \\ 0.29 & 0.04 & -0.17 \\ 0.27 & -0.23 & 0.04 \end{bmatrix}$$

The derivation of this matrix inverse has been quite long and time-consuming, but you need to understand this underlying method before learning how to do the calculations on a spreadsheet. However, first let us work through another example from first principles to make sure that you understand each stage of the analysis. This time we will start with a 2×2 matrix.

Example 15.16

Find the inverse matrix $\mathbf{A}^{-1}$ for matrix $\mathbf{A} = \begin{bmatrix} 20 & 5 \\ 6 & 2 \end{bmatrix}$

Solution

Because there are only four elements, the cofactor corresponding to each element of $\mathbf{A}$ will just be the element in the opposite corner, with the sign $(-1)^{i+j}$. Therefore, the corresponding cofactor matrix will be

$$\mathbf{C} = \begin{bmatrix} 2 & -6 \\ -5 & 20 \end{bmatrix}$$

The adjoint is the transpose of the cofactor matrix and so

$$\mathbf{AdjA} = \begin{bmatrix} 2 & -5 \\ -6 & 20 \end{bmatrix}$$

The determinant of the original matrix $\mathbf{A}$ is easily calculated as

$$|\mathbf{A}| = 20 \times 2 - 5 \times 6 = 40 - 30 = 10$$

The inverse matrix is thus

$$\mathbf{A}^{-1} = \frac{\mathbf{AdjA}}{|\mathbf{A}|} = \frac{\begin{bmatrix} 2 & -5 \\ -6 & 20 \end{bmatrix}}{10} = \begin{bmatrix} 0.2 & -0.5 \\ -0.6 & 2 \end{bmatrix}$$

Derivation of the matrix inverse formula

You can just take the above formula for the matrix inverse as given and there is no need for you to work through the proof of this result for the general case. However, we can show how the inverse formula can be derived for the case of a 2×2 matrix.

Assume that we wish to invert the matrix $\mathbf{A} = \begin{bmatrix} a & b \\ c & d \end{bmatrix}$

This inverse can be specified as $\mathbf{A}^{-1} = \begin{bmatrix} e & f \\ g & h \end{bmatrix}$

where e, f, g and h are numbers that the inverse formula will calculate.

Multiplying a square non-singular matrix by its inverse will give the identity matrix. Thus

$$\mathbf{AA}^{-1} = \begin{bmatrix} a & b \\ c & d \end{bmatrix}\begin{bmatrix} e & f \\ g & h \end{bmatrix} = \begin{bmatrix} ae + bg & af + bh \\ ce + dg & cf + dh \end{bmatrix} = \begin{bmatrix} 1 & 0 \\ 0 & 1 \end{bmatrix} = \mathbf{I}$$

From the calculations for each of the elements of $\mathbf{I}$ we get the four simultaneous equations

$$ae + bg = 1 \tag{1}$$

$$af + bh = 0 \tag{2}$$

$$ce + dg = 0 \tag{3}$$

$$cf + dh = 1 \tag{4}$$

The values of the elements of the inverse matrix e, f, g and h in terms of the values of the elements of the original matrix can now be solved by the substitution method.

From (1)

$$ae = 1 - bg$$

and so

$$e = \frac{(1 - bg)}{a} \tag{5}$$

Substituting the result (5) into (3) gives

$$\frac{c(1 - bg)}{a} + dg = 0$$

$$c - cbg + dga = 0$$

$$g(ad - bc) = -c$$

$$g = \frac{-c}{ad - bc} \tag{6}$$

Substituting the expression for g in (6) into (5) gives

$$e = \left(1 - \frac{b(-c)}{ad - bc}\right)\frac{1}{a} = \left(\frac{ad - bc + bc}{ad - bc}\right)\frac{1}{a} = \left(\frac{ad}{ad - bc}\right)\frac{1}{a} = \frac{d}{ad - bc} \qquad (7)$$

Using the same substitution method, you can check for yourself that the other two elements of the inverse matrix will be

$$f = \frac{-b}{ad - bc} \qquad (8)$$

and

$$h = \frac{a}{ad - bc} \qquad (9)$$

Since the values for e, f, g and h that are derived in (6), (7), (8) and (9) all contain the same term $\frac{1}{ad-bc}$ this can be written as a scalar multiplier so that

$$\mathbf{A}^{-1} = \begin{bmatrix} e & f \\ g & h \end{bmatrix} = \frac{1}{ad - bc}\begin{bmatrix} d & -b \\ -c & a \end{bmatrix} \qquad (10)$$

This checks out with the general inverse formula since for matrix $\mathbf{A} = \begin{bmatrix} a & b \\ c & d \end{bmatrix}$

The determinant is $|\mathbf{A}| = ad - bc$

The cofactor matrix $\mathbf{C} = \begin{bmatrix} d & -c \\ -b & a \end{bmatrix}$ and so the adjoint is $\mathbf{AdjA} = \begin{bmatrix} d & -b \\ -c & a \end{bmatrix}$

Substituting these results into (10) gives the inverse formula

$$\mathbf{A}^{-1} = \frac{\mathbf{AdjA}}{|\mathbf{A}|}$$

Using Excel for matrix inversion

Although you need to understand the rationale behind the matrix inversion process, for any actual computations involving a 3rd order or larger matrix, it is quicker to use Excel rather than do the calculations manually.

To invert a matrix using the Excel MINVERSE formula:

- Enter the matrix that you wish to invert.
- Highlight cells where inverted matrix will go (same dimension as original matrix).
- Enter in formula bar at top of screen:

  =MINVERSE (*cell range of matrix to be inverted*)

  instead of entering actual cell references for the matrix to be inverted, if you prefer you can use the mouse to mark out the matrix to be inverted. Do this after you have typed in the left bracket in the formula bar and then, when the required area is enclosed within the dotted lines, type in the right bracket.
- Hold down the Ctrl <u>and</u> Shift keys together and press ENTER. The programme will put curved brackets { } round the formula automatically and the inverted matrix should be calculated in the cells that you have chosen.

Test Yourself, Exercise 15.7

1. Derive the inverse matrix $\mathbf{A}^{-1}$ when $\mathbf{A} = \begin{bmatrix} 25 & 15 \\ 10 & 8 \end{bmatrix}$

2. For the matrix $\mathbf{A} = \begin{bmatrix} 5 & 0 & 2 \\ 3 & 4 & 5 \\ 2 & 1 & 2 \end{bmatrix}$ derive the cofactor matrix $\mathbf{C}$, the adjoint matrix $\mathbf{AdjA}$ and the inverse matrix $\mathbf{A}^{-1}$ by manual calculation.

3. Use Excel to derive the matrix inverse $\mathbf{A}^{-1}$ for
$$\mathbf{A} = \begin{bmatrix} 4 & 6 & 2 & 3 \\ 10 & 5 & 7 & 20 \\ 0 & 2 & 1 & 5 \\ 4 & -3 & 4 & 12 \end{bmatrix}$$

15.8   Application of the matrix inverse to the solution of linear simultaneous equations

Although small sets of linear equations can be solved by other algebraic techniques, e.g. row operations, we will work through a simple example here to illustrate how the matrix method works before explaining how larger sets of linear equations can be solved using Excel.

Example 15.17

Use matrix algebra to solve for the unknown variables x_1, x_2 and x_3 given that

$$10x_1 + 3x_2 + 6x_3 = 76$$
$$4x_1 \qquad + 5x_3 = 41$$
$$5x_1 + 2x_2 + 2x_3 = 34$$

Solution

This set of simultaneous equations can be set up in matrix format as $\mathbf{Ax} = \mathbf{b}$ where

$$\mathbf{Ax} = \begin{bmatrix} 10 & 3 & 6 \\ 4 & 0 & 5 \\ 5 & 2 & 2 \end{bmatrix} \begin{bmatrix} x_1 \\ x_2 \\ x_3 \end{bmatrix} = \begin{bmatrix} 76 \\ 41 \\ 34 \end{bmatrix} = \mathbf{b}$$

To derive the vector of unknowns $\mathbf{x}$ using the matrix formulation $\mathbf{x} = \mathbf{A}^{-1}\mathbf{b}$ we first have to derive the matrix inverse $\mathbf{A}^{-1}$. The first step is to derive the cofactor matrix, which will be

$$\mathbf{C} = \begin{bmatrix} (0-10) & -(8-25) & (8-0) \\ -(6-12) & (20-30) & -(20-15) \\ (15-0) & -(50-24) & (0-12) \end{bmatrix} = \begin{bmatrix} -10 & 17 & 8 \\ 6 & -10 & -5 \\ 15 & -26 & -12 \end{bmatrix}$$

The adjoint matrix will be the transpose of the cofactor matrix and so

$$\mathbf{AdjA} = \mathbf{C}^{\mathrm{T}} = \begin{bmatrix} -10 & 6 & 15 \\ 17 & -10 & -26 \\ 8 & -5 & -12 \end{bmatrix}$$

The determinant of $\mathbf{A}$, expanding along the second row, will be

$$|\mathbf{A}| = \begin{vmatrix} 10 & 3 & 6 \\ 4 & 0 & 5 \\ 5 & 2 & 2 \end{vmatrix} = -4(6-12)+0-5(20-15) = 24-25 = -1$$

The matrix inverse will therefore be

$$\mathbf{A}^{-1} = \frac{\mathbf{AdjA}}{|\mathbf{A}|} = \frac{\begin{bmatrix} -10 & 6 & 15 \\ 17 & 10 & -26 \\ 8 & -5 & -12 \end{bmatrix}}{-1} = \begin{bmatrix} 10 & -6 & -15 \\ -17 & 10 & 26 \\ -8 & 5 & 12 \end{bmatrix}$$

To solve for the vector of unknowns $\mathbf{x}$ we calculate

$$\mathbf{x} = \mathbf{A}^{-1}\mathbf{b} = \begin{bmatrix} 10 & -6 & -15 \\ -17 & 10 & 26 \\ -8 & 5 & 12 \end{bmatrix}\begin{bmatrix} 76 \\ 41 \\ 34 \end{bmatrix} = \begin{bmatrix} (10\times76)-(6\times41)-(15\times34) \\ (-17\times76)+(10\times41)+(26\times34) \\ (-8\times76)+(5\times41)+(12\times34) \end{bmatrix}$$

$$= \begin{bmatrix} 760-246-510 \\ -1292+410+884 \\ -608+205+408 \end{bmatrix} = \begin{bmatrix} 4 \\ 2 \\ 5 \end{bmatrix} = \begin{bmatrix} x_1 \\ x_2 \\ x_3 \end{bmatrix}$$

You can check that these are the correct values by substituting them for the unknown variables x_1, x_2 and x_3 in the equations given in this problem. For example, substituting into the first equation gives

$$10x_1 + 3x_2 + 6x_3 = 10(4) + 3(2) + 6(5) = 40 + 6 + 30 = 76$$

Using Excel to solve simultaneous equations

A promise was made that if you worked through all this matrix inversion analysis then you would learn how to solve a large set of simultaneous linear equations in a few seconds. Now it's payback time. The example below shows how to solve a set of six simultaneous equations with six unknown variables using Excel. Once you have worked through this example and understood what is involved, it should take you less than a minute to solve similar examples using Excel to do the necessary matrix inversion and multiplication.

Example 15.18

Solve for the unknown variables x_1, x_2, x_3, x_4, x_5 and x_6 given that

$$
\begin{aligned}
4x_1 + x_2 + 2x_3 - 17x_4 - 5x_5 + 8x_6 &= 21 \\
8x_1 + 9x_2 + 23x_3 + 15x_4 + 11x_5 + 39x_6 &= 593 \\
24x_1 + 41x_2 + 9x_3 + 3x_4 \qquad\qquad + x_6 &= 317 \\
6x_1 + 5x_2 \qquad\quad - x_4 + 3x_5 - 7x_6 &= 35 \\
9x_1 + 11x_2 + 39x_3 + 23x_4 + 15x_5 \qquad\quad &= 678 \\
28x_1 + 49x_2 + 4x_3 + 5x_4 + 9x_5 + 7x_6 &= 391
\end{aligned}
$$

Solution

Enter the matrix of coefficients **A** and the vector of constant values **b** into Excel, as shown in Table 15.5. In this table the cells (A3:F8) are used for the **A** matrix and the **b** column vector is in cells (H3:H8) and so the rest of the instructions below use these cell references.

Create the inverse matrix $\mathbf{A}^{-1}$ by highlighting a 6×6 block of cells (A10:F15) and type in the formula =MINVERSE(A3:F8) making sure both the Cntrl and Shift keys are held down when this is entered.

To derive the vector of unknowns **x** by finding the product matrix $\mathbf{A}^{-1}\mathbf{b}$, highlight a 6×1 column of cells (H10:H15) and then type =MMULT(A10:F15, H3:H8) in the formula bar and hold down the Cntrl *and* Shift keys when you hit the return key.

The vector of unknown variables should be calculated in the six cells of this column. You can now just read off the solution values $x_1 = 5$, $x_2 = 2$, $x_3 = 12$, $x_4 = 1$, $x_5 = 8$ and $x_6 = 4$.

Note that most numbers in this table have been rounded to 5 dp. However, as this would have rounded some very small numbers down to zero they have been left in the exponent format displayed in Excel. For example the number $-1\text{E} - 17$ is -1 divided by 10^{17}.

Table 15.5

	A	B	C	D	E	F	G	H
1	Example 15.17							
2	A MATRIX							b
3	4	1	2	-17	-5	8		21
4	8	9	23	15	11	39		593
5	24	41	9	3	0	1		317
6	6	5	0	-1	3	-7		35
7	9	11	39	23	15	0		678
8	28	49	4	5	9	7		391
9	Inverse A^-1						A^-1*b =	x
10	-0.0453	0.08783	0.11969	0.32077	-0.0634	-0.1339	solution	5
11	0.02431	-0.0509	-0.0504	-0.1805	0.03194	0.08268	values	2
12	0.03398	-0.0162	-1E-17	-0.0512	0.03343	-1E-17		12
13	-0.0723	0.03416	0.06457	0.05788	-0.0253	-0.0591		1
14	0.03184	-0.0257	-0.1339	-0.0156	0.0331	0.11024		8
15	0.00247	0.02302	-5E-18	-0.0118	-0.0137	4.5E-18		4

Estimating the parameters of an economic model

One important use of the matrix method of solution for a set of unknown variables is in econometrics, where estimates of the parameters of an economic model are derived using observations of different values of the variables in the model. Normally, relatively large data sets are used to estimate parameters, and a stochastic (random) error term has to be allowed for. However, to explain the basic principles involved we will work with only three observations and assume no error term. This should help you to understand the more sophisticated models that you will encounter if you go on to study intermediate econometric analysis of multi-variable models.

Assume that y is a linear function of three exogenous variables x_1, x_2 and x_3 so that

$$y_i = \beta_1 x_{1i} + \beta_2 x_{2i} + \beta_3 x_{3i}$$

where the subscript i denotes the observation number and β_1, β_2 and β_3 are the parameters whose values we wish to find. There are three observations, which give the values shown below:

Observation number	y	x_1	x_2	x_3
1	240	10	12	20
2	150	5	8	15
3	300	12	18	20

How can these observations be used to estimate the parameters β_1, β_2, and β_3?

If the function $y_i = \beta_1 x_{1i} + \beta_2 x_{2i} + \beta_3 x_{3i}$ holds for all three observations (i.e. all three values of i) then there will be three simultaneous equations

$$240 = \beta_1 10 + \beta_2 12 + \beta_3 20 \tag{1}$$

$$150 = \beta_1 5 + \beta_2 8 + \beta_3 15 \tag{2}$$

$$300 = \beta_1 12 + \beta_2 18 + \beta_3 20 \tag{3}$$

These can be written in matrix format as $\mathbf{y} = \mathbf{X}\boldsymbol{\beta}$

where $\mathbf{y} = \begin{bmatrix} 240 \\ 150 \\ 300 \end{bmatrix}$, $\quad \mathbf{X} = \begin{bmatrix} 10 & 12 & 20 \\ 5 & 8 & 15 \\ 12 & 18 & 20 \end{bmatrix}$ $\quad$ and $\boldsymbol{\beta} = \begin{bmatrix} \beta_1 \\ \beta_2 \\ \beta_3 \end{bmatrix}$

Since $\qquad\qquad\qquad\qquad\qquad\qquad\qquad \mathbf{X}\boldsymbol{\beta} = \mathbf{y}$

multiplying both sides by inverse $\mathbf{X}^{-1}$ gives $\qquad \mathbf{X}^{-1}\mathbf{X}\boldsymbol{\beta} = \mathbf{X}^{-1}\mathbf{y}$

A matrix times its inverse gives the identity matrix. Thus $\quad \mathbf{I}\boldsymbol{\beta} = \mathbf{X}^{-1}\mathbf{y}$

and so the vector of parameters $\boldsymbol{\beta}$ will be $\qquad\qquad\quad \boldsymbol{\beta} = \mathbf{X}^{-1}\mathbf{y}$

Although we could now finish the calculations using Excel we will continue working through this problem manually. Note that the notation is different from that used in the previous section because we are trying to find the values of the coefficients rather than the values of the variables x_1, x_2 and x_3, which are already given in matrix $\mathbf{X}$. To find the matrix inverse $\mathbf{X}^{-1}$ we first find the cofactor matrix

$$\mathbf{C} = \begin{bmatrix} 160 - 270 & -(100 - 180) & 90 - 96 \\ -(240 - 360) & 200 - 240 & -(180 - 144) \\ 180 - 160 & -(150 - 100) & 80 - 60 \end{bmatrix} = \begin{bmatrix} -110 & 80 & -6 \\ 120 & -40 & 36 \\ 20 & -50 & 20 \end{bmatrix}$$

The adjoint matrix will then be the transpose of this cofactor matrix

$$\mathbf{AdjX} = \mathbf{C}^{\mathrm{T}} = \begin{bmatrix} -110 & 120 & 20 \\ 80 & -40 & -50 \\ -6 & 36 & 20 \end{bmatrix}$$

The determinant of matrix $\mathbf{X}$ can be calculated as

$$|\mathbf{X}| = 10(160 - 270) - 12(100 - 180) + 20(90 - 96) = -260$$

Inserting these values into the formula for the inverse matrix gives

$$\mathbf{X}^{-1} = \frac{\mathbf{AdjX}}{|\mathbf{X}|} = \frac{\begin{bmatrix} -110 & 120 & 20 \\ 80 & -40 & -50 \\ -6 & 36 & 20 \end{bmatrix}}{-260} = \begin{bmatrix} 0.42 & 0.46 & -0.08 \\ -0.3 & 0.15 & 0.19 \\ 0.02 & -0.14 & -0.08 \end{bmatrix}$$

Therefore the vector of coefficients is

$$\boldsymbol{\beta} = \mathbf{X}^{-1}\mathbf{y} = \begin{bmatrix} 0.42 & 0.46 & -0.08 \\ -0.3 & 0.15 & 0.19 \\ 0.02 & -0.14 & -0.08 \end{bmatrix} \begin{bmatrix} 240 \\ 150 \\ 300 \end{bmatrix} = \begin{bmatrix} 9.23 \\ 6.92 \\ 3.23 \end{bmatrix} = \begin{bmatrix} \beta_1 \\ \beta_2 \\ \beta_3 \end{bmatrix}$$

To check the parameters in vector $\boldsymbol{\beta}$ have been calculated correctly, we can insert the values computed above into the first of the set of three simultaneous equations in this example. Thus, from equation (1)

$$y_1 = \beta_1 10 + \beta_2 12 + \beta_3 20 = 9.23(10) + 6.92(12) + 3.23(20) = 240$$

and so the calculated value of 240 for y_1 is correct (allowing for rounding error).

Test Yourself, Exercise 15.8

(You can solve questions 1 and 2 manually but Excel should be used for the others.)

1. Use the matrix inverse method to find the unknowns x and y when

   $$4x + 6y = 68$$
   $$5x + 20y = 185$$

2. Use matrix algebra to solve for x_1, x_2 and x_3 given that

   $$3x_1 + 4x_2 + 3x_3 = 60$$
   $$4x_1 + 10x_2 + 2x_3 = 104$$
   $$4x_1 + 2x_2 + 4x_3 = 60$$

3. Assume that demand for good (Q) depends on its own price (P), income (M) and the price of a substitute good (S) according to the demand function

   $$Q_i = \beta_1 P_i + \beta_2 M_i + \beta_3 S_i$$

where β_1, β_2, and β_3 are parameters whose values are not yet known and the subscript i denotes the observation number. Three observations of the amount Q demanded when P, M and S take on different values are shown below:

Observation number	P	M	S	Q
1	6	5	5	4
2	8	8	6	6.4
3	5	6	4	5.1

Find the values of β_1, β_2, and β_3 by setting up the relevant system of simultaneous equations in matrix format and solving using the inverse matrix.

Use the vector of parameters $\boldsymbol{\beta}$ that you have found to predict the value of Q when P, M and S take the values 7, 9 and 10, respectively, by vector multiplication.

4. Assume that the quantity demanded of oil (Q) depends on its own price (P), income (M), the price of the substitute fuel gas (G), the price of the complement good cars (C), population (N) and average temperature (T) according to the demand function

$$Q_i = \beta_1 P_i + \beta_2 M_i + \beta_3 G_i + \beta_4 C_i + \beta_5 N_i + \beta_6 T_i$$

where β_1, β_2, β_3, β_4, β_5 and β_6 are parameters whose value is not yet known and the subscript i denotes the observation number for Q and the explanatory variables.

Six observations of Q when P, M, G, C, N and T take on different values are:

Observation number	P	M	G	C	N	T	Q
1	15	80	12.5	5	4,000	18	6.980
2	20	95	14	8	4,100	17.4	6.919
3	28	108	11	6	4,150	19.2	4.522
4	35	112	16.2	7.5	4,230	18.3	4.659
5	36	110	16	8	4,215	18.9	4.082
6	30	103	14.5	5.8	4,220	19.2	4.981

Find the values of parameters β_1, β_2, β_3, β_4, β_5 and β_6 by setting up the relevant system of simultaneous equations in matrix format and solving using the inverse matrix.

Employing the vector of parameters $\boldsymbol{\beta}$ that you have found, use vector multiplication to predict the value of Q if the explanatory variables take the values shown below

P	M	G	C	N	T
41	148	23	8.2	4,890	21.2

15.9 Cramer's rule

Cramer's rule is another method of using matrices for solving sets of simultaneous equations, but it finds the values of unknown variables one at a time. This means that it can be quicker and easier to use than the matrix inversion method if you only wish to find the value of one unknown variable. This speed of manual calculation advantage is not that important if you can use an Excel spreadsheet for matrix inversion and multiplication. However, Cramer's rule is still useful in economics. Those of you who go on to study more advanced mathematical economics will use Cramer's rule to derive predictions from some multi-variable economic models specified in algebraic format.

We already know that a set of n simultaneous equations involving n unknown variables $x_1, x_2, \ldots, x_n$ and n constants values can be specified in matrix format as

$\mathbf{Ax} = \mathbf{b}$ where $\mathbf{A}$ is an $n \times n$ matrix of parameters,
$\qquad\qquad\qquad\quad \mathbf{x}$ is an $n \times 1$ vector of unknown variables and
$\qquad\qquad\qquad\quad \mathbf{b}$ is an $n \times 1$ vector of constant values.

Cramer's rule says that the value of any one of the unknown variables x_i can be found by substituting the vector of constant values $\mathbf{b}$ for the ith column of matrix $\mathbf{A}$ and then dividing the determinant of this new matrix by the determinant of the original $\mathbf{A}$ matrix.

Thus, if the term $\mathbf{A}_i$ is used to denote matrix $\mathbf{A}$ with column i replaced by the vector $\mathbf{b}$ then Cramer's rule gives

$$x_i = \frac{|\mathbf{A}_i|}{|\mathbf{A}|}$$

Example 15.19

Find x_1 and x_2 using Cramer's rule from the following set of simultaneous equations

$$5x_1 + 0.4x_2 = 12$$
$$3x_1 + 3x_2 = 21$$

Solution

These simultaneous equations can be represented in matrix format as

$$\mathbf{Ax} = \begin{bmatrix} 5 & 0.4 \\ 3 & 3 \end{bmatrix} \begin{bmatrix} x_1 \\ x_2 \end{bmatrix} = \begin{bmatrix} 12 \\ 21 \end{bmatrix} = \mathbf{b}$$

Using Cramer's rule to find x_1 by substituting the vector $\mathbf{b}$ of constants for column 1 in matrix $\mathbf{A}$ gives

$$x_1 = \frac{|\mathbf{A}_1|}{|\mathbf{A}|} = \frac{\begin{vmatrix} 12 & 0.4 \\ 21 & 3 \end{vmatrix}}{\begin{vmatrix} 5 & 0.4 \\ 3 & 3 \end{vmatrix}} = \frac{36 - 8.4}{15 - 1.2} = \frac{27.6}{13.8} = 2$$

In a similar fashion, by substituting vector $\mathbf{b}$ for column 2 in matrix $\mathbf{A}$ we get

$$x_2 = \frac{|\mathbf{A}_2|}{|\mathbf{A}|} = \frac{\begin{vmatrix} 5 & 12 \\ 3 & 21 \end{vmatrix}}{\begin{vmatrix} 5 & 0.4 \\ 3 & 3 \end{vmatrix}} = \frac{105 - 36}{15 - 1.2} = \frac{69}{13.8} = 5$$

15.10 Second-order conditions and the Hessian matrix

Matrix algebra can help derive the second-order conditions for optimization exercises involving any number of variables. To explain how, first consider the second-order conditions for unconstrained optimization with only two variables encountered in Chapter 10.

If one tries to find a maximum or minimum for the two variable function $f(x, y)$ then the FOC (first-order conditions) for both a maximum and a minimum require that

$$\frac{\partial f}{\partial x} = 0 \quad \text{and} \quad \frac{\partial f}{\partial y} = 0$$

SOC (second-order conditions) require that

$$\frac{\partial^2 f}{\partial x^2} < 0 \quad \text{and} \quad \frac{\partial^2 f}{\partial y^2} < 0 \quad \text{for a maximum}$$

$$\frac{\partial^2 f}{\partial x^2} > 0 \quad \text{and} \quad \frac{\partial^2 f}{\partial y^2} > 0 \quad \text{for a minimum}$$

and, for both a maximum and a minimum

$$\left(\frac{\partial^2 f}{\partial x^2}\right)\left(\frac{\partial^2 f}{\partial y^2}\right) > \left(\frac{\partial^2 f}{\partial x \partial y}\right)^2$$

These second-order conditions can be expressed more succinctly in matrix format. For clarity the abbreviated format for specifying second-order partial derivatives is also used, e.g. f_{xx} represents $\partial^2 f/\partial x^2$, f_{xy} represents $\partial^2 f/\partial x \partial y$, etc.

The Hessian matrix

The Hessian matrix contains all the second-order partial derivatives of a function, set out in the format shown in the following examples.

For the two variable function $f(x, y)$ the Hessian matrix will be

$$\mathbf{H} = \begin{bmatrix} f_{xx} & f_{xy} \\ f_{yx} & f_{yy} \end{bmatrix}$$

The Hessian will always be a square matrix with equal numbers of rows and columns. The *principal minors* of the Hessian matrix are the determinants of the matrices found by starting with the first element in the first row and then expanding by adding the next row and column. For any 2×2 Hessian there will therefore only be the two principal minors

$$|\mathbf{H}_1| = |f_{xx}| \quad \text{and} \quad |\mathbf{H}_2| = \begin{vmatrix} f_{xx} & f_{xy} \\ f_{yx} & f_{yy} \end{vmatrix}$$

Note that the second-order principal minor is the determinant of the Hessian matrix itself.

The *second-order conditions for a maximum and minimum* can now be specified in terms of the values of the determinants of these principal minors.

SOC for a *maximum* require $|\mathbf{H}_1| < 0$ and $|\mathbf{H}_2| > 0$ (Hessian is *negative definite*)

SOC for a *minimum* require $|\mathbf{H}_1| > 0$ and $|\mathbf{H}_2| > 0$ (Hessian is *positive definite*)

(The terms 'negative definite' and 'positive definite' are used to describe Hessians that meet the requirements specified.)

We can show that these requirements correspond to the second-order conditions for optimization of a two variable function that were set out in full above. For a *maximum* these SOC require

$$f_{xx} < 0, f_{yy} < 0 \tag{1}$$

and

$$f_{xx}f_{yy} > (f_{xy})^2 \tag{2}$$

From the Hessian matrix and its principal minors we can deduce that

$$|\mathbf{H}_1| < 0 \quad \text{means that} \quad f_{xx} < 0 \tag{3}$$

$$|\mathbf{H}_2| > 0 \quad \text{means that} \quad f_{xx}f_{yy} - f_{xy}f_{yx} > 0 \tag{4}$$

Given that for any pair of cross partial derivatives $f_{xy} = f_{yx}$ then (4) becomes

$$f_{xx}f_{yy} > (f_{xy})^2$$

and so condition (2) is met.

In (2) the term $(f_{xy})^2 > 0$ since any number squared will be greater than zero. Therefore it must be true that

$$f_{xx}f_{yy} > 0 \tag{5}$$

As we have already shown in (3) that $f_{xx} < 0$ then it must follow from (5) that

$$f_{yy} < 0$$

(a negative value must be multiplied by another negative value if the product is positive). Therefore SOC (1) also holds.

Thus we have shown that the matrix formulation of second-order conditions corresponds to the second-order conditions for optimization of a two variable function that we are already familiar with.

Returning to the price discrimination analysis considered in Chapter 10, we can now solve some problems using standard optimization techniques and check second-order conditions using the Hessian.

Example 15.20

A firm has the production function TC $= 120 + 0.1q^2$ and sells its output in two separate markets with demand functions

$$q_1 = 800 - 2p_1 \quad \text{and} \quad q_2 = 750 - 2.5p_2$$

Find the profit-maximizing output and sales in each market, using the Hessian to check second-order conditions for a maximum.

Solution

From the two demand schedules we can derive

$$p_1 = 400 - 0.5q_1 \qquad \text{TR}_1 = 400q_1 - 0.5q_1^2 \qquad \text{MR}_1 = 400 - q_1$$

$$p_2 = 300 - 0.4q_2 \qquad \text{TR}_2 = 300q_2 - 0.4q_2^2 \qquad \text{MR}_2 = 300 - 0.8q_2$$

Given that total output $q = q_1 + q_2$ then

$$\text{TC} = 120 + 0.1q^2 = 120 + 0.1(q_1 + q_2)^2$$
$$= 120 + 0.1q_1^2 + 0.2q_1q_2 + 0.1q_2^2$$

Therefore

$$\pi = \text{TR}_1 + \text{TR}_2 - \text{TC}$$
$$= 400q_1 - 0.5q_1^2 + 300q_2 - 0.4q_2^2 - 120 - 0.1q_1^2 - 0.2q_1q_2 - 0.1q_2^2$$
$$= 400q_1 - 0.6q_1^2 + 300q_2 - 0.5q_2^2 - 120 - 0.2q_1q_2$$

FOC for a maximum require

$$\frac{\partial \pi}{\partial q_1} = 400 - 1.2q_1 - 0.2q_2 = 0 \quad \text{therefore} \quad 400 = 1.2q_1 + 0.2q_2 \qquad (1)$$

$$\frac{\partial \pi}{\partial q_2} = 300 - q_2 - 0.2q_1 = 0 \qquad \text{therefore} \quad 300 = 0.2q_1 + q_2 \qquad (2)$$

To find the optimum values that satisfy the FOC, the simultaneous equations (1) and (2) can be set up in matrix format as

$$\mathbf{Aq} = \begin{bmatrix} 1.2 & 0.2 \\ 0.2 & 1 \end{bmatrix} \begin{bmatrix} q_1 \\ q_2 \end{bmatrix} = \begin{bmatrix} 400 \\ 300 \end{bmatrix} = \mathbf{b}$$

Using Cramer's rule to solve for the sales in each market gives

$$q_1 = \frac{\begin{vmatrix} 400 & 0.2 \\ 300 & 1 \end{vmatrix}}{\begin{vmatrix} 1.2 & 0.2 \\ 0.2 & 1 \end{vmatrix}} = \frac{400 - 60}{1.2 - 0.04} = \frac{340}{1.16} = 293.1$$

$$q_2 = \frac{\begin{vmatrix} 1.2 & 400 \\ 0.2 & 300 \end{vmatrix}}{\begin{vmatrix} 1.2 & 0.2 \\ 0.2 & 1 \end{vmatrix}} = \frac{360 - 80}{1.2 - 0.04} = \frac{280}{1.16} = 241.4$$

To check the second-order conditions we return to the first-order partial derivatives and then find the second-order partial derivatives and the cross partial derivatives. Thus, from

$$\frac{\partial \pi}{\partial q_1} = 400 - 1.2q_1 - 0.2q_2 \qquad \text{and} \qquad \frac{\partial \pi}{\partial q_2} = 300 - q_2 - 0.2q_1$$

we get

$$\frac{\partial^2 \pi}{\partial q_1^2} = -1.2 \qquad \frac{\partial^2 \pi}{\partial q_1 \partial q_2} = -0.2 \qquad \frac{\partial^2 \pi}{\partial q_2^2} = -1 \qquad \frac{\partial^2 \pi}{\partial q_2 \partial q_1} = -0.2$$

The Hessian matrix is therefore

$$\mathbf{H} = \begin{bmatrix} \pi_{11} & \pi_{12} \\ \pi_{21} & \pi_{22} \end{bmatrix} = \begin{bmatrix} -1.2 & -0.2 \\ -0.2 & -1 \end{bmatrix}$$

and the determinants of the principal minors are

$$|\mathbf{H}_1| = -1.2 < 0$$

and

$$|\mathbf{H}_2| = \begin{vmatrix} -1.2 & -0.2 \\ -0.2 & -1 \end{vmatrix} = 1.2 - 0.04 = 1.16 > 0$$

As $|\mathbf{H}_1| < 0$ and $|\mathbf{H}_2| > 0$ the Hessian is negative definite. Therefore SOC for a maximum are met.

3rd order Hessians

For a three variable function $y = f(x_1, x_2, x_3)$ the Hessian will be the 3×3 matrix of second-order partial derivatives

$$\mathbf{H} = \begin{bmatrix} f_{11} & f_{12} & f_{13} \\ f_{21} & f_{22} & f_{23} \\ f_{31} & f_{32} & f_{33} \end{bmatrix}$$

and the determinants of the three principal minors will be

$$|\mathbf{H}_1| = |f_{11}| \quad |\mathbf{H}_2| = \begin{vmatrix} f_{11} & f_{12} \\ f_{21} & f_{22} \end{vmatrix} \quad |\mathbf{H}_3| = \begin{vmatrix} f_{11} & f_{12} & f_{13} \\ f_{21} & f_{22} & f_{23} \\ f_{31} & f_{32} & f_{33} \end{vmatrix}$$

The SOC conditions for unconstrained optimization of a three variable function are:

(a) For a Maximum $|\mathbf{H}_1| < 0$, $|\mathbf{H}_2| > 0$ and $|\mathbf{H}_3| < 0$ (Hessian is Negative definite)

(b) For a Minimum $|\mathbf{H}_1|$, $|\mathbf{H}_2|$ and $|\mathbf{H}_3|$ are all > 0 (Hessian is Positive definite)

Example 15.21

A multiplant monopoly produces the quantities q_1, q_2 and q_3 in the three plants that it operates and faces the profit function

$$\pi = -24 + 839q_1 + 837q_2 + 835q_3 - 5.05q_1^2 - 5.03q_2^2 - 5.02q_3^2$$
$$- 10q_1q_2 - 10q_1q_3 - 10q_2q_3$$

Find the output levels in each of its three plants q_1, q_2 and q_3 that will maximize profit and use the Hessian to check that second-order conditions are met.

Solution

Differentiating this π function with respect to q_1, q_2 and q_3 and setting equal to zero to find the optimum values where the first-order conditions are met, we get:

$$\pi_1 = 839 - 10.1q_1 - 10q_2 - 10q_3 = 0 \tag{1}$$
$$\pi_2 = 837 - 10q_1 - 10.06q_2 - 10q_3 = 0 \tag{2}$$
$$\pi_2 = 835 - 10q_1 - 10q_2 - 10.04q_3 = 0 \tag{3}$$

These conditions can be rearranged to get

$$839 = 10.1q_1 + 10q_2 + 10q_3$$
$$837 = 10q_1 + 10.06q_2 + 10q_3$$
$$835 = 10q_1 + 10q_2 + 10.04q_3$$

These simultaneous equations can be specified in matrix format and solved by the matrix inversion method to get the optimum values of q_1, q_2 and q_3 as 42, 36.6 and 4.9, respectively. (The full calculations are not set out here as the objective is to explain how the Hessian is used to check second-order conditions, but you can check these answers using Excel if you are not sure how these values are calculated.)

Differentiating (1), (2) and (3) again we can derive the Hessian matrix of second-order partial derivatives

$$\mathbf{H} = \begin{bmatrix} \pi_{11} & \pi_{12} & \pi_{13} \\ \pi_{21} & \pi_{22} & \pi_{23} \\ \pi_{31} & \pi_{32} & \pi_{33} \end{bmatrix} = \begin{bmatrix} -10.1 & -10 & -10 \\ -10 & -10.06 & -10 \\ -10 & -10 & -10.04 \end{bmatrix}$$

The determinants of the three principal minors will therefore be

$$|\mathbf{H}_1| = |\pi_{11}| = -10.1$$

$$|\mathbf{H}_2| = \begin{vmatrix} \pi_{11} & \pi_{12} \\ \pi_{21} & \pi_{22} \end{vmatrix} = \begin{vmatrix} -10.1 & -10 \\ -10 & -10.06 \end{vmatrix} = 101.606 - 100 = 1.606$$

$$|\mathbf{H}_3| = \begin{vmatrix} \pi_{11} & \pi_{12} & \pi_{13} \\ \pi_{21} & \pi_{22} & \pi_{23} \\ \pi_{31} & \pi_{32} & \pi_{33} \end{vmatrix} = \begin{vmatrix} -10.1 & -10 & -10 \\ -10 & -10.06 & -10 \\ -10 & -10 & -10.04 \end{vmatrix} = -0.1242$$

(You can check the H_3 determinant calculations using the Excel MDETERM function.)
This Hessian is therefore negative definite as

$$|\mathbf{H}_1| = -10.1 < 0, \quad |\mathbf{H}_2| = 1.606 > 0, \quad |\mathbf{H}_3| = -0.1242 < 0$$

and so the second-order conditions for a maximum are met.

Higher order Hessians

Although you will not be asked to use the Hessian to tackle any problems in this text that involve more than three variables, for your future reference the general SOC conditions that apply to a Hessian of any order are:

(*a*) *Maximum*
   Principal minors alternate in sign, starting with $|\mathbf{H}_1| < 0$ (Negative definite)
   Thus a principal minor $|\mathbf{H}_i|$ of order i should have the sign $(-1)^i$
(*b*) *Minimum*
   All principal minors $|\mathbf{H}_i| > 0$ (Positive definite)

Test Yourself, Exercise 15.10

1. A firm that sells in two separate markets has the profit function

    $$\pi = -120 + 245q_1 - 0.3q_1^2 + 120q_2 - 0.4q_2^2 - 0.18q_1q_2$$

    where q_1 and q_2 sales in the two markets. Find the profit maximizing sales in each market, using the Hessian to check second-order conditions for a maximum.
2. Find the values of q_1 and q_2 that will maximizing the profit function

    $$\pi = -12 + 152q_1 - 0.25q_1^2 + 196q_2 - 0.2q_2^2 - 0.1q_1q_2$$

    and check that second-order conditions are met using the Hessian matrix.
3. If a firm producing three products faces the profit function

    $$\pi = -73 + 242q_1 + 238q_2 + 238q_3 - 8.4q_1^2 - 8.25q_2^2$$
    $$- 8.1q_3^2 - 16q_1q_2 - 16q_1q_3 - 16q_2q_3$$

Find the amounts of the three products q_1, q_2 and q_3 that will maximize profit and use the Hessian to check that second-order conditions are met.

4. A monopoly operates three plants with total cost schedules

$$TC_1 = 40 + 0.1q_1 + 0.04q_1^2 \quad TC_2 = 18 + 3q_2 + 0.02q_2^2$$

$$TC_3 = 30 + 4q_3 + 0.01q_3^2$$

and faces the market demand schedule

$$p = 250 - 2q \quad \text{where } q = q_1 + q_2 + q_3$$

Set up the profit function and then use it to determine how much the firm should make in each plant to maximize profit, using the Hessian to check that second-order conditions are met.

15.11   Constrained optimization and the bordered Hessian

In Chapter 11, the solution of constrained optimization problems using the Lagrange multiplier method was explained, but the explanation of how to check if second-order conditions for constrained optimization are met was put on hold. Now that the concept of the Hessian has been covered, we are ready to investigate how the related concept of the *bordered* Hessian can help determine if the second-order conditions are met when the Lagrange method is used.

If second-order partial derivatives are taken for a Lagrange constrained optimization objective function and put into a matrix format this will give what is known as the bordered Hessian.

For example, to maximize a utility function $U(X_1, X_2)$ subject to the budget constraint

$$M - P_1 X_1 - P_2 X_2 = 0$$

The Lagrange equation will be

$$G = U(X_1, X_2) + \lambda(M - P_1 X_1 - P_2 X_2)$$

Taking first-order derivatives and setting equal to zero we get the first-order conditions:

$$G_1 = U_1 - \lambda P_1 = 0 \tag{1}$$

$$G_2 = U_2 - \lambda P_2 = 0 \tag{2}$$

$$G_\lambda = M - P_1 X_1 - P_2 X_2 = 0 \tag{3}$$

These are used to solve for the optimum values of X_1 and X_2 when actual values are specified for the parameters.

Differentiating (1), (2) and (3) again with respect to X_1, X_2 and λ gives the bordered Hessian matrix of second-order partial derivatives

$$\mathbf{H_B} = \begin{bmatrix} U_{11} & U_{12} & -P_1 \\ U_{21} & U_{22} & -P_2 \\ -P_1 & -P_2 & 0 \end{bmatrix}$$

You can see that the bordered Hessian $\mathbf{H_B}$ has one more row and one more column than the ordinary Hessian. In this, and most other constrained maximization examples that you will encounter, the extra row and column each contain the negative of the prices of the variables in the constraint.

Although it is possible to use the Lagrange method to tackle constrained optimization problems with several constraints, we will only consider problems with one constraint here. The second-order conditions for optimization of a Lagrangian with one constraint require that for:

Maximization

- If there are two variables in the objective function (i.e. $\mathbf{H_B}$ is 3×3) then the determinant $|\mathbf{H_B}| > 0$.
- If there are three variables in the objective function (i.e. $\mathbf{H_B}$ is 4×4) then the determinant $|\mathbf{H_B}| < 0$ **and** the determinant of the naturally ordered principal minor of $|\mathbf{H_B}| > 0$. (The naturally ordered principal minor is the matrix remaining when the first row and column have been eliminated from $\mathbf{H_B}$.)

Minimization

- If there are two variables in the objective function the determinant $|\mathbf{H_B}| < 0$.
- If there are three variables in objective function then the determinant $|\mathbf{H_B}| < 0$ and the determinant of the naturally ordered principal minor of $|\mathbf{H_B}| < 0$.

Example 15.22

An individual has the utility function $U = 4X^{0.5}Y^{0.5}$ and can buy good X at £2 a unit and good Y at £8 a unit. If their budget is £100, find the combination of X and Y that they should purchase to maximize utility and check that second-order conditions are met using the bordered Hessian matrix.

Solution

The Lagrange function is

$$G = 4X^{0.5}Y^{0.5} + \lambda(100 - 2X - 8Y)$$

Differentiating and setting equal to zero to get the FOC for a maximum

$$G_X = 2X^{-0.5}Y^{0.5} - 2\lambda = 0 \tag{1}$$

$$G_Y = 2X^{0.5}Y^{-0.5} - 8\lambda = 0 \tag{2}$$

$$G_\lambda = 100 - 2X - -8Y = 0 \tag{3}$$

From (1)

$$X^{-0.5}Y^{0.5} = \lambda$$

From (2)

$$0.25X^{0.5}Y^{-0.5} = \lambda$$

Therefore

$$X^{-0.5}Y^{0.5} = 0.25X^{0.5}Y^{-0.5}$$

Multiplying both sides by $4X^{0.5}Y^{0.5}$

$$4Y = X \tag{4}$$

Substituting (4) into (3)

$$100 - 2(4Y) - 8Y = 0$$
$$Y = 6.25$$

and thus from (4) $\quad X = 25$

Differentiating (1), (2) and (3) again gives the bordered Hessian of second-order partial derivatives

$$\mathbf{H_B} = \begin{bmatrix} U_{XX} & U_{XY} & -P_X \\ U_{YX} & U_{YY} & -P_Y \\ -P_X & -P_Y & 0 \end{bmatrix} = \begin{bmatrix} -X^{-1.5}Y^{0.5} & X^{-0.5}Y^{-0.5} & -2 \\ X^{-0.5}Y^{-0.5} & -X^{0.5}Y^{-1.5} & -8 \\ -2 & -8 & 0 \end{bmatrix}$$

$$= \begin{bmatrix} -0.02 & 0.08 & -2 \\ 0.08 & -0.32 & -8 \\ -2 & -8 & 0 \end{bmatrix}$$

The determinant of this bordered Hessian, expanding along the third row is

$$|\mathbf{H_B}| = -2(-0.64 - 0.64) + 8(0.16 + 0.16) = 2.56 + 2.56 = 5.12 > 0$$

and so the second-order conditions for a maximum are satisfied.

To illustrate the use of the bordered Hessian to check the second-order conditions required for constrained optimization involving three variables, we shall just consider an example without any specific format for the objective function.

Example 15.23

If a firm is attempting to maximize output $Q = Q(x, y, z)$ subject to a budget of £5000 where the prices of the inputs x, y and z are £8, £12 and £6, respectively, what requirements are there for the relevant bordered Hessians to ensure that second-order conditions for optimization are met?

Solution

The Lagrange objective function will be

$$G = Q(x, y, z) + \lambda(5000 - 8x - 12y - 6z)$$

As there are three variables in the objective function and $\mathbf{H_B}$ is 4×4 then the second-order conditions for a maximum require that the determinant of the bordered Hessian of second-order partial derivatives $|\mathbf{H_B}| < 0$. Therefore

$$|\mathbf{H_B}| = \begin{vmatrix} Q_{xx} & Q_{xy} & Q_{xz} & -8 \\ Q_{yx} & Q_{yy} & Q_{yz} & -12 \\ Q_{zx} & Q_{zy} & Q_{zz} & -6 \\ -8 & -12 & -6 & 0 \end{vmatrix} < 0$$

As $\mathbf{H_B}$ is 4×4 the second-order conditions for a maximum also require that the determinant of the naturally ordered principal minor of $\mathbf{H_B} > 0$. Thus, when the first row and column have been eliminated from $\mathbf{H_B}$, this problem also requires that

$$\begin{vmatrix} Q_{yy} & Q_{yz} & -12 \\ Q_{zy} & Q_{zz} & -6 \\ -12 & -6 & 0 \end{vmatrix} > 0$$

Constrained optimization with any number of variables and constraints

All the constrained optimization problems that you will encounter in this text have only one constraint and usually do not have more than three variables in the objective function. However, it is possible to set up more complex Lagrange functions with many variables and more than one constraint.

Second-order conditions requirements for optimization for the general case with m variables in the objective function and r constraints are that the naturally ordered *border preserving* principal minors of dimension m of $\mathbf{H_B}$ must have the sign

$$(-1)^{m-r} \quad \text{for a maximum}$$
$$(-1)^{r} \quad \text{for a minimum}$$

'Border preserving' means not eliminating the borders added to the basic Hessian, i.e. the last column and the bottom row, which typically show the prices of the variables.

These requirements only apply to the principal minors of order $\geq (1 + 2r)$. For example, if the problem was to maximize a utility function $U = U(X_1, X_2, X_3)$ subject to the budget constraint $M = P_1X_1 + P_2X_2 + P_3X_3$ then, as there is only one constraint, $r = 1$. Therefore we would just need to consider the principal minors of order greater than three since

$$(1 + 2r) = (1 + 2) = 3$$

As the full-bordered Hessian in this example with three variables is 4th order then only $\mathbf{H_B}$ itself plus the first principal minor need be considered, as this is the only principal minor with order equal to or greater than 3.

The second-order conditions will therefore require that for a maximum:
For the full bordered Hessian $m = 4$ and so $|\mathbf{H_B}|$ must have the sign

$$(-1)^{m-r} = (-1)^{4-1} = (-1)^3 = -1 < 0$$

and the determinant of the 3rd order naturally ordered principal minor of $|\mathbf{H_B}|$ must have the sign

$$(-1)^{m-r} = (-1)^{3-1} = (-1)^2 = +1 > 0$$

These are the same as the basic rules for the three variable case stated earlier.

The last set of problems, below, just require you to use the bordered Hessian to check that second-order conditions for optimization are met in some numerical examples with a small number of variables to familiarize you with the method. Those students who go on to study further mathematical economics will find that this method will be extremely useful in more complex constrained optimization problems.

Test Yourself, Exercise 15.11

1. A firm has the production function $Q = K^{0.5}L^{0.5}$ and buys input K at £12 a unit and input L at £3 a unit and has a budget of £600. Use the Lagrange method to find the input combination that will maximize output, checking that second-order conditions are satisfied by using the bordered Hessian.
2. A firm operates with the production function $Q = 25K^{0.5}L^{0.4}$ and buys input K at £20 a unit and input L at £8 a unit. Use the Lagrange method to find the input combination that will minimize the cost of producing 400 units of Q, using the bordered Hessian to check that second-order conditions are satisfied.
3. A consumer has the utility function $U = 20X^{0.5}Y^{0.4}$ and buys good X at £10 a unit and good Y at £2 a unit. If their budget constraint is £450, what combination of X and Y will maximize utility? Check that second-order conditions are satisfied by using the bordered Hessian.
4. A consumer has the utility function $U = 4X^{0.75}Y^{0.25}$ and can buy good X at £12 a unit and good Y at £2 a unit. Find the combination of X and Y that they should purchase to minimize the cost of achieving a utility level of 580 and check that second-order conditions are met using the bordered Hessian matrix.

Answers

Chapter 2

2.1 1. 3,555   2. 865   3. 92,920   4. 23

2.2 1. 919   2. 225   3. 164   4. 627
      5. 440   6. 101

2.3 1. 840   2. 17   3. 172   4. 122
      5. £13,800   6. 598   7. £176

2.4 1. $\dfrac{73}{168}$ 2. $\dfrac{101}{252}$ 3. $4\dfrac{4}{5}$ 4. $\dfrac{19}{30}$

      5. $2\dfrac{13}{21}$ 6. $4\dfrac{37}{60}$ 7. $14\dfrac{1}{12}$ 8. $3\dfrac{12}{13}$

      9. $18\dfrac{39}{40}$ 10. $1\dfrac{1}{2}$

2.5 1. (a) $\dfrac{1}{3}$ (b) $\dfrac{5}{7}$ (c) $1\dfrac{2}{5}$ (d) 3   (e) 11

      2. 1   3. (a) 1 (b) 5   4. $15, 4\dfrac{1}{3}, 2\dfrac{1}{5}, 1\dfrac{2}{7}, \dfrac{7}{9}, \dfrac{5}{11}, \dfrac{3}{13}, \dfrac{1}{15}$

2.6 1. 36.914   2. 751.4   3. 435.1096   4. 36,082
      5. 0.09675   6. 610   7. 140
      8. (a) 0.1   (b) 0.001   (c) 0.000001
      9. (a) 0.452   (b) 2.431   (c) 0.075   (d) 0.002

2.7 1. -2 2. -24 3. -33 4. 0.45   5. 0.35
      6. -117 7. -330 8. 3600

      9. $\dfrac{-157}{140}$ 10. $\dfrac{17}{16}$

2.8 1. 0.25   2. 123   3. 6   4. 64
      5. 11.641754   6. 531,441   7. 0.0015328   8. 36
      9. -618.47021 10. 25.000655

2.9   1. ± 25 2. 2   3. 0.2   4. 7   5. 2.4494897

        6. 96   7. 10

        8. 5.2780316   9. 0.03423   10. 87.977857

2.10   1. 270,818.98   2. 220.9478   3. 2.8563×10^9 4. 1.5728×10^8

        5. 1.2683   6. 16,552,877   7. 93.696376   8. 4.38228

        9. 5.1331868

Chapter 3

3.1   1. (a) $0.01x$ (b) $0.5x$ (c) $0.5x$ 2. $0.01rx + 0.5wx + 0.5mx$

        3. (a) $\dfrac{x}{12}$ (b) $\dfrac{xp}{12}$ 4. (a) $0.1x$ kg   (b) $0.3x$ kg   (c) $x(0.1m + 0.3p)$

        5. $0.5w + 0.25$ 6. Own example   7. $10.5x + 6y$ 8. $3q - 6000$

3.2   1. 456   2. 77.312   3. $r + z, 9\%$ 4. Own example

        5. 1.094   6. £465.58   7. £2,100   8. (a) $99 + 0.78M$ (b) £2,166

3.3   1. $30x + 4$ 2. $24x - 18y + 7xy - 12$ 3. $6x + 5y - 650$

        4. $9H - 120$

3.4   1. $6x^2 - 24x$ 2. $x^2 + 4x + 9$ 3. $2x^2 + 6x + xy + 3y$

        4. $42x^2 - 16y^2 - 34xy + 6y$ 5. $33x + 2y - 20y^2 + 62xy - 21$

        6. $120 + 2x + 54y + 40z - x^2 + 6y^2 + xy + 4xz + 8yz$

        7. $200q - 2q^2$ 8. $13x + 11y$ 9. $8x^2 + 60x + 76$

        10. $4,000 + 150x$

3.5   1. $(x + 4)^2$ 2. $(x - 3y)^2$ 3. Does not factorize

        4. Does not factorize 5. Own example

3.6   1. $3x + 7 - 20x^{-1}$ 2. $x + 9$ 3. $4y + x + 12$ 4. $200x^{-1} + 21$

        5. $179x$ 6. $2(x + 3) + 4 - x - 3 - x + 2 = 9$ 7. Own example

3.7   1. 4   2. $\frac{1}{11}$ 3. 7   4. 14   5. 82   6. 20 p   7. 33%

        8. 40p   9. £3,062.50   10. 4 m   11. 26

3.8   1. $\frac{1}{n}\sum_{i=1}^{n} H_i$, 173.7 cm   2. 35   3. 60

        4. $\sum_{i=1}^{n} 6,000(0.9)^{i-1}$, 16,260 tonnes   5. (a) $\frac{1}{n}\sum_{i=1}^{n} R_i$, £4,425

        (b) $\frac{1}{3}\sum_{n-3}^{n-1} R_i$, £4,933   6. 13.25%, 8.2

3.9   1. (a) $\leq$ (b) $<$ (c) $\geq$ (d) $>$ 2. (a) $>$ (b) $\geq$ (c) $>$ (d) $>$

        3. (a) $Q_1 < Q_2$ (b) $Q_1 = Q_2$ (c) $Q_1 > Q_2$ 4. $P_2 > P_1$

Chapter 4

4.1   1. (a) Quantity demanded depends on the price of tea, average exp., etc.

(b) Q_t dependent, all others independent.

(c) $Q_t = 99 - 6P_t - 0.5Y_t + 0.8A + 1.2N + 1.4Pc$

(suggested number assumes tea is an inferior good)

2. (a) 202 (b) 7 (c) 6, $x \geq 0$ 3. Yes; no

4.2   1. $°F = 32 + 1.8°C$ 2. $P = 2,400 - 2Q$

3. It is not monotonic, e.g. TR $= 200$ when $q = 5$ or 10

4. $T = (0.0625X - 25)^2$; no   5. Own example

4.3   (Answers to 1 to 5 give intercepts on axes)

1. $x = -12, y = 6$ 2. $x = 3\frac{1}{3}, y = -40$ 3. $P = 60, Q = 300$

4. $P = 150, Q = 750$ 5. $K = 24, L = 40$ 6. Goes through origin only

7. Goes through $(Q = 0, TC = 200)$ and $(Q = 10, TC = 250)$

8. Horizontal line at TFC $= 75$ 9. Own example

10. (a) and (d); both slope upwards and have positive intercepts on P axis

4.4   1. $Q = 90 - 5P$; 50; $Q \geq 0, P \geq 0$ 2. $C = 30 + 0.75Y$

3. By £20 to £100   4. $P = 12 - 0.015Q$ 5. £6,440

4.5   1. 3.75, 0.75, 0.375, -0.75 2. $P = 12, Q = 40$; £4.50; 10   3. (a) 2/3 (b) 3

4. (c) (i) (a) (ii) (b)   5. (a), (d)   6. APC $= 400Y^{-1} + 0.5 > 0.5 = $ MPC

7. (a) 0.263 (b) 0.714 (c)1.667   8. Own example

4.6   1. -1.5 (a) becomes -1 (b) becomes -1 (c) no change (d) no change

2. $K = 100, L = 160, P_K = £8, P_L = £5$

3. Cost £520 > budget; P_L reduced by £10 to £30

4. (a) -10 (b) -1 (c) -0.1 (d) -0.025 (e) 0   5. No change

6. Height £120, base 12, slope $-10 = -($wage$)$ 7. Own example

4.7   1. Sketch graphs   2. Sketch graphs   3. Steeper

4. Like $y = x^{-1}$; £260   5. Own example

4.8   1. Sketch graphs   2. Own example   3. $\pi = 50x - 100 - 0.4x^3$; inverted U

4. (a) $40 = 3250q^{-1}$

(b) Original firms' π per unit $= £27.50$ but new firms' AC $= £170 > $ price

4.9   Plot Excel graphs

4.10   1. (a) $16L^{-1}$ (b) 0.16   (c) constant

2. (a) $57,243.34L^{-15}$ (b) 57.243   (c) constant

3. (a) $322.54L^{-1}$ (b) 3.2254   (c) increasing

4. (a) $3,125L^{-1.25}$ (b) 9.882   (c) increasing

5. (a) $23,415,916L^{-1.6667}$ (b) $10,868.71$ (c) decreasing

6. (a) $4,093.062L^{-1.7714}$ (b) 1.173 (c) decreasing

4.11   1. MR $= 33.33 - 0.00667Q$ for $Q \geq 500$

2. MR $= 76 - 0.222Q$ for $Q \geq 22.5$

3. MR $= 80 - 0.555Q$ for $Q \geq 562.5$

4. MC $= 30 + 0.0714Q$ for $Q \geq 56$

5. MC $= 56 + 0.1333Q$ for $Q \geq 30$

6. MC $= 3 + 0.0714Q$ for $Q \geq 59$

Chapter 5

5.1   1. $q = 40$, $p = 6$ 2. $x = 67$, $y = 17$ (approximately)   3. No solution exists

5.2   1. $q = 118$, $p = 256$ 2. (a) $q = 80$, $p = 370$ (b) q falls to 78, p rises to 376

3. Own example   4. (a) 40 (b) rises to 50   5. $x = 2.102$, $y = 62.25$

5.3   1. $A = 24$, $B = 12$ 2. 200   3. $x = 190$, $y = 60$

5.4   1. $x = 30$, $y = 60$ 2. $A = 6$, $B = 36$ 3. $x = 25$, $y = 20$

5.5   1. $x = 24$, $y = 14.4$, $z = 19.2$ 2. $x = 4$, $y = 6$, $z = 4$

3. $A = 6$, $B = 22$, $C = 2$ 4. $x = 17$, $y = 4$, $z = 8$

5. $A = 82.5$, $B = 35$, $C = 6$, $D = 9$

5.6   1. $q = 500$, $p = 275$ 2. $K = 17.5$, $L = 16$, $R = 10$

3. (a) p rises from £8 to £10   (b) p rises to £9

4. $Y = £3,750$ m; government deficit £150 m

5. $Y = £1,625$ m; balance of payments deficit £15 m

6. $L = 80$, $w = 52$

5.7   1. $p = 184 + 0.2a$, $q = 43.2 + 0.06a$, $p = 216$, $q = 52.8$

2. $p = 84 + 0.2t$, $q = 32 - 0.4t$, $p = 85$, $q = 30$

3. $p = 122.4 + 0.2t$, $q = 13.8 - 0.1t$, $p = 123.4$, $q = 13.3$

4. (a) Y $= 100/(0.25 + 0.75t)$, Y $= 250$ (b) Y $= 110/(0.25 + 0.75t)$, Y $= 275$

5. $p = (4200 + 3800v)/(9 + 5v)$,

    $q = (750 - 50v)/(9 + 5v)$

    $p = 494.30$, $q = 76.94$

5.8   1. $q_1 = 60$, $q_2 = 80$, $p_1 = £10$, $p_2 = £8$

2. $q_1 = 40$, $q_2 = 50$, $p_1 = £6$, $p_2 = £4$

3. $p_1 = £8.75$, $q_1 = 60$, $p_2 = £6.10$, $q_2 = 550$

4. £81 for extra 65 units

5. £7.50 for extra 25 units

6. $q_1 = 48$, $q_2 = 39$, $p_1 = £12$, $p_2 = £8.87$

7. (a) 190 units (b) £175 for extra 75 units

5.9   1. $q_1 = 180$, $q_2 = 200$, $p = £39$ 2. $q_1 = 1,728$, $q_2 = 780$, $p = £190.70$

3. $q_1 = 1,510$, $q_2 = 1,540$, $q_A = 800$, $q_B = 2,250$, $P_A = £500$, $P_B = £625$

4. $q_1 = 160$, $q_2 = 600$, $q_A = 293\frac{1}{3}$, $q_B = 266\frac{2}{3}$, $q_C = 200$, $P_A = £95$,

$$P_B = £80, \ P_c = £60$$

5. $q_1 = 15.47$, $q_2 = 27.34$, $q_3 = 26.17$, $p = £14.20$

5A.1   1. 8.4 of A, 4.64 of B (tonnes); (a) no change (b) no B, 12.16 of A

2. $A = 13$, $B = 27$ 3. 12 of A, 5 of B   4. 22.5 of A, 7.5 of B

5. 6 of A, 32 of B   6. Own example

7. 13.64 of A, 21.82 of B; £7092; surplus 2.72 of R, 22.72 of mix additive

8. Produce 15 of A, 21 of B   9. 30 of A, B = 0

10. Objective function parallel to first constraint

11. 24,000 shares in X, 18,000 shares in Y, return £8,640

12. Own example

5A.2   1. $C = 70$ when $A = 1$, $B = 1.5$, slack in $x = 30$ 2. $A = 3$, $B = 0$

3. $Q = 2.5$, $R = 1.5$; excesses 62.5 mg of B, 27.5 mg of C

4. 10 of A, 5 of B; space for 50 extra loads of X

5. Zero R, 15 tonnes of T; G exceeds by 45 kg

6. 100 of A, 40 of B   7. Own example

5A.3   1. 2 of A, 1 of B   2. 7.5 of X and 15 of Y (tonnes)   3. 8

4. Own example

Chapter 6

6.1   1. 2 or 3   2. 10 or 60   3. When $x = 2$ 4. 0.5   5. 9

6.2   1. 10 or -12.5

2. £16.353. (a) 1.01 or 98.99 (b) 11.27 or 88.73 (c) no solution exists

4. Own example

6.3   1. $x = 15$, $y = 15$ or $x = -3$, $y = 249$

2. $x = 1.75$, $y = 3.15$ or $x = -1.53$, $y = 20.97$ 3. 16.4

4. $q_1 = 3.2$, $q_2 = 4.8$, $p_1 = £136$, $p_2 = £96$

5. $p_1 = £15$, $q_1 = 80$, $p_2 = £8.50$, $q_2 = 70$

6.4   1. 52   2. 1069   3. 10

Chapter 7

7.1   1. £4,630.50   2. £314.70   3. £17,623.16

       4. £744.71   5. £40,441.40   6. £5,030.03

7.2   1. £43,747.41; 12.68%   2. £501,159.74; 7.44%   3. (a) APR 11.35%

       4. £2,083.61; 19.25%   5. £625; 5.09%   6. 19.28%

       7. 0.01467%   8. £494,531.25; 4.5%

7.3   1. £6,301.69   2. £355.89   3. No, $A = £9,106.27$

       4. £6,851.65   5. (a) £9,638.58   (b) £11,579.83 (c) £13,318.15

       6. 5   7. 5.27 years   8. 12.1 years   9. 5.45 years   10. 3.42 years

       11. 10.7%   12. 9.5%   13. 7.5%   14. 0.8%   15. 10.3%   16. 8.4%

       17. (b) as PV $= £5,269.85$

7.4   1. (a) £90.75   (b) $-£100.07$ (c) $-£474.01$ (d) £622.86   (e) £1,936.87

         (f) £877.33   (g) £791.25   (h) £992.16

       2. B, PV $= £6,569.10$ 3. (a) All viable (b) A best, NPV $= £6,824.68$

       4. Yes, NPV $= £7,433.56$ 5. Yes, NPV $= £4,363.45$

       6. (a) Yes, NPV $= £610.02$ (b) no, NPV $= -£522.30$

       7. B, NPV $= £856.48$

7.5   1. $r_A = 20\%, r_B = 41.6\%, r_C = 20\%, r_D = 20\%$;

       B consistently best, but others have same IRR with different NPV ranking

       2. (a) A, $r_A = 21.25\%, r_B = 20.42\%$ (b) B, $NPV_B = £2,698.94$,

       $NPV_A = £2,291.34$ 3. IRR $= 16.93\%$

7.6   1. (a) 2.5, 781.25, 50,857.3   (b) 3, 121.5, 14,762   (c) 1.4, 10.756, 139.6

         (d) 0.8, 19.66, 267.8   (e) 0.75, 0.57, 9.06

       2. 5,741 (to nearest whole unit)

       3. A, £1,149.32; B, £2,980.91; C, £45,216.47

       4. Yes, NPV $= £3,774.71$ 5. £4,149.20

7.7   1. (a) $k = 1.5$, not convergent   (b) $k = 0.8$, converges on 600

         (c) $k = -1.5$, not convergent   (d) $k = \frac{1}{3}$ converges on 54

         (e) converges on 961.54   (f) not convergent

       2. £3,076.92   3. Yes, NPV $= £50,000$

       4. (a) £240,000   (b) £120,000   (c) £80,000   (d) £60,000   5. £3,500

7.8   1. £152.59   2. £197.38   3. £191.46

       4. £794.66   5. (a) 14.02% (b) 26.08% (c) 23.86% (d) 14.71%

       6. Loan is marginally better deal (PV of payments $= £6,348.33 + £1,734$

         deposit $= £8,082.33$, less than cash price by £12.67)

7.9   1. 6.82 years   2. After 15.21 years   3. 4%

4. Yes, sum of infinite GP = 1,300 million tonnes   5. 4.85%

Chapter 8

8.1   1. $36x^2$ 2. 192   3. 21.6   4. $260x^4$ 5. Own example

8.2   1. $3x^2 + 60$ 2. 250   3. $-4x^{-2} - 4$ 4. 1
5. $0.2x^{-3} + 0.6x^{-0.7}$ 6. Own example

8.3   1. $120 - 6q, 20$ 2. 25   3. 14,400   4. £200   5. Own example

8.4   1. 7.5   2. $12q^2 - 40q + 60$
3. (a) $1.5q^2 - 6q + 25$ (b) $0.5q^2 - 3q + 25 + 20q^{-1}$ (c) $q - 3 - 20q^{-2}$
4. MC constant at 0.8   5. Own example

8.5   1. 4   2. (a) 80   (b) 158.33   (c) 40 or 120   3. 6

8.6   1. $50 - \frac{2}{3}q$ 2. 900   3. $24 - 1.2q^2$

8.7   1. 0.8   2. Proof   3. 0.16667   4. 1

8.8   1. £77.50   2. Own example   3. Rise, maximum TY when $t = £39$

8.9   1. (a) 0.8   (b) 4,400   (c) 5   (d) 120   (e) Yes, both 940

Chapter 9

9.1   1. 62.5   2. 150   3. (a) 500 (b) 600 (c) 300   4. 50

9.2   1. 1,200, max.   2. 25, max.   3. 4,096, max.   4. 4, not max.

9.3   1. 6, min.   2. 14.4956. min.   3. 0, min.   4. 3, not min.
5. No stationary point exists

9.4   1. (a)   MC $= 2q^2 - 28q + 222$, min. when $q = 7$, MC $= 124$
(b)   AVC $= \frac{2}{3}q^2 - 14q + 222$, min. when $q = 10.5$, AVC $= 148.5$
(c)   AFC $= 50q^{-1}$, min. when $q \to \infty =$, AFC $\to 0$
(d)   TR $= 200q - 2q^2$, max. when $q = 50$, TR $= 5,000$
(e)   MR $= 200 - 4q$, no turning point, end-point max. when $q = 0$
(f) $\pi = -\frac{2}{3}q^3 + 12q^2 - 22q - 50$, max. when $q = 11, \pi = 272.67$
$\left(\pi \text{ min. when } q = 1, \pi = -60\frac{2}{3}\right)$
2. Own example   3. (a) 16   (b) 8   (c) 12
4. No turning point but end–point min. when $q = 0$
5. No turning point but end–point min. when $q = 0$
6. Max. when $x = 63.33$, no minimum

9.5   1. π max. when $q = 4$ (theoretical min. when $q = -1.67$ not realistic)

2. (a) Max. when $q = 10$ (b) no min. exists

3. π max. when $q = 12.67$, gives $\pi = -48.8$ 4. 5,075 when $q = 10$ 5. 27.6 when $q = 37$

9.6   1. 15 orders of 400   2. 560   3. 480 every 4.5 months   4. 140

9.7   1.   (i) (a) $q = 90 - 0.2t$, $p = 270 + 0.4t$ (b)&(c) $q = 90$, $p = 270$

(ii) (a) $q = 250 - 1.25t$, $p = 125 + 0.375t$ (b)&(c) $q = 250$, $p = 125$

(iii) (a) $q = 25 - 0.9615t$, $p = 160 + 0.385t$ (b)&(c) $q = 25$, $p = 160$

(there is no tax impact for (b) and (c) in all cases)

2. $q = 100$, $p = 380$ (no tax impact)

Chapter 10

10.1   1. (a) $3 + 8x$, $16 + 4z$ (b) $42x^2z^2$, $28x^3z$ (c) $4z + 6x^{-3}z^3$, $4x - 9x^{-2}z^2$

2. $MP_L = 4.8K^{0.4}L^{-0.6}$, falls as L increases

3. $MP_K = 12K^{-0.7}L^{0.3}R^{0.4}$, $MP_L = 12K^{0.3}L^{-0.7}R^{0.4}$, $MP_R = 16K^{0.3}L^{0.3}R^{-0.6}$

4. $MP_L = 0.7$, does not decline as L increases   5. No   6. $1.2x_j^{-0.7}$

10.2   1. (a) 0.228   (b) falls to 0.224   (c) inferior as $\partial q / \partial m < 0$ (d) elasticity with respect to $p_s = 0.379$ and so a 1% increase in both prices would cause a percentage rise in q of $0.379 - 0.228 = 0.151\%$

2. (a) Yes, MU_A and MU_B will rise at first but then fall;

(b) no, MU_A falls but MU_B continually rises, therefore law not obeyed;

(c) yes, both MU_A and MU_B continually fall.

3. No, MU will never reach zero for finite values of A or B.

4. 3,738.46; balance of payments changes from 4.23 deficit to 68.85 surplus.

5. $25 + 0.6q_1^2 + 2.4q_1q_2$ 6. 0.45; 1.81818; 55

10.3   1. $-2K^{0.6}L^{-1.5}$, $2.4K^{-0.4}L^{-0.5}$ 2. $Q_{LL} = 6.4$, MP_L function has constant slope; $Q_{LK} = 35 + 2.8K$, position of MP_L will rise as K rises; $Q_{KK} = 2.8L$, MP_K has constant slope, actual value varies with L; $Q_{KL} = 35 + 2.8K$, increase in L will increase MP_K, effect depends on level of K.

3. $TC_{11} = 0.008q_3^2$, $TC_{22} = 0$, $TC_{33} = 0.008q_1^2$

$TC_{12} = 1.2q_3 = TC_{21}$, $TC_{23} = 9 + 1.2q_1 = TC_{32}$

$TC_{31} = 0.016q_1q_3 + 1.2q_2 = TC_{13}$

10.4   1. $q_1 = 12.46$; $q_2 = 36.55$ 2. $p_1 = 97.60$, $p_2 = 101.81$

3. $q_1 = 0$, $q_2 = 501.55$ (mathematical answer gives $q_1 = -1,292.24$, $q_2 = 1,701.77$ so rework without market 1)

4. £575.81 when $q_1 = 47.86$ and $q_2 = 39.01$ 5. $q_1 = 266.67$, $q_2 = 333.33$

6. $q_1 = 1,580.2$, $q_2 = 1,791.8$ 7. $K = 2,644.2$, $L = 3,718.5$

8. £29,869.47 when $K = 1,493.47$ and $L = 2,489.12$

9. Because max. $\pi = £18,137.95$ when $K = 2,176.5$ and $L = 2,015.22$

10. $K = 10,149.1, L = 9,743.1$

10.5   1. (a) $12K^{-0.4}L^{0.4}dK + 8K^{0.6}L^{-0.6}dL$

(b) $14.4K^{-0.7}L^{0.2}R^{0.4}dK + 9.6^{0.3}L^{-0.8}R^{0.4}dL + 19.2K^{0.3}L^{0.2}R^{-0.6}dR$

(c) $(4.8K^{-0.2} + 1.6KL^2)dK + (3.5L^{-0.3} + 1.6K^2L)dL$

2. (a) Yes (b) no, surplus (c) no, surplus

3. $40x^{-0.6}z^{-0.45} + 12x^{0.4}z^{-0.7}$

4. $\dfrac{\partial Q_A}{\partial P_A} + \dfrac{\partial Q_A}{\partial M}\dfrac{dM}{dP_A}$

Chapter 11

11.1   1. $K = 12.6, L = 21$ 2. $K = 500, L = 2,500$ 3. $A = 6, B = 4$

4. 141.42 when $K = 25, L = 50$ 5. Own example

6. (a) $K = 1,000, L = 50$ (b) $K = 400, L = 20$

7. 1,950 when $K = 60, L = 120$ 8. $L = 241, K = 201$, TC $= £3,617$

11.2   See answers to 11.1

11.3   1. See answers to 11.1   2. $L = 38.8, K = 20.7$, TC $= £3,104.50$

3. $C_1 = £480,621, C_2 = £213,609$

4. $L = 19.04, K = 8.18$, TC $= £1,145.30$

11.4   1. $x = 30, y = 30, z = 90$ 2. 877.8 when $K = 15, L = 45, R = 13$

3. $x = 50, y = 100, z = 150$ 4. 79,602.1 when $x = 300, y = 300, z = 1,875$

5. $K = 26.7, L = 33.3, R = 8.9, M = 55.6$ 6. Own example

7. $L = 60, K = 45, R = 40$

Chapter 12

12.1   1. 9   2. Answer given   3. $3M(1+i)^2$ 4. $0.6x(3 + 0.6x^2)^{-0.5}$

5. $0.5(6 + x)^{-0.5}$ 6. $\text{MRP}_L = 60L^{-0.5} - 8, L = 16$ 7. 169 units

8. £8   9. 0.000868

12.2   1. $(6x + 7)^{-0.5}(39x^2 + 36.4x - 5.7)$ 2. 12

3. $76.5L^{-0.5}(0.5K^{0.8} + 3L^{0.5})^{-0.4}$ 4. 312.5   5. £190

6. Own example   7. (a) $-0.05(60 - 0.1q)^{-0.5}$

(b) rate of change of slope $= -0.0025(60 - 0.1q)^{-0.5} < 0$ when $q < 600$

(c) 400

12.3   1. $(24 + 6.4x - 4.5x^{1.5} - 3x^{2.5})(8 - 6x^{1.5})^{-1.5}$

2. $(18,000 + 360q)(25 + q)^{-1.5}$ 3. -0.113

4. $q = 1,333\frac{1}{3}$, $d^2TR/dq^2 = -0.00367$ 5. $L = 4.8$, $H = 7.2$

6. Adapt proof in text for MC and AC to $AVC = TVC(q)^{-1}$

12.4   1. (a) $12.5x^2 + C$ (b) $5x + 0.6x^2 + 0.05x^3 + C$ (c) $24x^5 - 15x^4 + C$
    (d) $42x + 18x^{-1} + C$ (e) $60x^{1.5} + 220x^{-0.2} + C$

2. (a) $4q + 0.05q^2$ (b) $42q - 9q^2 + 2q^3$ (c) $35q + 0.3q^3$
    (d) $62q - 8q^2 + 0.5q^3$ (e) $185q - 12q^2 + 0.3q^4$

12.5   1. (a) £750,000 (b) £81,750 (c) £250,000 (d) £67,750

2. £49,600   3. Own example

Chapter 13

13.1   1. 20   2. No production in period 4   3. (a) Unstable (b) stable

13.2   1. $P_t = 4 + 0.25(-2)^t$ 2. Stable, 118.54   3. 404.64

13.3   1. 2,790.625; yes   2. 39,946.789   3. 492.57   4. 1,848.259

13.4   1. 2,460.79   2. No, 1,976.67 < 1,980   3. $P_t^x = 562 - 63(0.83)^t$, 555.27

Chapter 14

14.1   1. 64.44 million   2. 61,062 units   3. 16.8 million tonnes
    4. Usage in million units: (a) 94.6, yes   (b) 137.6, yes   (c) 200.2, no
        (d) 291.31, no   5. 56,609 units   6. €31,308.07

14.2   1. 2%; 9.84 million; no   2. 9%, 401,767,300 barrels
    3. £122,197.54   4. 587

14.3   1. 0.48%   2. 2.05%; 3.49%   3. 0.83%, 621.43 million tones
    4. €6,446.39 million   5. 8.8%   6. 5.83%   7. 6.18%
    8. 9% discrete (equivalent to 8.62% continuous)

14.4   1. (a) $200e^{0.2t}$, 1477.81   (b) $45e^{1.2t}$, 7323965.61   (c) $14e^{-0.4t}$, 0.26
    (d) $40e^{1.32t}$, 21614597.49 (e) $128e^{-0.03t}$, 99.69   2. 10 %, 6.77

14.5   1. $-20e^{0.4t} + 200$, 52.22, unstable   2. $-19.2e^{-1.5t} + 32$, 31.99, stable
    3. $-20e^{-0.75t} + 120$, 119.53, stable   4. $75e^{0.08t} - 300$, -188.11, unstable

14.6   1. $7e^{-0.325t} + 30$, stable   2. $-7.25e^{-0.96t} + 26.25$, difference 0.01
    3. Yes, as predicted spot price is $27.56
    4. $32.54e^{-0.347t} + 17.46$, £23.20, 3 periods   5. $44.01

14.7   1. $25e^{-0.2t} + 180, 183.38$ 2. $-63.33e^{-0.195t} + 1583.33, 1574.32$

3. $12.22e^{-0.036t} + 2027.78, 2036.3$ 4. $-9.49e^{-0.226t} + 141.49, 140.495$

5. $-18.154e^{-0.176t} + 346.15, 343.015$

Chapter 15

15.1   1. $\begin{bmatrix} 2 & 8 & 23 \\ 3 & 5 & 26 \end{bmatrix}$ 2. Yes

3. a. yes, $\begin{bmatrix} 6 & 70 \\ 27 & 23 \end{bmatrix}$ b. no   c. yes, $\begin{bmatrix} 14 \\ 3 \\ 14 \\ -3 \\ 2 \end{bmatrix}$

4. $\begin{bmatrix} 1.4 & 0.6 & 0.2 & 0.8 \\ 1.2 & 0.6 & 1.6 & 0.5 \\ 0.8 & 0.24 & 0.4 & 0 \end{bmatrix}$

15.2   1. $\begin{bmatrix} 27 & 39 \end{bmatrix}$ 2. a. $\begin{bmatrix} 98 & 84 \\ 163 & 134 \end{bmatrix}$ b. $\begin{bmatrix} 71 & 24 & 13 \\ 68 & 8 & 17 \end{bmatrix}$

c. not possible   3. $\mathbf{PR} = \begin{bmatrix} 52 & 30 & 56 & 43 \\ 62 & 31.5 & 71 & 55 \end{bmatrix}$

15.3   1. a. $\begin{bmatrix} 17.5 & 45 & 19 \\ 61 & 130 & 84 \end{bmatrix}$ b. $\begin{bmatrix} 136 & 67.5 & 90 & 40 \\ 130 & 47.5 & 51 & 44 \\ 59 & 10 & 9 & 36 \end{bmatrix}$

c. $\begin{bmatrix} 460 & 579 & 299 & 400 & 2110 & 3181 \\ 291 & 1077 & 240 & 418 & 2155 & 1166 \\ 907 & 4505 & 400 & 1030 & 9932 & 7386 \\ 114 & 20 & 466 & 560 & 151 & 318 \\ 93 & 133 & 165 & 213 & 135 & -215 \end{bmatrix}$

15.4   1. a $\begin{bmatrix} 5 & 4 & 9 \\ 2 & 1 & 4 \\ 2 & 5 & 4 \end{bmatrix} \begin{bmatrix} x \\ y \\ z \end{bmatrix} = \begin{bmatrix} 95 \\ 32 \\ 61 \end{bmatrix}$ b. $\begin{bmatrix} 6 & 4 & 8 \\ 3 & 2 & 4 \\ 1 & -8 & 2 \end{bmatrix} \begin{bmatrix} x \\ y \\ z \end{bmatrix} = \begin{bmatrix} 56 \\ 28 \\ 34 \end{bmatrix}$

c. $\begin{bmatrix} 5 & 4 & 2 \\ 9 & 4 & 0 \\ 2 & 4 & 4 \end{bmatrix} \begin{bmatrix} x \\ y \\ z \end{bmatrix} = \begin{bmatrix} 95 \\ 32 \\ 61 \end{bmatrix}$ d. not possible

2. (b) and (c)   3. (b) not square, (c) rows linearly dependent

15.5   1. A. 2   B.0   C.56   D.137   E.119

15.6   1. a. -39 b.15   c. -4 d. 28   e. 50   f. 4

2.   A.636   B. -101 C. -4462

15.7   1. $\begin{bmatrix} 0.16 & -0.3 \\ -0.2 & 0.5 \end{bmatrix}$ 2. $\mathbf{C} = \begin{bmatrix} 3 & 4 & -5 \\ 2 & 6 & -5 \\ -8 & -19 & 20 \end{bmatrix}$

$\mathbf{AdjA} = \begin{bmatrix} 3 & 2 & -8 \\ 4 & 6 & -19 \\ -5 & -5 & 20 \end{bmatrix}$ $\mathbf{A}^{-1} = \begin{bmatrix} 0.6 & 0.4 & -1.6 \\ 0.8 & 1.2 & -3.8 \\ -1 & -1 & 4 \end{bmatrix}$

3. $\begin{bmatrix} -0.5 & 0.5 & -0.5 & -0.5 \\ 0.1075 & -0.0215 & 0.1505 & -0.0538 \\ 1.7742 & -1.3548 & 0.4839 & 1.6129 \\ -0.3978 & 0.2796 & 0.043 & -0.3011 \end{bmatrix}$

15.8   1. $x = 5, y = 8$ 2. $x_1 = 10, x_2 = 6, x_3 = 2$

3. $\beta_1 = -0.5, \beta_2 = 1, \beta_3 = 0.4, Q = 9.5$

4. $\beta_1 = -300, \beta_2 = 75, \beta_3 = 400, \beta_4 = -100, \beta_5 = 0.2, \beta_6 = 10, Q = 8370$

15.9   1. $x = 3, y = 7$ 2. 6   3. See 15.8 answers

15.10   1. $q_1 = 389.6, q_2 = 62.3$, max SOC met as $|H_1| = -0.6, |H_2| = 0.448$

2. $q_1 = 216.8, q_2 = 435.8$, max SOC met as $|H_1| = -0.5, |H_2| = 0.19$

3. $q_1 = 6.485, q_2 = 2.376, q_3 = 5.4$, max SOC met as $|H_1| = -16.8$,

$$|H_2| = 21.2, |H_3| = -16.8$$

4. $q_1 = 5.2, q_2 = 35.4, q_3 = 20.8$, max SOC met as $|H_1| = -4.08$,

$$|H_2| = 0.4832, |H_3| = -0.0225$$

15.11   1. $K = 25, L = 100$, max SOC met as $|H_B| = 0.72$

2. $K = 16, L = 32$, min SOC met as $|H_B| = -14.618$

3. $X = 25, Y = 100$, max SOC met as $|H_B| = 4.543$

4. $K = 121.93, L = 243.86$, min SOC met as $|H_B| = -0.945$

Symbols and terminology

Index